About this workbook

This workbook is an industrial grade program designed to cover the fundamentals as well as the more advanced applications of geometric tolerancing. It covers what you need to know in order to function and work as a professional in an organization utilizing geometric tolerancing. It is programmed to accompany the Geometric Dimensioning and Tolerancing Video Series. (ASME Y14.5M-1994 & ASME Y14.5.1M-1994) The video training program consists of 16 tapes that average 45 to 60 min in length and 16 workshop exercises. The first 5 units, plus unit 10, cover the fundamentals of geometric tolerancing. The remaining units thru unit 16 cover the more advanced material and include geometric applications.

The delivery schedule is flexible, but the video is usually completed in sixteen 1 1/2 to 2 hour units for a total program length of 24 to 30 hours. A typical unit consists of the participant viewing the video and following along in the workbook. Afterwards, a workshop exercise is completed. The solutions to the workshop exercises are presented, reviewed and discussed by a program leader. A leaders guide as well as a wood and plastic model set are available to assist the program leader. The model set is comprised of parts, assemblies and inspection equipment shown in this workbook.

The workbook is written in 16 units that follow along with the 16 video tapes. It is suggested the participant review the unit in the workbook prior to the delivery session. The participant should follow along and turn the pages in the workbook as topics are presented in the video program. The workbook will provide a good reference to the material and concepts covered. The intent of the video program is to lay the basic ground work so the participant may begin building knowledge on a solid foundation.

The workbook can also be used in a classroom lecture situation without the video series. It is written in a very simple and easy to understand language with extensive 3D isometric graphics. The workbook is designed as a simple practical hands-on approach to the subject. The workbook can be used for basic as well as advanced programs. The program leader will decide which material is relevant for a particular group. The remaining workbook material can be used as reference.

The workbook takes a global, practical approach to explaining how the geometrics system works. (see unit 2) Parts are shown in functional assemblies and are fully toleranced illustrating actual on-the-job applications. Inspection, gaging and verification procedures are also shown.

Make sure you read and understand the first five units very thoroughly. After you understand the basic principles, the remaining units can be read in any order that strikes your interest. Geometrics is a new language, and like any new language, it will take time and practice. The more you use it, the more you will understand it. Suggestions for improvement are welcome and encouraged.

Good Luck.

About the Author

Alvin G. Neumann is the President/Director of Technical Consultants Inc. (TCI). TCI is a training, consulting and management firm located in Longboat Key, Fl. It specializes in geometric tolerancing and related activities for customers in the U.S. and worldwide. Mr. Neumann has over thirty-two years experience working with small, medium and large corporations in product design, tooling, gaging, manufacturing, and assembly. He has over twenty years experience developing and implementing engineering, quality and manufacturing training programs.

Mr. Neumann has written many books on geometric dimensioning and tolerancing. He has also authored two extensive video training programs on the subject. Over the years, Mr. Neumann has provided consultation and has delivered thousands of geometric tolerancing training programs to industry.

He is a member and section sponsor on the American National ASME Y14.5M Dimensioning and Tolerancing and Y14.5.1M Mathematical Definitions subcommittee and chairman of the U.S. Technical Advisory Group to the ISO TC 10/SC5 international subcommittee on dimensioning and tolerancing. He is also active on the international Joint Harmonization Group.

Mr. Neumann has traveled worldwide presenting material related to geometric tolerancing. He is a senior member of the Society of Manufacturing Engineers, American Society of Mechanical Engineers and the American Society of Quality Control.

Acknowledgments

The author would like to express his appreciation and gratitude to the many professional associates and friends on the standards committees, schools, colleges, universities, companies and professional organizations, both nationally and internationally, who provided technical input and draft review of this workbook. Many thanks to Bill Haefele, my good friend, who provided technical input and many excellent graphics used in this text. Most importantly, I would like to acknowledge the support, patience and understanding from my wife, Kathy and two children Scott and Holly. I also thank Kathy for her final copy editing.

Credit is also gratefully given and acknowledgement made for certain references and definitions derived from the ASME Y14.5M-1994 and ASME Y14.5.1M-1994 standards published by American Society of Mechanical Engineers (ASME), New York, NY.

Dedicated to my Mother and Father

GEOMETRIC DIMENSIONING AND TOLERANCING WORKBOOK

Table of Contents
&
Program Outline

UNIT 1

INTRODUCTION, SYMBOLS AND TERMS

Introduction to Geometric Tolerancing
American National and International Standards
      American National Standards
      ISO Standards
Common Symbols
      Common Symbol Application
      Dimension Origin Symbol
      Statistical Tolerancing Symbol
      Radius, Controlled Radius
Feature Definition - With Size and Without Size
Features and Characteristics
Geometric Tolerance Zones
Geometric Characteristic Symbols
Feature Control Frame
      Common Symbols in a Feature Control Frame
Material Condition Modifiers - Definitions
Workshop Exercise 1.1

INTRODUCTION TO GEOMETRIC TOLERANCING

Geometric dimensioning and tolerancing (GD&T) is an international engineering language that is used on engineering drawings (blueprints) to describe product in three dimensions. GD&T uses a series of internationally recognized symbols rather than words to describe the product. These symbols are applied to the features of a part and provide a very concise and clear definition of design intent.

GD&T is a very precise mathematical language that describes the form, orientation and location of part features in zones of tolerance. These zones of tolerance are then described relative to a Cartesian coordinate system. The GD&T system has a strong mathematical base which is essential in today's computerized world.

This international engineering language is similar to any other language such as French, German, Japanese or Spanish. There are certain grammar and punctuation rules that must be learned and practiced. This is especially important because we want to avoid the possibility of slang creeping into the language. This slang could possibly lead to misunderstandings or incorrect interpretations of the language.

Like any language it takes a while to learn the GD&T system. People from various companies and backgrounds have various knowledge levels. People can apply and interpret the system on a scale of 1 to 10. There are some personnel that might be considered conversational, others can read it but not write it, and others might be experts. It takes time, practice and patience to thoroughly understand the system.

In the late 1930's a fellow by the name of Stanley Parker from the Royal Torpedo Factory in Scotland first realized that there were some problems using limit type tolerancing. He was one of the first people to devise the geometric tolerancing system. The geometric tolerancing system has been updated, refined and expanded since that time and is now recognized world wide as the method to use in defining product.

The latest American National Standards on the subject is the American National Standards Institute/American Society of Mechanical Engineers ASME Y14.5M-1994 Dimensioning and Tolerancing and ASME Y14.5.1M-1994 Mathematical Definitions of Dimensioning and Tolerancing. In the international arena there is the International Standards Organization ISO 1101:1983 and other associated series of standards on the subject.

The information in this book must be considered as advisory and is to be used at the discretion of the user. In some instances, figures show added detail for emphasis. In other instances, figures are incomplete by intent. Numerical values of dimensions and tolerances are illustrative only. Be sure to consult the above standards for additional specific information on the subject.

AMERICAN NATIONAL AND INTERNATIONAL STANDARDS

On the following pages are American National Standards and International Standards on the subject of dimensioning and tolerancing and product definition. The included list of standards is only a sample of the more common documents available. For a complete listing or catalog see the proper organization.

ISO (the International Organization for Standardization) is a world wide federation of national standards institutes (ISO member bodies). The work of developing International Standards is carried out through ISO technical committees. The American National Standards Institute (ANSI) is the primary source and official sales agent for ISO standards in the United States. ANSI is also a source for American National Standards and national standards from other countries.

American National Standards Institute
11 W. 42nd St.
New York, NY 10036
(212) 642-4953 or 642-4993

Additional sources for some American National Standards and other standards:

American Society of Mechanical Engineers
345 E. 47th St
New York, NY 10017
(800) 843-2763

Society of Automotive Engineers
400 Warrendale Dr.
Warrendale. PA 15096

Institute of Electrical and Electronics Engineers
Service Center
445 Hoes Lane
Piscataway, NJ 08854

AMERICAN NATIONAL STANDARDS RELATED TO PRODUCT DEFINITION AND DIMENSIONAL METROLOGY

Y14 series of standards typically related to product definition

ASME Y1.1-1989	Abbreviations
ANSI Y14.1-1980	Drawing Sheet Size and Format
ASME Y14.2M-1992	Line Conventions and Lettering
ASME Y14.3-1993	Multi View and Sectional View Drawings
ASME Y14.5M-1994	Dimensioning and Tolerancing
ASME Y14.5.1M-1994	Mathematical Definition of Dimensioning and Tolerancing Principles
ANSI Y14.6-1978	Screw Thread Representation
ANSI Y14.6aM-1981	Screw Thread Representation (Metric Supplement)
ASME Y14.7.1-1971	Gear Drawing Standards-Part 1, for Spur, Helical, Double Helical, and Rack
ASME Y14.7.2-1978	Gear and Spline Drawing Standards-Part 2, Bevel and Hypoid Gears
ASME Y14.8M-1989	Castings and Forgings
ANSI Y14.36-1978	Surface Texture Symbols
ANSI/IEEE 268-1982	Metric Practice

B89 series of standards typically related to metrology

ASME/ANSI B1.2-1983	Gages and Gaging for Unified Inch Screw Threads
ANSI B4.1-1967	Preferred Limits and Fits for Cylindrical Parts
ANSI B4.2-1978	Preferred Metric Limits and Fits
ANSI B4.3-1978	General Tolerances for Metric Dimensioned Product
ANSI B4.4M-1981	Inspection of Workpieces
ANSI B5.10-1981	Machine Tapers-Self Holding and Steep Taper Series
ANSI/ASME B46.1-1985	Surface Texture
ASME B89.1.2M-1991	Calibration of Gage Blocks by Contact Comparison
ASME B89.1.6M-1984	Measurement of Qualified Plain Internal Diameters for Use as Master Rings and Ring Gages
ASME B89.1.9M-1984	Precision Gage Blocks for Length Measurement
ASME B89.1.10M-1987	Dial Indicators (For Linear Measurements)
ASME B89.1.12M-1990	Methods for Performance Evaluation of Coordinate Measuring Machines
ANSI B89.3.1-1972	Measurement of Out-of-Roundness
ASME B89.3.4M-1985	Axes of Rotation, Methods for Specifying and Testing
ANSI B89.6.2-1973	Temperature and Humidity Environment for Dimensional Measurement
ASME B92.1-1970	Involute Splines and Inspection, Inch Version
ASME B92.2M-1980	Metric Module, Involute Splines
ANSI/ASME B94.6-1984	Knurling
ANSI B94.11M-1979	Twist Drills

Note: See ANSI or ASME for complete listing.

ISO STANDARDS RELATED TO DIMENSIONING AND TOLERANCING

ISO 129:1985(E)   Technical drawings - General principles, definitions, methods of execution and special indications

ISO 286-1:1988(E) ISO system of limits and fits. Part 1, Bases of tolerances, deviations and fits

ISO 286-2:1988(E) ISO system of limits and fits. Part 2, Tables of standard tolerance grades and limit deviations for holes and shafts

ISO 406:1987(E)   Technical drawings - Tolerancing of linear and angular dimensions

ISO 1101:1983(E)   Technical drawings - Geometrical tolerancing - Tolerancing of form, orientation, location and run-out - Generalities, definitions, symbols, indications on drawings

ISO 1660:1987(E)   Technical drawings - Dimensioning and tolerancing of Profiles

ISO 1829:1975(E)   Selection of tolerance zones for general purposes

ISO 2692:1988(E)   Technical drawings - Maximum material principle

Amd 1: 1992 (E)   Technical drawings - Least material principle

ISO 2768-1:1989(E) General tolerances- Part 1, Tolerances for linear and angular dimensions without individual tolerance indications.

ISO 2768-2:1989(E) General tolerances- Part 2, Geometrical tolerances for features without individual tolerance indications

ISO 3040:1990(E)   Technical drawings - Dimensioning and tolerancing - Cones

ISO 5458:1987(E)   Technical drawings - Geometrical tolerancing - Positional tolerancing

ISO 5459:1981(E)   Technical drawings - Geometrical tolerancing - Datums and datum-systems for geometrical tolerances

ISO 7083:1983(E)   Technical drawings - Symbols for geometrical tolerancing - Proportions and dimensions

ISO 8015:1985(E)   Technical drawings - Fundamental tolerancing principles

ISO 10578:1992(E) Technical drawings - Tolerancing of orientation and location - Projected tolerance zone

ISO 10579:1993(E) Technical drawings - Dimensioning and tolerancing - Non-rigid parts

Additional ISO Standards related to product definition
ISO 128:1982(E)   Technical drawings - General principles of presentation
ISO 273:1979(E)   Fasteners - Clearance holes for bolts and screws
ISO 1000:1992 (E) SI units and recommendations for the use of their multiples and of certain other units
ISO 1302:1992(E)   Technical drawings - Method of indicating surface texture

ISO Technical Reports
ISO TR5460:1985   Technical Drawings - Geometrical tolerancing - Tolerancing of form, orientation, location and run-out - Verification principles and methods - Guidelines

Note: A majority of this information can also be found in Handbook #12

COMMON SYMBOLS

Shown below are the most common symbols that are used with geometric tolerancing and other related dimensional requirements on engineering drawings. Note the comparison with the ISO standards. Most of the symbology is identical. There are a few symbols that are used in the ASME Y14.5M, 1994 standard that are being proposed for the ISO standards. The symbols marked with an "X" are new or revised from the previous Y14.5M, 1982 standard.

	SYMBOL	ASME Y14.5M	ISO
	FEATURE CONTROL FRAME	⊕ ⌀.030 Ⓜ A B C	⊕ ⌀.030 Ⓜ A B C
	DIAMETER	⌀	⌀
	SPHERICAL DIAMETER	S⌀	S⌀
	AT MAXIMUM MATERIAL CONDITION	Ⓜ	Ⓜ
	AT LEAST MATERIAL CONDITION	Ⓛ	Ⓛ
	REGARDLESS OF FEATURE SIZE	NONE	NONE
	PROJECTED TOLERANCE ZONE	Ⓟ	Ⓟ
X	FREE STATE	Ⓕ	Ⓕ
X	TANGENT PLANE	Ⓣ	Ⓣ (proposed)
X	STATISTICAL TOLERANCE	⟨ST⟩	NONE
X	RADIUS	R	R
X	CONTROLLED RADIUS	CR	NONE
	SPHERICAL RADIUS	SR	SR
	BASIC DIMENSION (theoretically exact dimension in ISO)	50	50
X	DATUM FEATURE	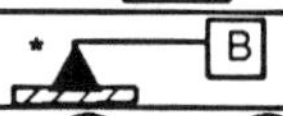 * ▲ B	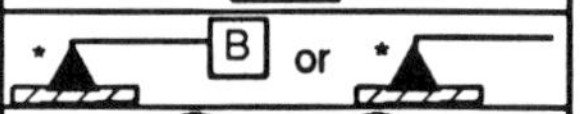 * ▲ B or * ▲
	DATUM TARGET	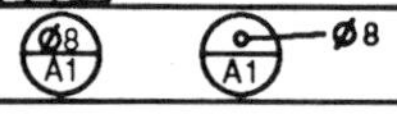 ⌀8/A1 ⊙/A1—⌀8	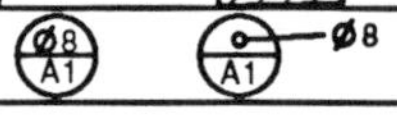 ⌀8/A1 ⊙/A1—⌀8
	TARGET POINT	✕	✕
	DIMENSION ORIGIN	 ⊕→	 ⊕→
	REFERENCE DIMENSION (auxiliary dimension in ISO)	(50)	(50)
	NUMBER OF PLACES	8X	8X
	COUNTERBORE/SPOTFACE	⊔	⊔ (proposed)
	COUNTERSINK	⌵	⌵ (proposed)
	DEPTH/DEEP	↧	↧ (proposed)
	SQUARE	□	□
	ALL AROUND	 ⌒○	NONE
	DIMENSION NOT TO SCALE	150	150
	ARC LENGTH	⌒150	⌒150
X	BETWEEN	↔	NONE
	SLOPE	◁	◁
	CONICAL TAPER	▷	▷
	ENVELOPE PRINCIPLE	NONE (implied)	Ⓔ

*MAY BE FILLED OR NOT FILLED

COMMON SYMBOL APPLICATION

The drawing below is shown with many of the common symbols applied. The symbols are a universal method of specifying requirements without the use of notes or words. The symbols are designed to be very intuitive and look like the requirement that they are identifying.

Repetitive features such as holes, slots and tabs which are repeated often can be specified by stating the required number of features or places and an "X" and then followed by the requirement. A space is used between the "X" and the requirement as shown below. Where used with a basic dimension, the number of places and the X may be placed either inside or outside the basic dimension frame.

The symbol for diameter is a circle with a slash. The symbol for radius is the letter R. The symbol for square features is a square box. The symbol for counterbore and spotface are the same. A spotface will have no depth indicated while the counterbore symbol will always have a depth specified. The countersink symbol is shown as a 90° Vee. The deep or depth symbol will identify the depth of the indicated feature. There is no space between the value or other specified symbols.

SYMBOL APPLICATION

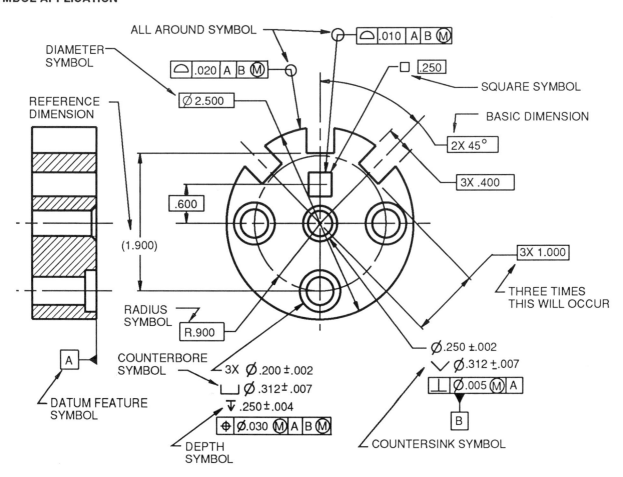

1.7

DIMENSION ORIGIN

The dimension origin is a means of indicating that a toleranced dimension between two features originates from one of the features. The high points of the surface indicated as the origin define a plane from where the dimension originates.

The example below illustrates a part where the origin is identified as the bottom surface. Without such indication the longer surface could have been selected as the origin, thus permitting a greater angular variation between surfaces.

The dimension origin concept is only a 2 dimensional control and is usually used in very simple applications. It does not establish a datum reference frame. If it is necessary to control more complex parts relative to a datum reference frame, a profile control is used.

THIS ON THE DRAWING

.510
.490

MEANS THIS

NOT THIS

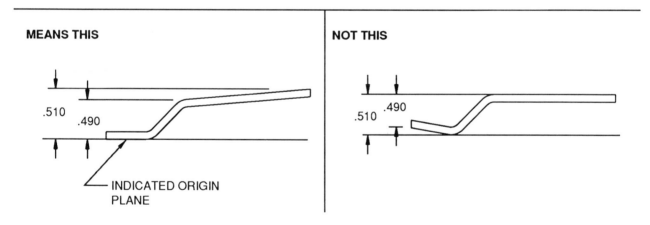

.510
.490

INDICATED ORIGIN
PLANE

.510 .490

STATISTICAL TOLERANCING SYMBOL

Often, tolerances are calculated on an arithmetic basis. Tolerances are assigned to individual features on a component by dividing the total assembly tolerance by the number of components and assigning a portion of this tolerance to each component. When tolerances are stacked up in this manner, the tolerance may become very restrictive or tight.

Statistical tolerancing is the assignment of tolerances to related components of an assembly on the basis of sound statistics. An example is: The assembly tolerance is equal to the square root of the sum of the squares of the individual tolerances.

Statistical tolerancing may be applied to features to increase tolerances and reduce manufacturing cost. To ensure compatability, the larger tolerance identified by the statistical tolerance symbol may only be used where appropriate statistical process control will be used. A note such as the one shown below shall be placed on the drawing.

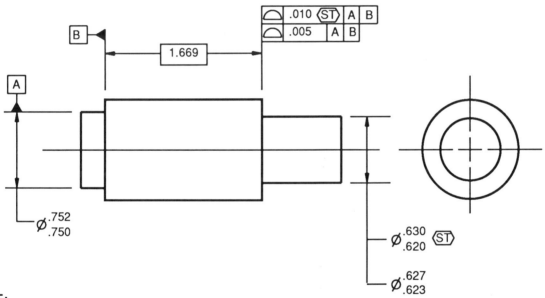

NOTE:

FEATURES IDENTIFIED AS STATISTICALLY TOLERANCED ⟨ST⟩ SHALL BE PRODUCED WITH STATISTICAL PROCESS CONTROLS, OR TO THE MORE RESTRICTIVE ARITHMETIC LIMITS.

In some cases, it may be desirable to state only the statistical tolerance and the arithmetic number will not be shown. In this case, a note such as the following must be placed on the drawing.

FEATURES IDENTIFIED AS STATISTICALLY TOLERANCED ⟨ST⟩ SHALL BE PRODUCED WITH STATISTICAL PROCESS CONTROLS.

For additional information on statistical tolerancing, see appropriate statistics or engineering design manuals.

RADIUS, CONTROLLED RADIUS

There are two types of radii tolerance that can be applied, the radius and controlled radius. The radius (R) tolerance is for general applications. The controlled radius (CR) is used when it is necessary to place further restrictions on the shape of the radius, as in high stress applications.

Note: This is a change from the previous editions of the Y14.5 standard. The definition of the tolerance zone for the former term tangent radius, previously noted by the symbol R, is now meant to apply to a controlled radius (symbol CR).

RADIUS, SYMBOL R

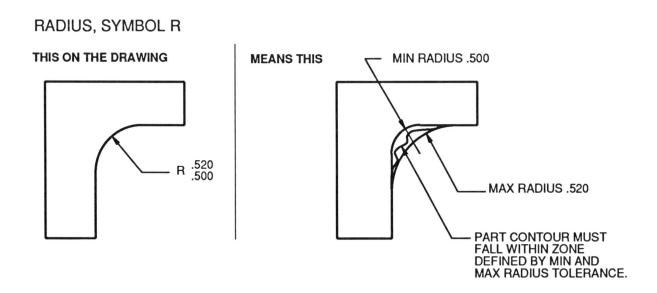

THIS ON THE DRAWING

R .520 / .500

MEANS THIS

MIN RADIUS .500

MAX RADIUS .520

PART CONTOUR MUST FALL WITHIN ZONE DEFINED BY MIN AND MAX RADIUS TOLERANCE.

CONTROLLED RADIUS, SYMBOL CR

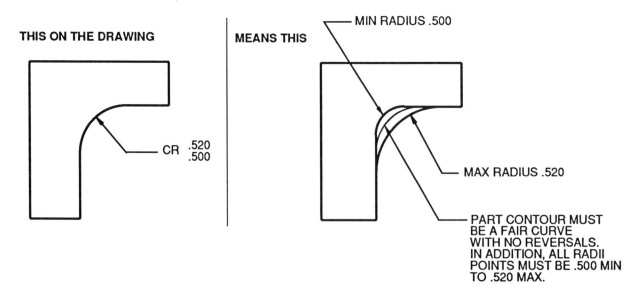

THIS ON THE DRAWING

CR .520 / .500

MEANS THIS

MIN RADIUS .500

MAX RADIUS .520

PART CONTOUR MUST BE A FAIR CURVE WITH NO REVERSALS. IN ADDITION, ALL RADII POINTS MUST BE .500 MIN TO .520 MAX.

FEATURE DEFINITION - WITH SIZE AND WITHOUT SIZE

Geometric tolerancing is a feature based system. A feature is a general term applied to a physical portion of a part such as a surface, pin, tab, hole or slot. Parts may have many features.

There are two types of features: Features with size and features without size.

Feature without size: **A plane surface**

Feature with size: **One cylindrical or spherical surface or a set of two opposed elements or opposed parallel surfaces, each of which is associated with a size dimension. (An axis, median plane or a center point can be derived from these features.)**

A feature control frame is attached or directed to these features to control their form, orientation and/or location.

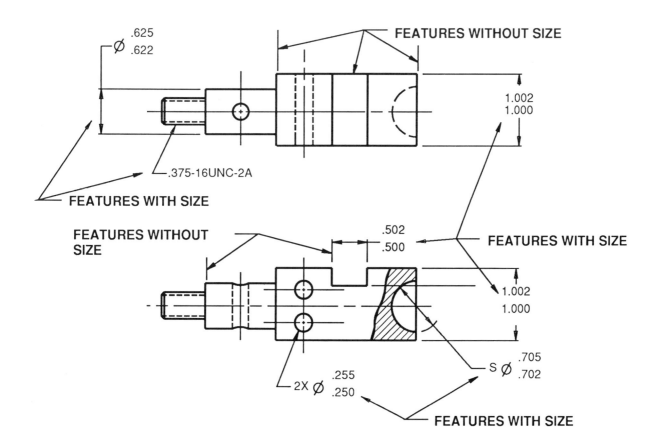

GEOMETRIC CHARACTERISTIC SYMBOLS

There are 14 geometric characteristic symbols. These characteristic symbols are placed in the first compartment of a feature control frame. They define the specification that is required.

The characteristics are grouped together in categories. The major categories are form, orientation, location, runout and profile tolerances. They can also be grouped together relative to their use such as: for individual features, for related features and for individual or related features.

GEOMETRIC CHARACTERISTIC SYMBOLS

	TYPE OF TOLERANCE	CHARACTERISTIC	SYMBOL
FOR INDIVIDUAL FEATURES	FORM	STRAIGHTNESS	—
		FLATNESS	▱
		CIRCULARITY (ROUNDNESS)	○
		CYLINDRICITY	⌭
FOR INDIVIDUAL OR RELATED FEATURES	PROFILE	PROFILE OF A SURFACE	⌓
		PROFILE OF A LINE	⌒
FOR RELATED FEATURES	ORIENTATION	ANGULARITY	∠
		PERPENDICULARITY	⊥
		PARALLELISM	//
	LOCATION	POSITION	⊕
		CONCENTRICITY	◎
		SYMMETRY	≡
	RUNOUT	CIRCULAR RUNOUT	↗ *
		TOTAL RUNOUT	↗↗ *

*ARROWHEADS MAY BE FILLED OR NOT FILLED

1.12

FEATURES AND CHARACTERISTICS

Geometric tolerancing is a feature based system. Parts are composed of features. Geometric characteristics are applied to the features to control their variation. The characteristics are divided in three main categories, form, orientation and location. The form tolerances are the lowest category and refine size. The orientation tolerances are used with basic angles, not basic linear dimensions. If used on surfaces they also control form. The location tolerances are the highest category and are always associated with basic linear dimensions. Position locates the axis or median plane of features and also controls orientation. Profile is the most powerful characteristic of all and locates surfaces. It will control orientation and form as well. The characteristics shown below are the most common. The other characteristics not listed are variations of these basic controls.

FORM TOLERANCES

ORIENTATION TOLERANCES

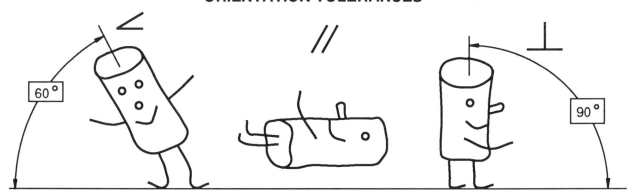

LOCATION TOLERANCES

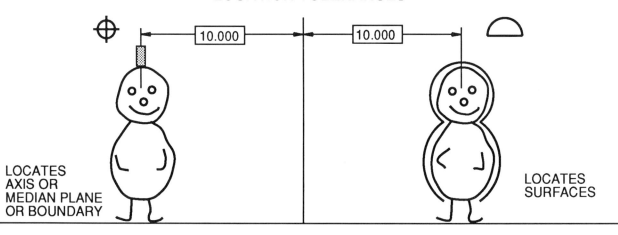

GEOMETRIC TOLERANCE ZONES

Geometric tolerancing is a three dimensional, mathematically based tolerancing system. Parts are composed of features that have no relationship to each other. In order to establish a relationship between the features, a datum reference frame DRF (coordinate system) is established on the part. The feature's location or orientation is defined with "theoretically exact" or basic dimensions relative to the DRF. Feature control frames point to the features and identify the constraint requirements for the identified features.

DATUM REFERENCE FRAME OR CARTESIAN COORDINATE SYSTEM

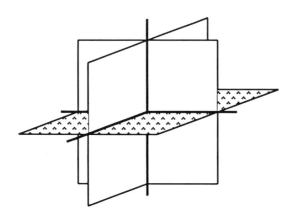

The form, orientation and location of features is controlled with zones of tolerance that are created with feature control frames. These zones of tolerance can be 3D or 2D. Some common shapes of tolerance zones can be seen below.

GEOMETRIC TOLERANCING ZONES CAN BE 3D OR 2D.

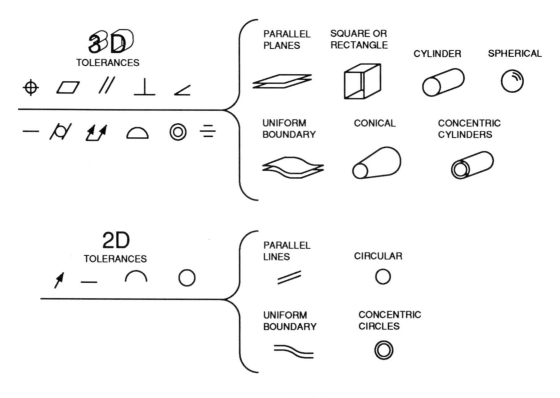

FEATURE CONTROL FRAME

The feature control frame is probably the most important symbol in the geometric tolerancing system. It states the requirements or instructions for the features to which it is attached. As its name implies, the feature control frame controls features. Each feature control frame will state only one requirement or one message. There is only one set-up or only one gage for each feature control frame. If there are two requirements for a feature, it will require two feature control frames.

To help interpret a feature control frame, it is important to note that there are always two words that are implied to precede a feature control frame. The two words are "This feature" as in "This feature flat" or "These features positioned."

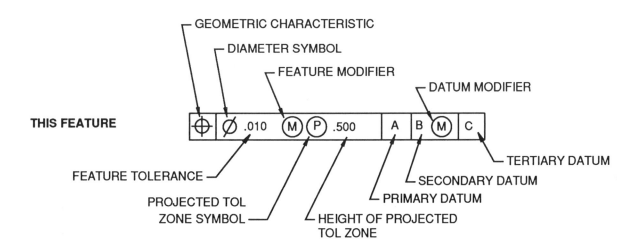

The first compartment of a feature control frame will always contain one of the 14 geometric characteristic symbols. There can never be two geometric characteristic symbols placed in a feature control frame. If there are two requirements, we must use two feature control frames. The geometric characteristic symbol in the first compartment will stipulate the requirement for the feature such as: this feature must be flat, or this feature must be parallel, etc.

The second compartment of a feature control frame will always contain the total tolerance for the feature. The feature tolerance is always specified as a total tolerance. It is never a plus/minus value.

If the tolerance is preceded by a diameter symbol ∅, the tolerance will be a diameter or cylindrical shaped tolerance zone as in the location of a hole. If the tolerance is preceded by a spherical diameter symbol S∅, the tolerance will be a spherical shaped zone as in the location of a ball or sphere. If there is no symbol preceding the tolerance, the default shape tolerance is parallel planes or a total wide zone as in the position of a slot or a profile of a surface.

Following the feature tolerance a feature modifier such as MMC Ⓜ or LMC Ⓛ can be specified. This will occur if the feature being controlled is a feature of size such as a hole, slot, tab or pin. If the feature being controlled is a feature of size and no modifier is specified, the default is RFS. (See modifier rules for more information.) If the feature being controlled has no size, such as a plane surface, then no modifier can be specified.

Also, in the second compartment, following the tolerance and specified modifier, can be found additional symbols, if they are applicable. These symbols can be for the projected tolerance zone, free state, tangent plane and statistical tolerance.

The third and following compartments of a feature control frame contain the specified datums, if datums are applicable. For example, a form tolerance like flatness or straightness will not allow specified datums. The datums are specified in their order of importance such as primary, secondary and tertiary. The alphabetical order of the datums has no significance. The significance is their order of precedence in the feature control frame. (See datum section for more information.)

If any of the datums are features of size, a datum modifier such as MMC Ⓜ or LMC Ⓛ may be specified. If the specified datum feature has size and no modifier is specified, the default is RFS. (See modifier rules for more information.)

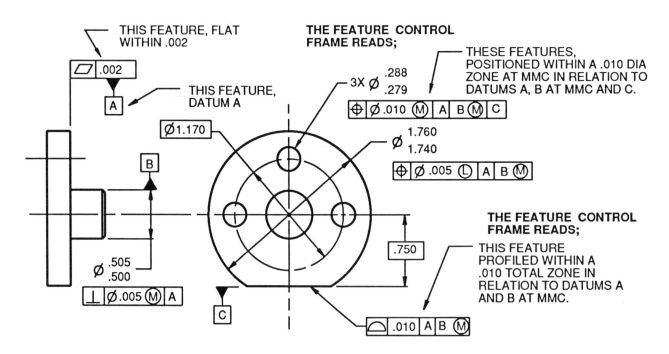

The placement of a feature control frame is very important. If the feature control frame is attached or directed to a surface, it controls that surface as in a flatness or profile control. If the feature control frame is attached to or associated with a feature of size, then it controls the axis or median plane of that feature as in a position callout.

Common symbols that can be found in a feature control frame.

TERM	SYMBOL
FEATURE CONTROL FRAME	⊕ ∅ .010 A B C
DIAMETER	∅
SPHERICAL DIA	S∅
MAXIMUM MATERIAL COND	Ⓜ
LEAST MATERIAL COND	Ⓛ
PROJECTED TOL ZONE	Ⓟ
FREE STATE	Ⓕ
TANGENT PLANE	Ⓣ
STATISTICAL TOLERANCE	⟨ST⟩
REGARDLESS OF FEATURE SIZE*	Ⓢ

* The RFS symbol has been eliminated from the ASME Y14.5M, 1994 standard. It is applicable in earlier versions.

All of the above symbols can be used inside a feature control frame. The symbols for projected tolerance zone, free state, tangent plane, and statistical tolerance always follow the material condition modifier. The minimum height of the projected tolerance zone can be specified in the feature control frame following the projected tolerance zone symbol. Sample feature control frames are shown below.

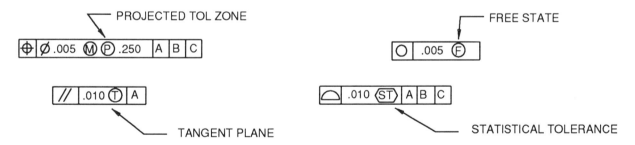

FORMER PRACTICE: In the ANSI Y14.5, 1982 standard, the datum feature symbol was shown as a rectangular box with two dashes. This practice was changed in the ASME Y14.5, 1994 standard to the new symbol shown below. This change was enacted to bring the United States closer in line with international practices.

DATUM FEATURE SYMBOL

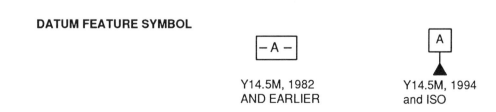

1.17

MATERIAL CONDITION MODIFIERS - DEFINITIONS

One of the most important concepts in geometric tolerancing is the use of material condition modifiers. It is essential that the reader understand the definitions and examples of the modifiers shown below.

In geometric tolerancing there is often a need to refer to particular feature of size at its largest size, smallest size or regardless of feature size. The terms maximum material condition (MMC), least material condition (LMC) and regardless of feature of size (RFS) allow us to do this.

These terms can only be used when referring to features of size such as holes, slots tabs, pins, etc. The terms have no meaning when applied to non-features of size such as plane surfaces. The application or implication of these material condition modifiers inside the feature control frame can have a substantial effect on the tolerance. See additional discussion on modifiers later in text.

MAXIMUM MATERIAL CONDITION - Abbreviation (MMC) Symbol Ⓜ
The condition where the feature contains the maximum material within the stated limits of size - for example, the largest pin or the smallest hole.

LEAST MATERIAL CONDITION -Abbreviation (LMC) Symbol Ⓛ
The condition where the feature contains the least material within the stated limits of size - for example, the smallest pin or largest hole.

REGARDLESS OF FEATURE SIZE - Abbreviation (RFS)
The term used to indicate that a geometric tolerance applies at any increment of size of the feature within its size tolerance.
Note: In the current ASME Y14.5M, 1994 standard there is no symbol for RFS. (Unless otherwise specified, all geometric tolerances are implied RFS). In the previous editions of the Y14.5 standard the Ⓢ symbol was used for RFS.

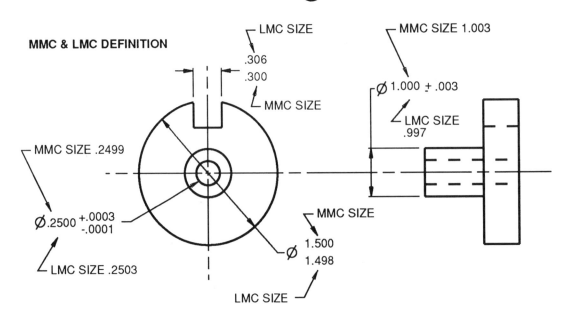

WORKSHOP EXERCISE 1.1

1. What is the name and date of the current American National Standard on dimensioning and tolerancing?

ASME   Y14.5m-1994

2. What is the name and date of the current American National Standard on mathematical definitions of dimensioning and tolerancing?

ASME Y14.5.1m-1994

3. In the International Standards Organization (ISO) there are many documents that cover dimensioning and tolerancing. What is the number of the ISO standard that covers the generalities of geometric tolerancing? - Tolerancing of form, orientation, location and runout.

1101

On the drawing below find and label the following symbols.

4. Feature control frame
5. Reference dimension
6. Datum feature symbol
7. Basic dimension
8. Diameter symbol
9. Radius symbol

10. Countersink symbol
11. Counterbore symbol
12. Square symbol
13. Two places designation
14. Depth symbol
15. All around symbol

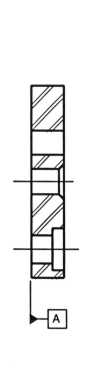
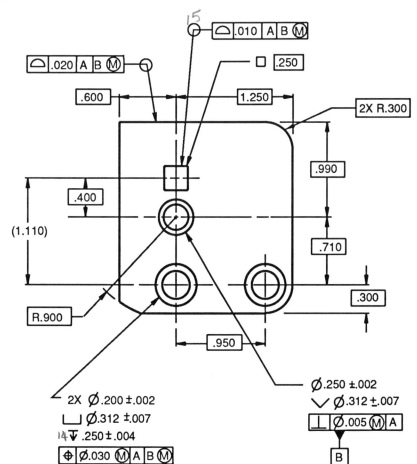

WORKSHOP EXERCISES 1.1

Sketch the proper symbol for the following terms.

16. Statistical tolerance

17. Free state (F)

18. Projected tolerance zone (P)

19 Maximum material condition (M)

20. Regardless of feature size (former symbol)
(S)

21. Least material condition (L)

22. Controlled radius CR

23. Tangent Plane

24. Spherical radius SR

25. In the chart below, sketch the proper geometric symbol next to the geometric characteristic term. Also, identify the categories or type of tolerance of the geometric characteristics such as: form, orientation, runout, location and profile.

GEOMETRIC CHARACTERISTIC SYMBOLS

	TYPE OF TOLERANCE	CHARACTERISTIC	SYMBOL
FOR INDIVIDUAL FEATURES		STRAIGHTNESS	
		FLATNESS	
		CIRCULARITY (ROUNDNESS)	
		CYLINDRICITY	
FOR INDIVIDUAL OR RELATED FEATURES		PROFILE OF A SURFACE	
		PROFILE OF A LINE	
FOR RELATED FEATURES		ANGULARITY	
		PERPENDICULARITY	
		PARALLELISM	
		POSITION	
		CONCENTRICITY	
		SYMMETRY	
		CIRCULAR RUNOUT	
		TOTAL RUNOUT	

26. Label the parts of the feature control frame. Choose from the terms below.

Feature Tolerance

Feature Modifier

Primary Datum

Datum Modifier

Secondary Datum

Tertiary Datum

Diameter Symbol

Geometric Characteristic

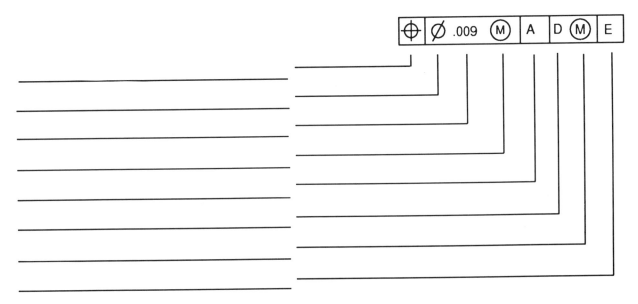

27. Next to the dimensions, label the maximum material condition (MMC) and least material condition of the size features below.

MMC & LMC DEFINITION

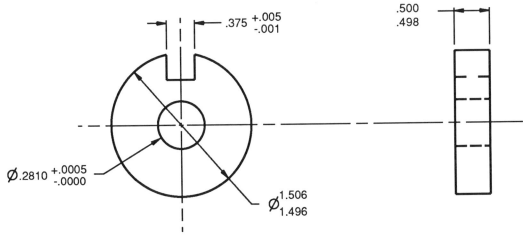

1.21

UNIT 2

HOW THE GEOMETRIC SYSTEM WORKS

LIMIT TOLERANCING APPLIED TO AN ANGLE BLOCK

In order to understand the geometric tolerancing system, it is important to understand the interpretations of the limit or plus/minus system of tolerancing. Limit tolerancing is basically a two dimensional tolerancing system.

Consider the limit tolerancing applied to the angle block drawing. When the part is drawn on a print it usually looks very good, the lines are straight the corners are square and the hole is round. The product designer, using their drafting or CAD equipment, will draw the picture very straight and square as shown in the top illustration. This part is then produced in the manufacturing process. Since all manufacturing equipment has inherent errors and is not perfect, it stands to reason that the imperfect equipment will not build perfect parts. We can not build anything square or perfect; the part will have inherent variations - the corners are no longer square, the surfaces are no longer straight or flat.

Visually, the angle block looks straight and square. The manufacturing variations are so small that they are usually undetectable with the human eye. However, when an inspector looks at these parts with very precise measuring instruments (micrometers, indicators and CMM's) that can detect variations of .001 or .0001 of an inch or better, the angle block might look like the part in the lower illustration.

Notice that the produced angle block is not square in the front view or the side view. The surfaces are warped and are not flat. The hole that was put in is not square to any surface, and the hole itself is not round. It is at this point that the limit system of tolerance breaks down. The plus/minus tolerances are a two dimensional system; the actual parts are three dimensional. The plus/minus dimensions are simply high and low limits. They are not oriented to specified datums. (See Taylor Principle for more detailed information.)

There are implied datums shown on the product drawing. The dimensions appear to be originating from the left corner. The corner on the produced part is not square and the surfaces are warped and not flat. The problem is: How do we orient the part to begin making the measurements to evaluate the limit tolerances?

As you can see, it is not clear exactly how to verify the product specifications. In the measurement process all measurements should be made square to avoid measurements that are like a parallelogram. The problem here is that the part is not square, and everyone can measure the part differently and arrive at different answers.

Limit tolerances usually do not have an origin or any orientation or location relative to datums. The datums are usually implied. In addition, if the plus/minus dimensioned features do not have opposing elements, there is often confusion on where to start the measurements. Should we start on the long side and measure the short side, or should we start on the short side and measure the long side?

LIMIT TOLERANCING APPLIED TO AN ANGLE BLOCK

THIS ON THE DRAWING

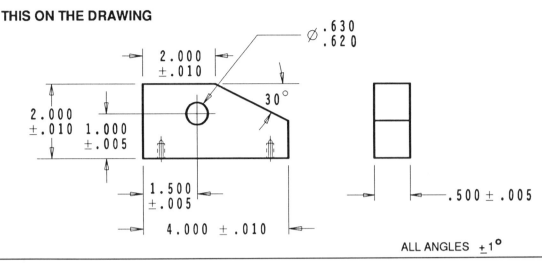

$\emptyset \begin{smallmatrix}.630\\.620\end{smallmatrix}$

2.000 ±.010

30°

2.000 ±.010   1.000 ±.005

1.500 ±.005

4.000 ± .010

.500 ± .005

ALL ANGLES ±1°

MEANS THIS ON PRODUCED PART

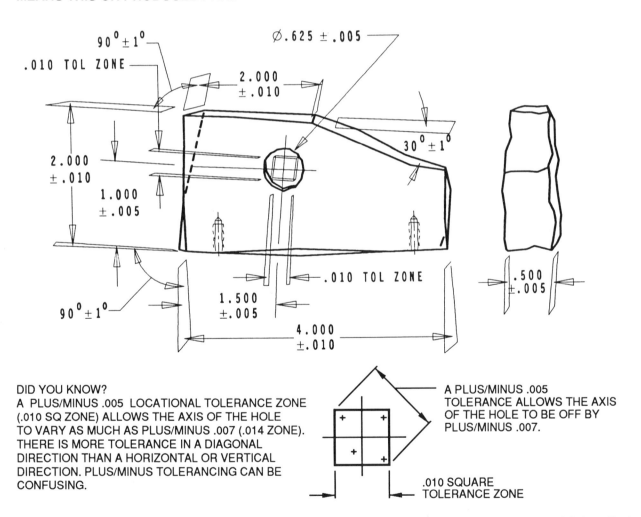

90°±1°

∅.625 ±.005

.010 TOL ZONE

2.000 ±.010

30°±1°

2.000 ±.010

1.000 ±.005

.010 TOL ZONE

.500 ±.005

90°±1°

1.500 ±.005

4.000 ±.010

DID YOU KNOW?
A PLUS/MINUS .005 LOCATIONAL TOLERANCE ZONE (.010 SQ ZONE) ALLOWS THE AXIS OF THE HOLE TO VARY AS MUCH AS PLUS/MINUS .007 (.014 ZONE). THERE IS MORE TOLERANCE IN A DIAGONAL DIRECTION THAN A HORIZONTAL OR VERTICAL DIRECTION. PLUS/MINUS TOLERANCING CAN BE CONFUSING.

A PLUS/MINUS .005 TOLERANCE ALLOWS THE AXIS OF THE HOLE TO BE OFF BY PLUS/MINUS .007.

.010 SQUARE TOLERANCE ZONE

Plus/minus tolerancing is a very complex two dimensional tolerancing system with implied datums. It is subject to various interpretations. The limits of size do not control the orientation or locational relationship between features. In addition, a 90 degree angle applies where centerlines and lines depicting features are shown at right angles. The tolerance on these angles is usually found in the title block or in a note on the drawing.

The limit dimensions on the drawing are shown at 90 degrees and are implied to be square to each other. This implied angle tolerance is usually a standard tolerance that is specified in the title block. The implied angle between the plus/minus dimensions applies in the front view as well as the side and top views. Limit tolerancing is very complex and is very difficult to stack up and predict with any confidence. Most of our modern engineering, manufacturing and quality systems all work square or relative to a coordinate system.

Plus/minus tolerancing is basically a caliper or micrometer type measurement. It works very well for individual features of size (see rule 1- Taylor Principle) but does not control the relationship between individual features very well. This does not mean that limit tolerancing cannot be used. It is just important that we realize its limitations and problems. Having a clear understanding of limit tolerancing helps us understand what geometric tolerancing will do for us.

Another problem with limit tolerancing on our angle block is the way in which the .620 - .630 hole is located. First of all, the hole is located from implied datums. Since the datum reference frame is perfectly square and the part may be produced within tolerance but out of square, the order of precedence of the implied datums is not clear. This can pose major problems. The solution is described in detail later as you study the concept on datum reference frame (DRF).

The second problem is that the limit tolerancing that locates the hole will establish a square tolerance zone within which the hole axis must lie. This square tolerance zone allows the axis of the hole to be off more in a diagonal direction than it can be off in a horizontal or vertical direction. In fact, with a little right angle trigonometry, it can be seen that a plus/minus .005 tolerance (.010 total zone) actually allows the axis of the hole to vary within plus/minus .007 (.014 total zone).

As you can see, the plus/minus tolerance system does present problems in clearly defining the product. Limit tolerancing worked for us in the past with a lot of verbal communication and when parts are built in one plant or one country. Things have changed dramatically in the past few years. Parts are now built and shipped all over the world. Computers have also entered the scene. Engineers and designers use computer aided design (CAD), manufacturing uses computer numerical control (CNC) machines, and quality control uses coordinate measuring machines (CMM's) to inspect and verify parts. Parts must be described in a three dimensional mathematical language to ensure clear and concise communication of information relating to product definition. This is why we need geometric tolerancing.

GEOMETRIC TOLERANCING APPLIED TO AN ANGLE BLOCK

In the following illustration the same angle block that was defined with plus/minus tolerancing is now shown with geometric tolerancing applied. Notice that datums A, B and C have been applied to features on the part establishing a X, Y and Z cartesian coordinate system. All of the dimensions radiate from the coordinate system and are enclosed in a "box". These dimensions are called basic dimensions and have no tolerance. Tolerance zones are established at the basic locations through the use of a feature control frame. The basic dimensions locate the tolerance zones and not the actual features. The actual features of the part must then lie within these established tolerance zones.

HOW THE GEOMETRIC TOLERANCING SYSTEM WORKS.

Use plus/minus or limit tolerancing to define size and form of features.

Establish a datum reference or coordinate system on the part.

Basic dimensions define the part relative to a coordinate system and are theoretically exact.
The location of features holes, slots, tabs and surfaces are shown on a drawing with basic dimensions. The basic dimensions are enclosed in a box and have no tolerance. The basic dimensions are considered to be theoretically exact and define the location of tolerance zones. The designated features must fall within the tolerance zones. The size and shape of the tolerance zone and its relation to a coordinate system can be found inside the feature control frame. Another way to look at the basic dimensions is that they locate or define the theoretically exact location of the tolerance zones. The features must then lie within the tolerance zones defined in the feature control frame.

The two main or most often used geometric characteristics are position and profile of a surface.
Position locates features of size such as holes, slots, tabs and pins. Profile of a surface locates surfaces.

Position tolerance is used to locate the hole on the angle block. On the angle block part the size of the hole is defined with a plus/minus tolerance. The plus/minus tolerance is well suited for defining the size of individual features having opposed elements. A very detailed definition of size can be found later in the text in rule #1 (Taylor Principle).

The location of the hole is defined in the feature control frame below the size tolerance. The position characteristic in the first compartment of the feature control frame states the hole is positioned. In the second compartment of the feature control frame, the diameter symbol precedes the tolerance indicating that the tolerance is diameter or cylindrical shaped. This zone extends for the full length or depth of the hole. The axis of the .610 - .620 hole must fall within the .010 diameter tolerance zone.

The .010 diameter zone locates the axis of the hole the same amount in any direction. This is a much clearer specification than limit tolerancing, which allows the location of hole more tolerance in the diagonal direction than the horizontal or vertical direction. The .010 round positional zone can be compared to a bullseye at a rifle or archery range. The manufacturing process will aim for the center or basic dimensions and afterwards chart the variation off center to insure the goal has been accomplished.

The position tolerance zone defines the capability of the process or the manufacturing equipment necessary to produce the part within limits. Generally, the tighter the tolerance, the better the processing equipment. For example, if the position tolerance on the hole is .050, a drill press might be used to locate the hole. If the position tolerance is .010, a bridgeport might be used. If the position tolerance is .0005 a jig bore might be used. A position tolerance is not any more important than a plus/minus tolerance. It is simply a clearer method for communicating product requirements.

Application of MMC modifier concept.
A way to allow additional manufacturing tolerance is with the circle M or maximum material condition (MMC) modifier that follows the position tolerance. This modifier allows additional positional tolerance (up to .020 total) as the feature departs from maximum material condition or a .620 size hole. The product designer will allow this condition if the hole is used for clearance applications.

If the hole is to clear a mating pin, the hole may move off location within a .010 diameter if it is produced at its smallest size. The hole, however, may move more off location as it increases in size. Since the hole can increase in size by a total of .010 in diameter, the positional tolerance may increase in diameter by .010 providing a maximum of .020 diameter position tolerance. This MMC concept allows manufacturing the flexibility to adjust the processing on the hole size and location as necessary to achieve maximum process capability. The MMC concept is described in detail in the effect of modifiers section later in this text.

In the last three compartments of the position specification, the hole is related to the cartesian coordinate system through the specification of datums. The datums are specified in an order of precedence relating the imperfect produced part to the theoretical datum reference frame or cartesian coordinate system.

Profile of a surface is used to locate the surfaces on the part.
On the upper illustration of the geometrically toleranced angle block, a feature control frame with a profile characteristic entered in the first compartment is used to locate the surfaces between A and B. Basic dimensions radiating from the datum reference frame are used to define the theoretical or basic surface. The .020 total profile

GEOMETRIC TOLERANCING APPLIED TO AN ANGLE BLOCK

Geometric tolerancing is a very clear and concise three dimensional mathematical language for communicating product definition. In the example below, the angle block hole and surfaces are clearly defined with geometric tolerancing. The form and orientation tolerances (zones are not shown on "means this" figure) on the datum features control the stability of the imperfect part relative to the theoretically perfect datum reference frame.

THIS ON THE DRAWING

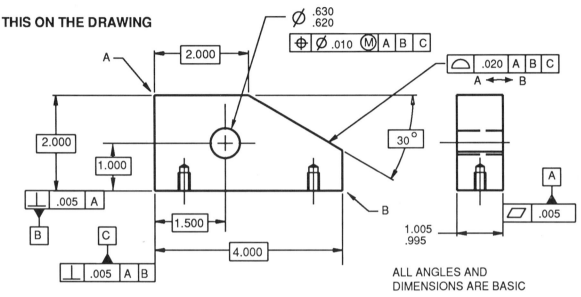

ALL ANGLES AND
DIMENSIONS ARE BASIC

**MEANS THIS ON
PRODUCED PART**

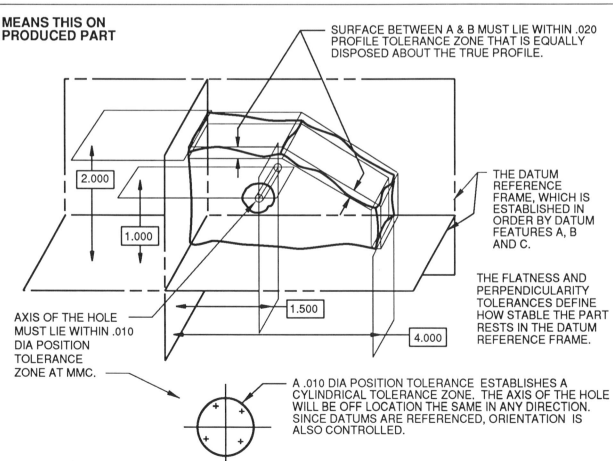

SURFACE BETWEEN A & B MUST LIE WITHIN .020 PROFILE TOLERANCE ZONE THAT IS EQUALLY DISPOSED ABOUT THE TRUE PROFILE.

THE DATUM REFERENCE FRAME, WHICH IS ESTABLISHED IN ORDER BY DATUM FEATURES A, B AND C.

THE FLATNESS AND PERPENDICULARITY TOLERANCES DEFINE HOW STABLE THE PART RESTS IN THE DATUM REFERENCE FRAME.

AXIS OF THE HOLE MUST LIE WITHIN .010 DIA POSITION TOLERANCE ZONE AT MMC.

A .010 DIA POSITION TOLERANCE ESTABLISHES A CYLINDRICAL TOLERANCE ZONE. THE AXIS OF THE HOLE WILL BE OFF LOCATION THE SAME IN ANY DIRECTION. SINCE DATUMS ARE REFERENCED, ORIENTATION IS ALSO CONTROLLED.

tolerance specified in the second compartment of the feature control frame defines a three dimensional uniform boundary (.010 either side) within which the actual surface must lie. The symbol A arrow B under the feature control frame specifies the area and distance to which the profile specification applies.

The indicated surface of the angle block must then lie within the specified profile boundary. Since the surface must fall within this 3D boundary, the location, orientation and form of the surface are also controlled.

The profile tolerance is not any more important than a plus/minus specification. Profile is just another method for communicating product specifications very clearly. The importance or "tightness" of the tolerance will determine the manufacturing and verification process. If surfaces are unimportant, a large profile tolerance is specified. See the profile section for more detailed information.

In the last three compartments of the profile specification, the surface is related to the cartesian coordinate system through the specification of datums. The datums are specified in an order of precedence relating the imperfectly produced part to the theoretical datum reference frame or cartesian coordinate system.

Form and orientation tolerances establish the stability of the imperfect part to the datum reference frame.
If the surfaces on the angle block that are identified as datum features have irregularities in form or orientation, the part will "rock" or be unstable when matched to the perfect reference frame. In the upper illustration, the unstablity of the part relative to the datum reference frame is controlled by the form and orientation tolerances applied to the datum surfaces.

On the geometrically toleranced angle block, notice the flatness (form tolerance) and two perpendicularity requirements (orientation tolerances) specified on the datums. These tolerances will control or limit the instability of the imperfect part relative to the theoretically perfect datum reference frame. These tolerances are selected by a designer to match functional requirements.

GEOMETRIC TOLERANCING APPLIED TO AN ANGLE BLOCK - 2D VIEW

Geometric tolerancing is a very clear and concise three dimensional mathematical language for communicating product definition. A fully geometrically toleranced product drawing is shown in the top view. In the bottom view, the produced part is shown in the datum reference frame established by datum features A, B and C. The surfaces must lie within the specified tolerance zones.

THIS ON THE DRAWING

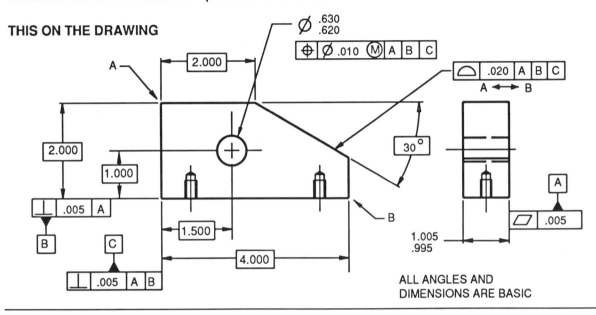

ALL ANGLES AND
DIMENSIONS ARE BASIC

MEANS THIS ON PRODUCED PART

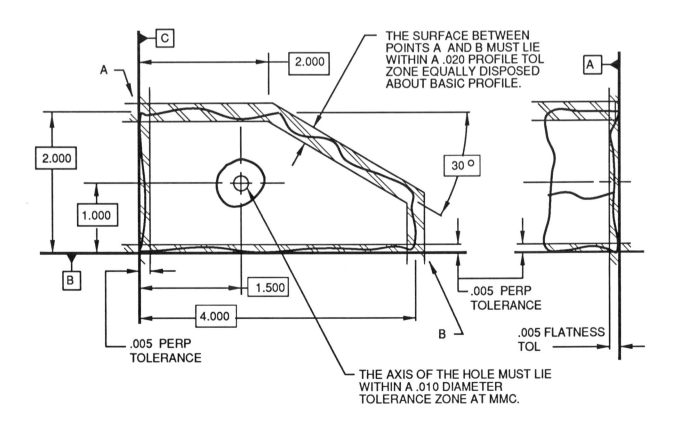

THE SURFACE BETWEEN POINTS A AND B MUST LIE WITHIN A .020 PROFILE TOL ZONE EQUALLY DISPOSED ABOUT BASIC PROFILE.

.005 PERP TOLERANCE

.005 FLATNESS TOL

.005 PERP TOLERANCE

THE AXIS OF THE HOLE MUST LIE WITHIN A .010 DIAMETER TOLERANCE ZONE AT MMC.

COMPARISON - GEOMETRIC TOLERANCING AND LIMIT TOLERANCING

The holes in the two parts below are clearance and are toleranced to fit over the .610 pin in the assembly on the following page. Notice that the location of the hole on the limit toleranced part can only have a maximum of plus/minus .0035 because of the problem with the square tolerance zone. The hole on the geometric toleranced part is allowed a .010 dia. zone and may have up to .020 dia because of the MMC modifier on the position call out. The geometric toleranced part has specified datums, rather than implied datums as used on the limit toleranced part. Geometric tolerancing provides more tolerance and clearly conveys design intent.

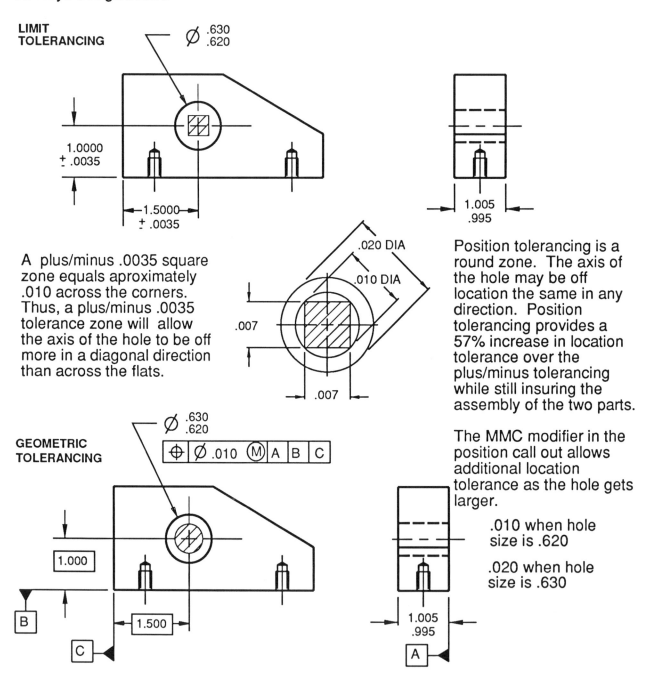

LIMIT TOLERANCING

Ø .630 / .620

1.0000 ± .0035

1.5000 ± .0035

1.005 / .995

A plus/minus .0035 square zone equals aproximately .010 across the corners. Thus, a plus/minus .0035 tolerance zone will allow the axis of the hole to be off more in a diagonal direction than across the flats.

.020 DIA

.010 DIA

.007

.007

Position tolerancing is a round zone. The axis of the hole may be off location the same in any direction. Position tolerancing provides a 57% increase in location tolerance over the plus/minus tolerancing while still insuring the assembly of the two parts.

The MMC modifier in the position call out allows additional location tolerance as the hole gets larger.

.010 when hole size is .620

.020 when hole size is .630

GEOMETRIC TOLERANCING

Ø .630 / .620

⊕ Ø .010 Ⓜ A B C

1.000

1.500

B

C

1.005 / .995

A

Geometric tolerancing uses basic dimensions, shown enclosed in a box. Basic dimensions are theoretically exact and have no tolerance. The position call out, directed at the hole, specifies a tolerance zone about the basic location within which the axis of the hole must lie.

ANGLE BLOCK ASSEMBLY

The angle block mounts in the base assembly shown below. The object of this discussion is to show how the part mounts in the assembly and compare the limit type tolerancing with geometric tolerancing. There are two drawings of the angle block in this unit. One is shown with limit tolerancing, and the other is shown with geometric tolerancing. If you study both drawings you will find that the limit toleranced part, with its many interpretations, may not interchange in the assembly.

The angle block mounts in the assembly, on the back face, bottom surface and left hand surface. The .620 - .630 diameter hole in the angle block is supposed to clear the .610 diameter pin shown in the base. In order for the two parts to interchange, the maximum the .620 dia hole in the angle block can be off location is .010 total.

The hole on the limit toleranced part must be toleranced at a maximum of plus/minus .0035 to insure that, across the corners of the square zone, the hole may only move a maximum of .010.

The hole on the geometric toleranced part is toleranced with a .010 dia zone. The MMC modifier in the position call out allows the axis of the hole up to a maximum of .020 position tolerance if the size of the hole is produced at its largest size of .630.

To carry the example further, on the geometric toleranced part, datums are used to define the exact order of how the part mounts in the assembly. Form and orientation tolerances are used to control the stability of the imperfect part in the theoretically perfect datum reference frame. Profile tolerancing is used to clearly define the outside part contour in 3D with a large profile tolerance. Geometric tolerancing replaces limit tolerancing for the location of features. (both holes and surfaces) Geometric tolerancing is a clear and concise method for defining design intent.

MATING PART BASE ASSEMBLY FOR ANGLE BLOCK

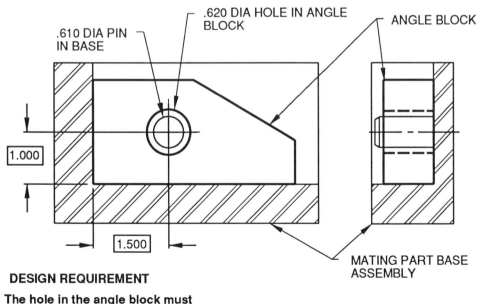

DESIGN REQUIREMENT

**The hole in the angle block must
clear the pin in the base**

The tolerances on the mating part base are ignored in this example and are shown theoretically exact to keep the problem simple. Later in the text, more complex examples will illustrate tolerancing applied to both parts in an assembly.

EFFECT OF MAXIMUM MATERIAL CONDITION - (MMC)

A geometric tolerance (size features only) may be applied on an MMC basis by placing the (M) symbol in the feature control frame following the feature tolerance. This will have a substantial effect on the allowable feature tolerance. The allowable feature tolerance is dependent on the actual mating size of the considered feature.

The allowable geometric tolerance for the feature applies when the feature is produced at its maximum material condition (smallest hole or largest pin). If the considered feature's size departs from its maximum material condition, an increase in the allowable geometric tolerance is permitted equal to the amount of the feature's departure from MMC.

Consider the part shown below. The (M) modifier in the feature control frame states that the features must be positioned within a .005 diameter tolerance zone when the features are at their maximum material condition. The maximum material condition (MMC) for the holes is .260 dia. If the features depart from the .260 dia. size, they can have additional position tolerance equal to the amount of their departure from MMC.

How much the hole can be off location depends on the actual mating size of the hole. If the hole is produced .001 larger in size, then the position tolerance can increase by .001. If the hole is produced .002 larger in size, then the position tolerance can increase by .002. The maximum available position on the hole is .013 dia as the size of the hole can only depart from MMC to LMC by .008 dia.

The tolerance is always one for one as the hole size is a diameter and the position tolerance is also a diameter zone. The MMC principle applies to each hole independently. Each hole may have a different position tolerance depending on the actual size of each hole. The MMC concept is the opposite of the LMC concept.

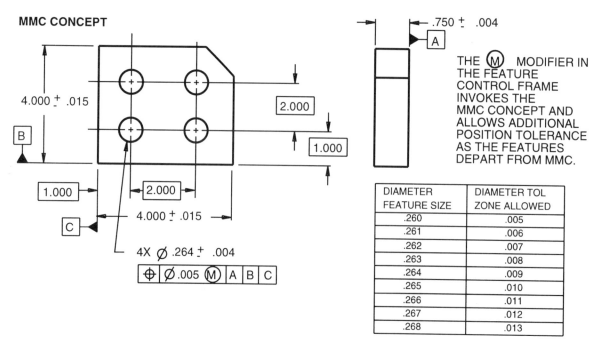

MMC CONCEPT

4.000 ± .015

B

2.000

1.000

1.000

2.000

4.000 ± .015

C

4X ⌀ .264 ± .004

⊕ | ⌀.005 Ⓜ | A | B | C

.750 ± .004

A

THE Ⓜ MODIFIER IN THE FEATURE CONTROL FRAME INVOKES THE MMC CONCEPT AND ALLOWS ADDITIONAL POSITION TOLERANCE AS THE FEATURES DEPART FROM MMC.

DIAMETER FEATURE SIZE	DIAMETER TOL ZONE ALLOWED
.260	.005
.261	.006
.262	.007
.263	.008
.264	.009
.265	.010
.266	.011
.267	.012
.268	.013

EFFECT OF LEAST MATERIAL CONDITION - (LMC)

A geometric tolerance (size features only) may be applied on an LMC basis by placing the Ⓛ symbol in the feature control frame following the feature tolerance. This will have a substantial effect on the allowable feature tolerance. The allowable feature tolerance is dependent on the actual mating size of the considered feature.

The allowable geometric tolerance for the feature applies when the feature is produced at its least material condition (smallest pin or largest hole). If the considered feature's size departs from its least material condition, an increase in the allowable geometric tolerance is permitted equal to the amount of the feature's departure from LMC.

Consider the part shown below. The Ⓛ modifier in the feature control frame states that the features must be positioned within a .005 diameter tolerance zone when the features are at their least material condition. The least material condition (LMC) for the holes is .268 dia. If the features depart from the .268 dia. size, they can have additional position tolerance equal to the amount of their departure from LMC.

How much the hole can be off location depends on the actual mating size of the hole. If the hole is produced .001 smaller in size, then the position tolerance can increase by .001. If the hole is produced .002 smaller in size, then the position tolerance can increase by .002. The maximum available position on the hole is .013 dia as the size of the hole can only depart from LMC to MMC by .008 dia.

The tolerance is always one for one, as the hole size is a diameter and the position tolerance is also a diameter zone.The LMC principle applies to each hole independently. Each hole may have a different position tolerance depending on the actual size of each hole.   The LMC concept is the opposite of the MMC concept.

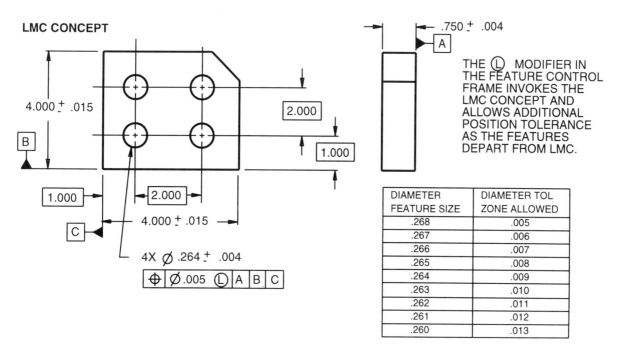

LMC CONCEPT

THE Ⓛ MODIFIER IN THE FEATURE CONTROL FRAME INVOKES THE LMC CONCEPT AND ALLOWS ADDITIONAL POSITION TOLERANCE AS THE FEATURES DEPART FROM LMC.

DIAMETER FEATURE SIZE	DIAMETER TOL ZONE ALLOWED
.268	.005
.267	.006
.266	.007
.265	.008
.264	.009
.263	.010
.262	.011
.261	.012
.260	.013

EFFECT OF REGARDLESS OF FEATURE SIZE - RFS

Geometric tolerances (size features only) are implied on an RFS basis by implication. The modifier rule #2 states that unless otherwise specified, all geometric tolerances are by default implied to apply RFS. Since all geometric tolerances apply at RFS, there is no need for a RFS symbol and it has been eliminated in the ASME Y14.5M, 1994 standard. Note: In the past edition of the Y14.5 standard, the RFS symbol (S) was specified for position tolerances. (See modifier rules for more information.)

If a geometric tolerance is by implication applied RFS, the specified allowable geometric tolerance is independent of the actual size of the considered feature. The allowable geometric tolerance is limited to the specified value regardless of the actual size of the feature.

Consider the part shown below. Since no modifier is specified in the feature control frame following the feature tolerance, this tolerance is implied to apply RFS. The features must be positioned within a .005 dia regardless of their feature size. This means that regardless of the actual mating size of the features being positioned, they have a .005 dia position tolerance and no more. If the holes get large or small, the position tolerance remains .005 dia. The RFS condition is more restrictive than the MMC or LMC concept.

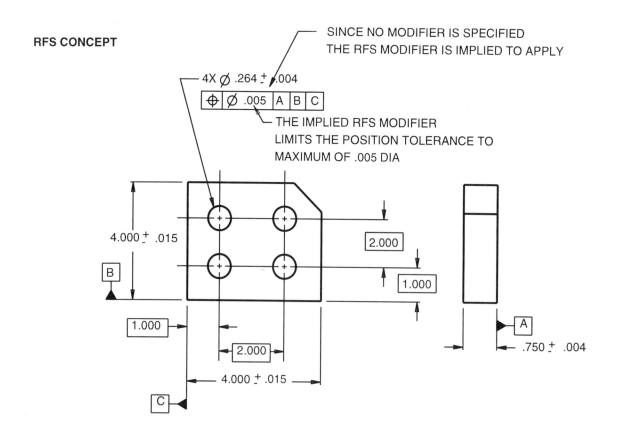

2.14

EFFECT OF MODIFIERS WITH ZERO TOLERANCE

There are many cases where features of size may be controlled with a zero geometric tolerance. If a feature is specified with a zero geometric tolerance, it must be modified on an MMC or LMC basis. The RFS concept is not applicable with zero geometric tolerance. The amount of allowable geometric tolerance is totally dependent on the actual mating size of the feature.

The example below illustrates holes positioned with zero tolerance at MMC. The allowable position tolerance for the holes is dependent on the actual mating size of the holes. The (M) modifier in the feature control frame states that the features must be positioned within a .000 diameter tolerance zone when the features are at their maximum material condition. The (MMC) for the holes is .255 dia. If the features depart from the .255 dia. size, they can have additional position tolerance equal to the amount of their departure from MMC. See chart below.

ZERO TOLERANCING AT MMC

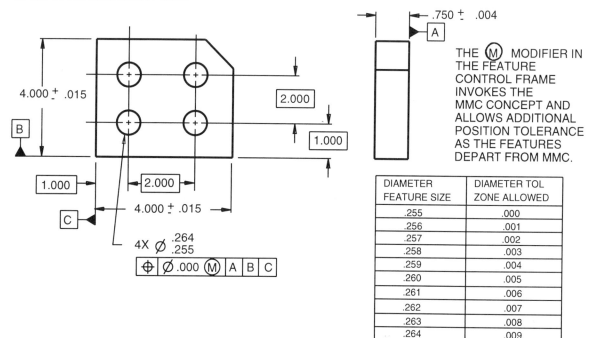

THE (M) MODIFIER IN THE FEATURE CONTROL FRAME INVOKES THE MMC CONCEPT AND ALLOWS ADDITIONAL POSITION TOLERANCE AS THE FEATURES DEPART FROM MMC.

DIAMETER FEATURE SIZE	DIAMETER TOL ZONE ALLOWED
.255	.000
.256	.001
.257	.002
.258	.003
.259	.004
.260	.005
.261	.006
.262	.007
.263	.008
.264	.009

APPLICABILITY OF MATERIAL CONDITION MODIFIERS

In using geometric tolerancing, the application or implication of material condition modifiers within the feature control frame can have a substantial effect on the stated tolerance or datum reference. The material condition modifiers are maximum material condition (MMC) Ⓜ, least material condition (LMC) Ⓛ and regardless of feature of size (RFS).

The application of these modifier symbols within the feature control frame is limited to features that have size and datums that have size. (holes, slots tabs, pins, etc.) If modifiers are applied to non-size features or non-size datum references, (surfaces) they have no meaning. If no modifier is specified in the feature control frame for features of size and datum references, the default condition is implied to be RFS. See chart below for applicability.

SYMBOL	TYPE OF TOLERANCE	APPLICABILITY OF FEATURE MODIFIER	APPLICABILITY OF DATUM MODIFIER
▱	FLATNESS	NO	NOT APPLICABLE
—	STRAIGHTNESS (LINE ELEMENT)	NO	
—	STRAIGHTNESS AXIS OR MEDIAN PLANE	YES	
○	CIRCULARITY	NO	
⌭	CYLINDRICITY	NO	
⊥	PERPENDICULARITY	YES IF FEATURES HAVE SIZE	YES IF DATUM REFERENCE HAS SIZE
∥	PARALLELISM		
∠	ANGULARITY		
⊕	POSITION	YES	
⌒	PROFILE OF A SURFACE	NO	
⌒	PROFILE OF A LINE	NO	
↗↗	TOTAL RUNOUT	NO	NOT APPLICABLE
↗	CIRCULAR RUNOUT	NO	
◎	CONCENTRICITY	NO	
≡	SYMMETRY	NO	

MODIFIER RULES - APPLICABILITY OF MMC, LMC & RFS

The applicability of the modifiers MMC, LMC and RFS in the feature control frame is only allowed for features of size and datums of size. These are features or datum features whose axis or median plane is controlled by or referenced to a geometric specification. In the current ASME Y14.5M-1994 standard there is a rule, often referred to as rule #2, that governs the applicability of modifiers in the feature control frame. This rule states:

Current ASME Y14.5M-1994 and ISO - Rule #2 (old rule #3 eliminated)

RFS applies for all geometric tolerances with respect to the individual tolerance, datum reference, or both, where no modifying symbol is specified. MMC and LMC must be specified where required.

Datum feature symbol
Y14.5M-1994 and ISO.

Since all geometric tolerances are implied RFS, there is no need for an RFS (S) symbol. The RFS symbol does not exist in the ISO standard and has been eliminated in the current ASME/ANSI Y14.5M-1994 standard.

The geometric controls of symmetry and concentricity and their specified datums always apply at RFS. The specified datums with runout controls also apply at RFS. Due to the nature of these controls, they can not be modified at MMC or LMC.

There is an alternative practice in the ASME Y14.5M-1994 standard, referred to as rule #2a. This alternative rule states:

(Rule #2a) For a tolerance of position, RFS (S) may be specified in the feature control frame with respect to the individual tolerance, datum reference, or both, as applicable.

This alternative rule #2a is not recommended as it is contrary to international practice. The modifier rule in the ISO standard is identical to rule 2 in the ASME Y14.5M-1994 standard, which states, that unless otherwise specified all geometric controls are RFS where applicable.

MODIFIER RULES - FORMER PRACTICE

The rules for the applicability of modifiers for position have undergone many changes in past editions of the Y14.5 standard. The rules for all other geometric tolerances, except position, have remained the same implied RFS where applicable.

The changes for modifiers on position have taken place to bring the United States closer in line with international practices. In the Y14.5-1973 standard, the U.S. standard implied MMC on position, and the ISO standard implied RFS on position. The interpretations were directly opposite. A decision was made to make the Y14.5 standard compatible with ISO. Since the difference between the two standards was so great, it was decided to make the change in stages.

The ANSI Y14.5M-1982 standard required modifiers to be specified on position. The current ASME Y14.5M-1994 standard specifies that position tolerance and all geometric tolrances are implied RFS. The ISO standard, just as the current U.S. standard, implies RFS to all geometric tolerances where no modifier is specified.

Former practice - ANSI Y14.5M-1982

   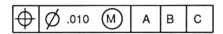

Rule #2 - MMC, LMC and RFS must be specified for individual tolerances and datum references for all position tolerances.

Rule #3 - RFS applies for individual tolerances and datums on all other geometric tolerances. MMC and LMC must be specifed where it is required.

$$\boxed{-A-}$$ Datum feature symbol
Y14.5M-1982 and earlier.

Former practice - ANSI Y14.5M-1973

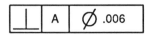

Rule #2 - MMC applies for individual tolerances and datum references for position. RFS and LMC must be specified where it is required.

Rule #3 - RFS applies for individual tolerances and datums on all other geometric tolerances. MMC and LMC must be specifed where it is required.

RULES FOR SCREW THREADS, GEARS AND SPLINES

Screw threads

Each tolerance of orientation, position or datum reference for a screw thread applies to the axis of the thread derived from the pitch cylinder. Where an exception to this practice is necessary, the specific feature of the screw thread (such as MAJOR DIA or MINOR DIA) shall be stated under the feature control frame or adjacent to the datum feature symbol, as applicable.

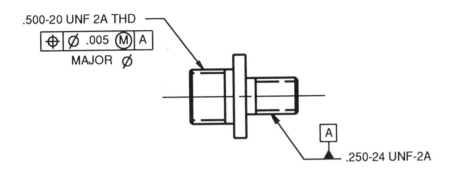

Unless otherwise specified, on screw threads, all geometric tolerances and datum references apply to the pitch diameter.

Gears and splines

Each tolerance of orientation, position or datum reference specified for features other than screw threads, such as gears and splines, must designate the specific feature of the gear or spline to which each applies (such as MAJOR DIA, PITCH DIA, PD, OR MINOR DIA). This information is stated under the feature control frame or under the datum feature symbol, as applicable.

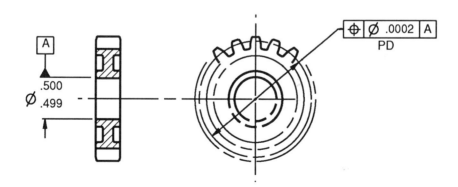

All geometric tolerances and datum references specified for gears and splines must designate the specific feature to which it applies.

FEATURE MODIFIERS - WHEN TO USE MMC, LMC AND RFS.

The hole in the rectangular plate has been located with a position tolerance relative to datums A, B and C. There is a position callout for the hole. Since the hole is a feature of size, it requires the application of a feature modifier MMC, LMC or RFS. Modifier rule #2 states that, if no modifier is specified, the default condition is automatically RFS. If the designer wanted the MMC or LMC condition to apply, it must be specified following the feature tolerance in the feature control frame.

In the past, it seemed that everything was positioned at MMC, because it was easy to make a functional gage. The functional gage answer allowed us to get by without much thought on the subject. This theory does not make sense any longer. The designer should define the function of the part and think the application through. Parts should not be just blindly applied with MMC. Besides, with the proliferation of coordinate measuring machines, functional gaging is not as popular as it was in the past. The designer should not design product for a particular inspection method. We should design for function.

The question is: Under what conditions would the designer select the MMC, LMC or RFS modifier? As we might imagine, the selection of the feature modifiers are determined by the function of the part. On the following page three identical rectangular plates are shown mounted in the bottom left corner with a mating part. (The mating part will be considered perfect for these examples.) The datum selection for the three parts are identical. The selection of the datums is unimportant in this example as the concept being illustrated is the application of feature modifiers. The important thing to look at is the function of the hole being used in the three examples.

Figure A - The hole in the rectangular plate is designed to clear the fixed pin in the mating part. It is important to note that the position tolerance of the hole plus the size tolerance of the clearance hole will have an effect on the clearance. Clearance applications are usually MMC.

Figure B - The hole in the rectangular plate is designed to locate the pilot pin. It is not a press fit. It is a loose fit. It is important to note that the position tolerance of the hole in the rectangular plate plus the size tolerance on the hole will have an effect on the location of the pilot pin.

Figure C - The hole in the rectangular plate is designed to accept the interference fit pin. The location of the hole in the rectangular plate will, in effect, locate the pin as the pin is pressed and centered in the hole. It is important to note that the position tolerance of the hole will have an effect on the location of the pressed pin, but the size tolerance of the hole will have no effect on the location of the pin. Press fit applications are usually RFS.

Review the illustration and examples on the following pages.

WHEN TO USE MMC, LMC AND RFS

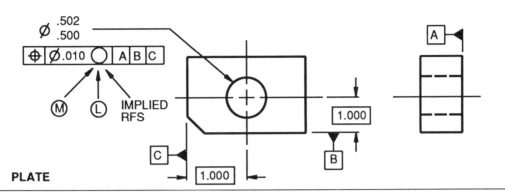

Ø .502
.500

⊕ Ø.010 Ⓜ A B C

Ⓜ Ⓛ IMPLIED RFS

PLATE

1.000

1.000

A

B

C

MMC AND LMC ARE USED WHEN THE SIZE OF THE FEATURE INTERACTS WITH LOCATION.

APPLICATION: CLEARANCE HOLE IN PLATE MUST CLEAR PIN IN BASE.

MMC IS USED
FOR CLEARANCE
APPLICATIONS

SYMBOL = Ⓜ

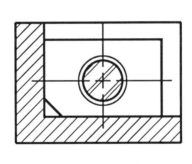
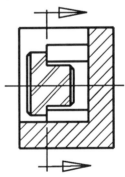

FIGURE A

APPLICATION: HOLE IN PLATE PROVIDES LOCATION FOR PILOT PIN.

LMC IS USED
FOR LOCATION
APPLICATIONS

SYMBOL = Ⓛ

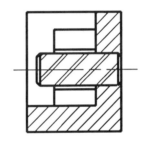

FIGURE B

RFS IS USED IF THE SIZE OF A FEATURE DOES NOT INTERACT WITH LOCATION.

APPLICATION: PIN IS PRESSED IN PLATE

RFS IS USED FOR PRESS
FITS

IMPLIED RFS IF NO
MODIFIER IS SPECIFIED

PAST PRACTICE SYMBOL

Ⓢ

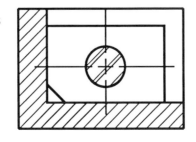
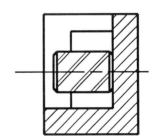

FIGURE C

MAXIMUM MATERIAL CONDITION APPLICATION

The maximum material condition (MMC) modifier is usually used for clearance type applications. The MMC modifier in the feature control frame states that as the feature departs from its MMC (MMC = smallest hole or largest pin) the amount of departure from MMC can be added to the position tolerance. In other words, the more the feature departs from MMC the more it can be off position.

MMC APPLICATION

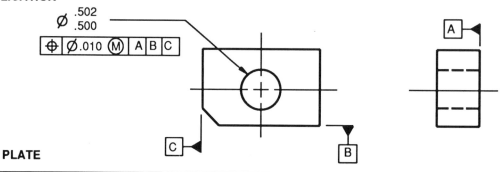

PLATE

MMC IS USED WHEN THE SIZE OF THE FEATURE INTERACTS WITH LOCATION.

APPLICATION: CLEARANCE HOLE IN PLATE MUST CLEAR PIN IN BASE

MMC IS USED
FOR CLEARANCE
APPLICATIONS

SYMBOL = Ⓜ

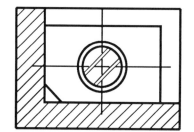

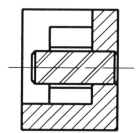

Consider the above example. This is a classic MMC application. The rectangular plate fits in the corner of the base part. There is a fixed pin in the base part that the hole in the rectangular plate must clear. All that is important in this example is that the hole will clear the pin. The hole does not locate the pin. The hole is designed to only clear the pin.

The position tolerance for the hole is applied at MMC. The MMC in the feature control frame for this example states that, if the size of the hole is produced at the smallest size of .500 dia, then there is .010 dia position available. The larger the hole is produced the more it can be off position - up to a maximum position of .012 dia.

Functionally, in this case, all we are trying to do is clear the pin. The hole cannot, of course, be produced too large as there is a tolerance on the size of the hole.

Another way to understand MMC is to imagine that there is a 3D solid inner boundary of .490 dia on the basic location that the hole must clear. There can never be any material in this area. The 3D solid is the MMC virtual condition and represents the worst condition of the mating part which, in this case, is the fixed pin that the hole is trying to clear.

TYPICAL MMC APPLICATION - BOLTED OR SCREWED ASSEMBLIES

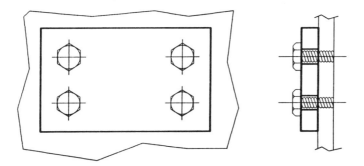

The simple example above illustrates a typical maximum material application. The holes in the top flat plate must mate and provide clearance with the threaded fasteners below. The clearance holes in the flat plate should be designed in such a way that the combination of the hole size and position will never cover up any portion of the tapped holes. Otherwise, the screws could not be inserted in the holes.

The fixed fastener formula is used to calculate the position on the holes and the MMC modifier would be applied to the tolerance. This allows additional position as the holes depart from MMC.

TYPICAL MMC APPLICATION - RIVETED ASSEMBLIES

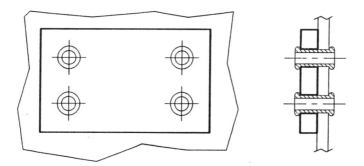

The example above is a riveted assembly. The holes in the top flat plate must clear the rivets. The clearance holes in the top plate should be designed in such a way that the combination of the hole size and location will always provide clearance so the rivets may be assembled with the mating part.

The floating fastener formula is used to calculate the position on the holes and the MMC modifier is applied to the tolerance. This allows additional position as the holes depart from MMC. For additional information on calculating positional tolerance for clearance holes, see the fixed and floating fastener formulas later in the text.

REGARDLESS OF FEATURE SIZE APPLICATION

The RFS modifier is usually used for press fits or centering applications. RFS is implied and no modifier is required. The feature control frame states that, regardless of the feature size, the hole must be positioned within the stated position tolerance. There is no additional positional tolerance allowed as the feature departs from MMC or LMC. In other words, the size of the feature has nothing to do with the location of the feature. The feature is located regardless of its size.

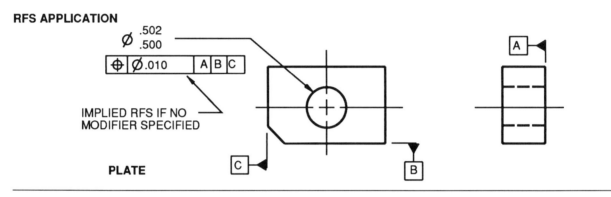

RFS IS USED IF THE SIZE OF A FEATURE DOES NOT INTERACT WITH LOCATION.

APPLICATION: PIN IS PRESSED IN PLATE

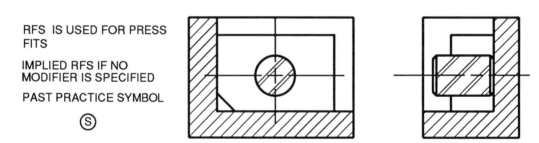

Consider the figure above. This is a classic RFS application. The rectangular plate fits in the corner of the mating part. There is a pin that is pressed in the hole in the rectangular plate. The size of the hole in the rectangular plate is designed to accept the press fit pin and to provide an interference fit. The position tolerance on the hole is designed to locate the hole which in turn will locate the pin.

The position tolerance for the hole in this case is applied at RFS. The implied RFS in the feature control frame for this example states that, regardless of the size of the hole, the hole must be located within the stated position tolerance.

Functionally this makes sense. If the hole were produced larger or smaller, the pin would just have a heavier or lighter press fit . The fit of the pin in the hole has nothing to do with the location of the hole. The press condition on the hole is one design consideration, and the location of the hole is another consideration. The design should not allow additional position as the hole gets larger or smaller.

LEAST MATERIAL CONDITION APPLICATION

The least material condition (LMC) modifier is usually used for location type applications that are not press fits but rather are loose fits. The LMC modifier in the feature control frame states that as the feature departs from its LMC (LMC = largest hole or smallest pin) the amount of departure from LMC can be added to the position tolerance. In other words, the more the feature departs from LMC the more it can be off position.

LMC APPLICATION

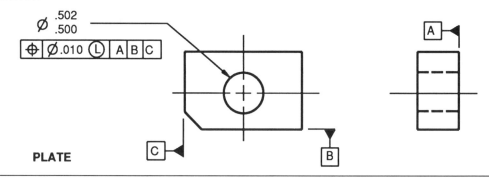

LMC IS USED WHEN THE SIZE OF THE FEATURE INTERACTS WITH LOCATION.

THE HOLE IN THE PLATE LOCATES THE PILOT PIN

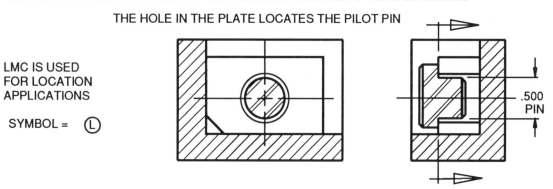

LMC IS USED
FOR LOCATION
APPLICATIONS

SYMBOL = Ⓛ

.500
PIN

Consider the figure above. This is a classic LMC application. The rectangular plate fits in the corner of the base. There is a pilot on the pin that is installed in the hole in the rectangular plate. The hole in the rectangular plate is designed to locate the pin. The size of the hole was designed to provide a loose fit with the mating pin, possibly for ease of assembly.

The position tolerance for the hole is applied at LMC. The LMC in the feature control frame states that if the size of the hole is produced at the largest size of .502 dia, then there is .010 dia position available. The smaller the hole is produced the more it can be off position up to a maximum position of .012 dia.

Functionally, this makes sense. If the designer is trying to locate the pin, there are two things that must be considered - the location of the hole and the size of the hole. The size of the hole will affect location because, if the size of the hole is produced at its largest size (LMC), the mating pin will rattle around in the hole more than if the hole is produced at its smallest size. In effect, the size of the hole will have an impact on the location of the hole.

Since the purpose of the hole in this plate is to locate the pin, the hole is positioned at LMC. The LMC states that, if the hole is produced at its largest size (LMC) of .502 dia, the only position tolerance available is the .010 dia stated in the feature control frame. The designer would reason that if the hole is produced at its largest size, then the size of the hole must be considered as location because the pin may rattle around in the large hole. If the hole is produced at LMC, the actual position of the pilot pin is the sum of the position tolerance plus the rattle of the pin in the hole.

Conversely, the LMC also states that, if the hole departs from its largest size (LMC) of .502 dia, the amount of departure can be added to the position tolerance. The designer would reason that if the hole departs from its largest size and is produced at its smallest size then the rattle of the pin in the hole will not have an affect on the location of the pin. Therefore, as the size of the hole gets smaller (departs from LMC), it may be off position equal to the amount of the hole's departure from LMC. If the hole is produced at the MMC, the actual position of the pin would include only the position tolerance and not the size of the hole because the pin will not rattle in the hole.

Another way to understand LMC, is to think of it as the opposite concept of MMC. There is a 3D solid outer boundary of .512 dia on the basic location that must always contain material. The 3D solid is the LMC virtual condition and represents the worst condition of the mating part which, in this case, is farthest the loose mating pin can ever be off the basic location.

TYPICAL LMC APPLICATION - SELF PILOTING WELD NUTS

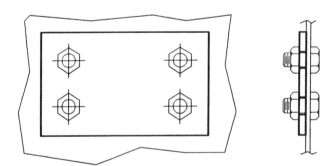

The simple example above illustrates a typical least material application. Weld nuts in the top flat plate must mate to provide assembly with the clearance holes below. The holes in the flat plate locate the pilot on the weld nut. The pilot on the weld nut is inserted in the holes and welded in place. The hole in the top plate will "fixture" or locate the weld nut in place.

In determining the final location of the weld nut, the designer must consider the size of the hole that accepts the pilot of the weld nut as well as the location of the hole. The size of the hole is important because, as the hole increases in size, the pilot on the weld nut will rattle around more in the hole and, in effect, increase the location.

The fixed fastener formula is used to calculate the position on the holes. The designer must also subtract from the formula the extra rattle or clearance of the weld nut pilot and the hole in the top flat plate. The tolerance on the holes in the top plate are modified at LMC. This allows additional position tolerance as the holes depart from LMC.

The designer would reason that, if the holes that located the pilot on the weld nut were smaller, then the weld nut would not rattle in the hole. If the weld nut rattle tolerance was decreased, then the position could be increased proportionately. Note: The holes in the bottom part are clearance and are modified at MMC.

TYPICAL LMC APPLICATION - SLIP FIT DRILL BUSHINGS

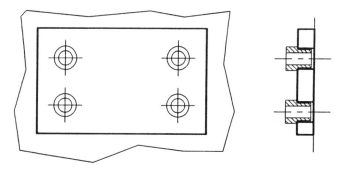

The example above illustrates slip fit drill bushings that are installed in a drill fixture plate. In this example the slip fit holes in the drill fixture plate are located at LMC. The holes in the drill plate will locate the drill bushings. The size of the holes in the drill fixture plate, as well as the position tolerance, will have an impact on the location of the drill bushings.

WORST CASE CALCULATIONS FOR LOCATION OF FEATURES

To ensure that the modifier concept is clear, let us do some simple worst case calculations comparing the MMC, LMC and RFS modifiers for locating loose fit mating features. The object will be to determine which modifier will locate the axis of the pilot pin as close to basic as possible while still providing maximum position tolerance. It is important to note that the diameter of the pilot pin is included in the following calculations.

To keep things simple, the size of the pilot pin is a basic .500 diameter. The tolerance of the pilot pin will not be considered. (In a "real life stack-up" the size tolerance of the pilot pin is also considered.)

THE LMC MODIFIER PROVIDES MAXIMUM POSITION TOLERANCE WHILE MINIMIZING ALLOWABLE STACK-UP OF THE AXIS OF THE PILOT PIN FROM THE BASIC LOCATION.

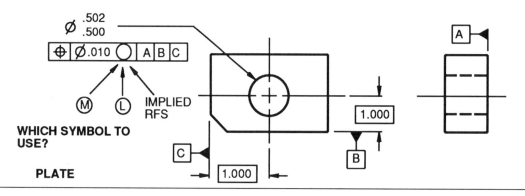

APPLICATION: HOLE IN PLATE PROVIDES LOCATION FOR PILOT PIN. A POSSIBLE REASON IS THAT THE HEAD OF THE PIN MAY NOT TOUCH THE SIDE WALLS OF THE BASE.

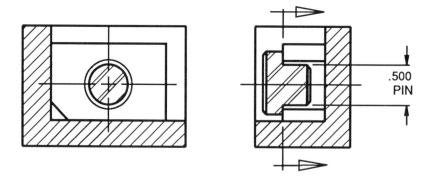

As you will see from the following calculations, the choice for the location of loose fit features is certainly the LMC modifier. In the following examples, LMC allows .012 position tolerance. A tolerance stack-up evaluation finds the axis of the pilot pin can be .012 dia off basic location. MMC allows .012 position tolerance. A tolerance stack-up evaluation finds the axis of the pilot pin can be .014 dia off basic location. RFS allows .010 position tolerance. A tolerance stack-up evaluation finds the axis of the pilot pin can be .012 dia off basic location. LMC allows as much position tolerance as MMC while locating the pilot pin as close as that of an RFS application.

If MMC is applied to the feature, it will allow a .012 dia max position tolerance for manufacturing. The resulting worst case stack-up of tolerances will allow the axis of the pilot pin to drift off the basic location by .014 dia. (Including the rattle of the pilot pin in the hole.)

Assume hole is produced at .502 dia and off location .012 dia.
Maximum position tolerance at MMC   .010 dia
Additional position tolerance as feature departs from MMC (bonus)   .002 dia
Rattle of a .500 pilot pin in a .502 hole   .002 dia
Total amount pilot pin axis can move off basic location   .014 dia

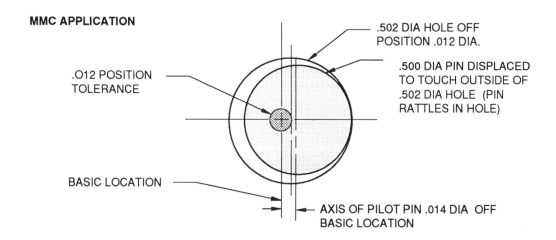

MMC APPLICATION

.012 POSITION TOLERANCE

BASIC LOCATION

.502 DIA HOLE OFF POSITION .012 DIA.

.500 DIA PIN DISPLACED TO TOUCH OUTSIDE OF .502 DIA HOLE (PIN RATTLES IN HOLE)

AXIS OF PILOT PIN .014 DIA OFF BASIC LOCATION

If RFS is applied to the feature, it will allow a .010 dia max position tolerance for manufacturing. The resulting stack-up of tolerances will allow the axis of the pilot pin to drift off the basic location by .012 dia. (Including the rattle of the pilot pin in the hole.)

Assume hole is produced at .502 dia and off location .010 dia.
Maximum position tolerance RFS   .010 dia
Additional position tolerance (bonus)   N/A
Rattle of a .500 pilot pin in a .502 hole   .002 dia
Total amount pilot pin axis can move off basic location   .012 dia

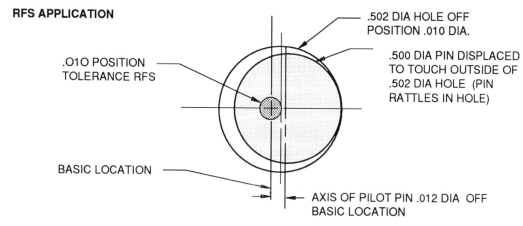

RFS APPLICATION

.010 POSITION TOLERANCE RFS

BASIC LOCATION

.502 DIA HOLE OFF POSITION .010 DIA.

.500 DIA PIN DISPLACED TO TOUCH OUTSIDE OF .502 DIA HOLE (PIN RATTLES IN HOLE)

AXIS OF PILOT PIN .012 DIA OFF BASIC LOCATION

If LMC is applied to the feature, it will allow a .012 dia max position tolerance for manufacturing. The resulting stack-up of tolerances will allow the axis of the pilot pin to drift off the basic location by .012 dia. (Including the rattle of the pilot pin in the hole.)

Case 1 - Assume .502 dia produced hole and off location .010 dia.

Maximum position tolerance at LMC	.010 dia
Additional position tolerance as feature departs from LMC (bonus)	N/A
Rattle of a .500 pilot pin in a .502 hole	.002 dia
Total amount pilot pin axis can move off basic location	**.012 dia**

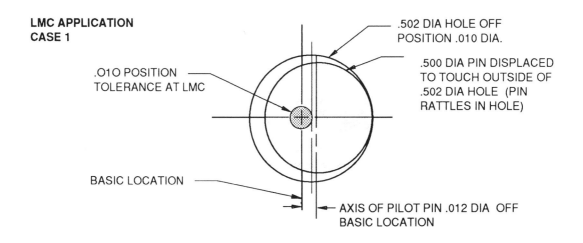

Case 2 - Assume .500 dia produced hole and off location .012 dia.

Case 2 - Assume .500 dia produced hole and off location .012 dia.

Maximum position tolerance at LMC	.010 dia
Additional position tolerance as feature departs from LMC	.002 dia
Rattle of a .500 pilot pin in a .500 hole	none
Total amount pilot pin axis can move off center	**.012 dia**

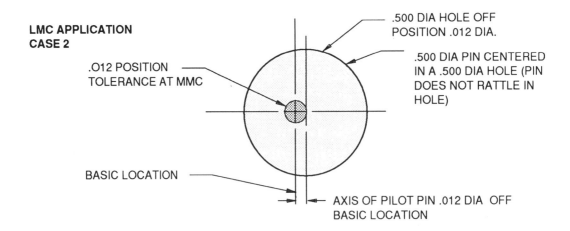

WORKSHOP EXERCISES 2.1

This exercise is designed to give you practical experience in correctly applying and interpreting geometric tolerancing symbology. If you have trouble applying the symbols per the following directions, page through the workbook looking for similar examples. The geometric tolerancing applied to an angle block example earlier in this unit is a good reference. Write clearly and neatly. The following directions apply to the angle plate drawing below.

1. Make the right hand face in the side view flat within .005. Also identify this surface as datum feature A.

2. Make the top surface in the plan view perpendicular within .005 to datum A. Also identify this surface as datum feature B.

3. Make the left hand surface in the plan view perpendicular within .005 to datum A (primary) and B (secondary). Also identify this surface as datum feature C.

4. Make all the dimensions basic, except of course, the size tolerances.
   Boxes or note saying all are basic.

5. Position the three holes within a diameter zone of .010 at MMC to datums A (primary) B (secondary) and C (tertiary).

6. Position the .250 diameter hole within a diameter zone of .008 RFS in relation to datum A (primary), datum B (secondary) and datum C (tertiary).

7. In the plan view, identify the far right hand corner on the top surface as point A. Also identify the lower corner on the left hand surface as point B. On the contoured surface between A and B, apply a profile tolerance of .020 total referenced to datum A (primary), B (secondary) and C (tertiary). Under the profile feature control frame state that the tolerance applies between A and B.

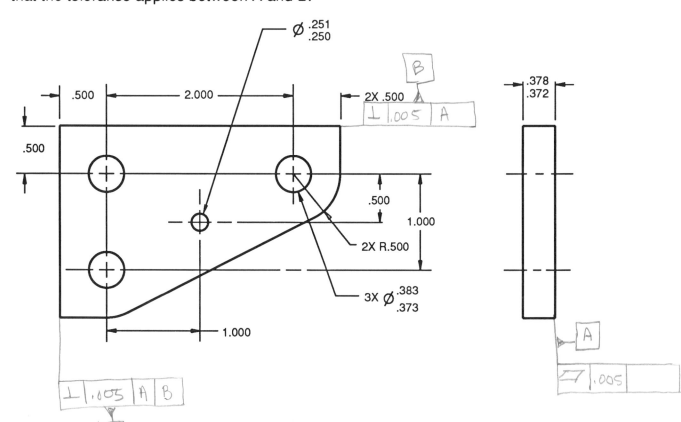

2.31

8. The produced part (angle plate) is shown below loaded in the datum reference frame. Shade in and label the geometric tolerance zones that were applied in the previous problem. Illustrate the tolerance zones very neatly and clearly. If you need help look at other examples in the workbook.

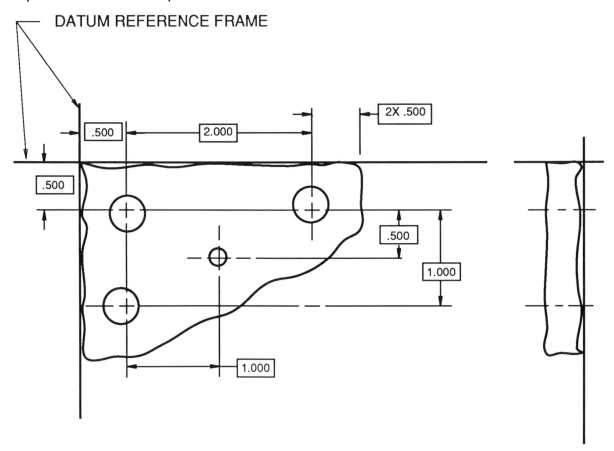

9. In the two feature control frames below there is no feature modifier present. In the current ASME Y14.5M, 1994 standard is the tolerance for the features implied to be MMC, LMC or RFS? What is the rule number? 2

| ⊕ | ⌀.030 | A | B Ⓜ |

| ⊥ | ⌀.005 | A |

10. In interpreting older drawings, the former ANSI Y14.5M, 1982 standard rules for the applicability of modifiers was different. What was the rule for position? What was implied for all other geometric characteristics?

MMC, LMC, RFS called out

all other implied RFS

11. On the drawing below, a position tolerance has been applied to the four holes in the plate. The MMC modifier has been placed in the feature control frame. The MMC modifier allows additional positional tolerance as the features depart from MMC. In the chart below, fill in the missing blanks to show the additional position tolerance as the features depart from MMC.

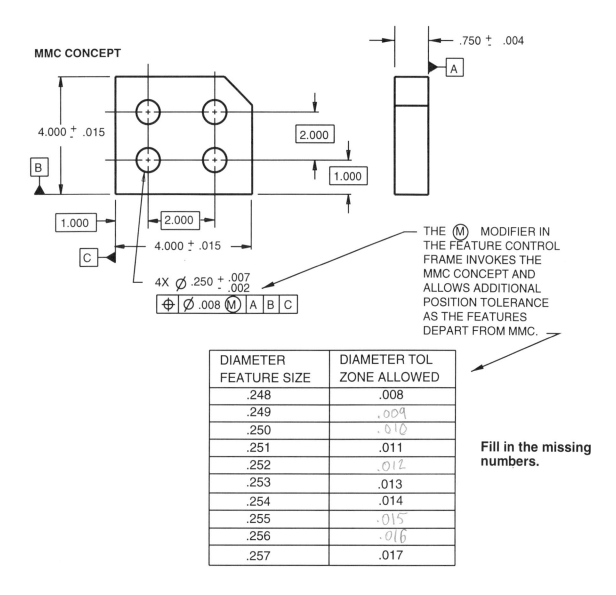

MMC CONCEPT

THE (M) MODIFIER IN THE FEATURE CONTROL FRAME INVOKES THE MMC CONCEPT AND ALLOWS ADDITIONAL POSITION TOLERANCE AS THE FEATURES DEPART FROM MMC.

DIAMETER FEATURE SIZE	DIAMETER TOL ZONE ALLOWED
.248	.008
.249	.009
.250	.010
.251	.011
.252	.012
.253	.013
.254	.014
.255	.015
.256	.016
.257	.017

Fill in the missing numbers.

UNIT 3

POSITION TOLERANCE VERIFICATION

POSITION TOLERANCING VERIFICATION

Geometric tolerancing is a very clear and concise three dimensional mathematical language for communicating product definition. This unit will cover some simple verification techniques for position tolerancing. The exercises are designed to provide practice with the MMC concept and position tolerancing. It will also cover the collection and reporting of data. Simple paper gaging techniques are also included.

THIS ON THE DRAWING

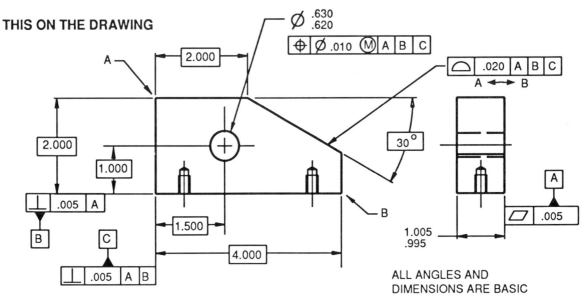

ALL ANGLES AND
DIMENSIONS ARE BASIC

**MEANS THIS ON
PRODUCED PART**

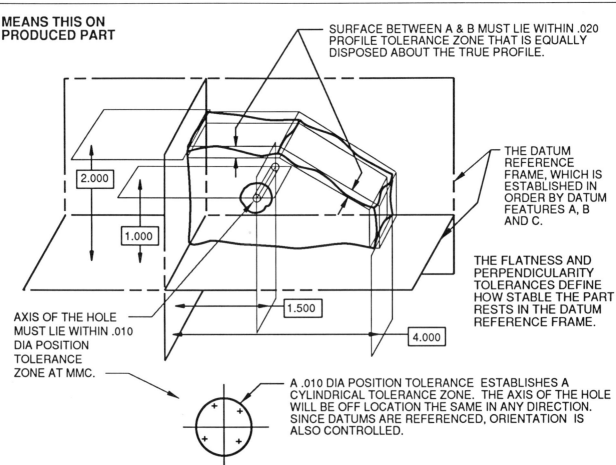

SURFACE BETWEEN A & B MUST LIE WITHIN .020
PROFILE TOLERANCE ZONE THAT IS EQUALLY
DISPOSED ABOUT THE TRUE PROFILE.

THE DATUM
REFERENCE
FRAME, WHICH IS
ESTABLISHED IN
ORDER BY DATUM
FEATURES A, B
AND C.

THE FLATNESS AND
PERPENDICULARITY
TOLERANCES DEFINE
HOW STABLE THE PART
RESTS IN THE DATUM
REFERENCE FRAME.

AXIS OF THE HOLE
MUST LIE WITHIN .010
DIA POSITION
TOLERANCE
ZONE AT MMC.

A .010 DIA POSITION TOLERANCE ESTABLISHES A
CYLINDRICAL TOLERANCE ZONE. THE AXIS OF THE HOLE
WILL BE OFF LOCATION THE SAME IN ANY DIRECTION.
SINCE DATUMS ARE REFERENCED, ORIENTATION IS
ALSO CONTROLLED.

POSITION - HOLE VERIFICATION AT MMC

The example below illustrates how to verify a position tolerance at MMC.

AS DRAWN

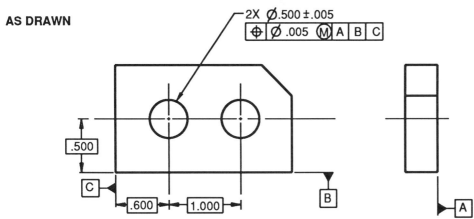

AS PRODUCED

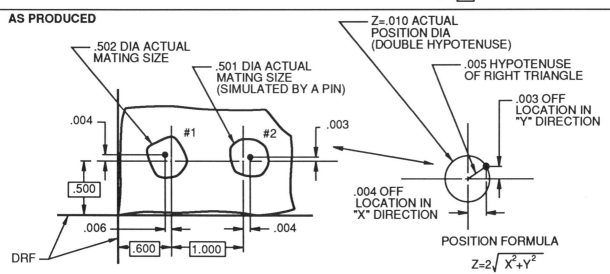

POSITION FORMULA

$$Z = 2\sqrt{X^2 + Y^2}$$

Position tolerance actual is calculated by using "X" and "Y" deviations in the position formula or position chart. In order to pass acceptance, position tolerance actual must be less than position tolerance allowed.

Allowed position tolerance is calculated by taking the difference between the hole MMC and hole actual size and adding it to the position tolerance stated in the feature control frame.

Enter MMC of hole

Enter produced hole size

Enter "X" deviation from basic location

Enter "Y" deviation from basic location

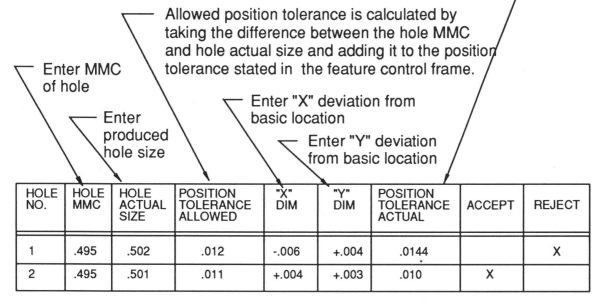

HOLE NO.	HOLE MMC	HOLE ACTUAL SIZE	POSITION TOLERANCE ALLOWED	"X" DIM	"Y" DIM	POSITION TOLERANCE ACTUAL	ACCEPT	REJECT
1	.495	.502	.012	-.006	+.004	.0144		X
2	.495	.501	.011	+.004	+.003	.010	X	

POSITION - HOLE VERIFICATION AT LMC

The example below illustrates how to verify a position tolerance at LMC.

AS DRAWN

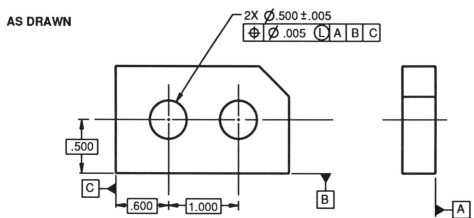

AS PRODUCED

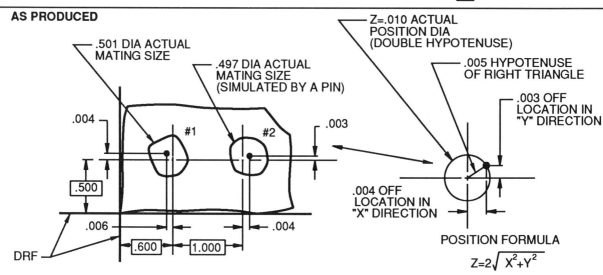

POSITION FORMULA

$$Z = 2\sqrt{X^2 + Y^2}$$

Position tolerance actual is calculated by using "X" and "Y" deviations in the position formula or position chart. In order to pass acceptance, position tolerance actual must be less than position tolerance allowed.

Allowed position tolerance is calculated by taking the difference between the hole LMC and hole actual size and adding it to the position tolerance stated in the feature control frame.

Enter LMC of hole

Enter produced hole size

Enter "X" deviation from basic location

Enter "Y" deviation from basic location

HOLE NO.	HOLE LMC	HOLE ACTUAL SIZE	POSITION TOLERANCE ALLOWED	"X" DIM	"Y" DIM	POSITION TOLERANCE ACTUAL	ACCEPT	REJECT
1	.505	.501	.009	-.006	+.004	.0144		X
2	.505	.497	.013	+.004	+.003	.010	X	

CLASS EXERCISE - HOLE VERIFICATION AT MMC

The top drawing is a part with position tolerance applied. The lower drawing is the produced part. Evaluate the dimensions on the produced part to verify conformance to the position tolerances. Use the chart below to record your calculations. Review the previous examples if you have any problems.

AS DRAWN

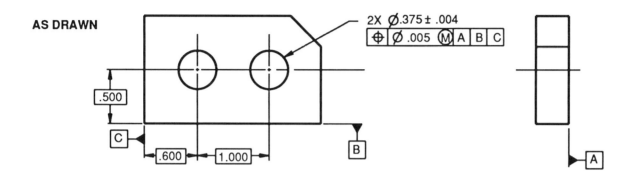

AS PRODUCED

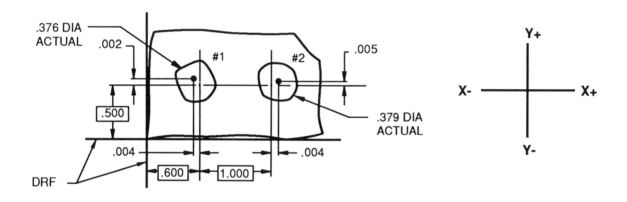

HOLE NO.	HOLE MMC	HOLE ACTUAL SIZE	POSITION TOLERANCE ALLOWED	"X" DIM	"Y" DIM	POSITION TOLERANCE ACTUAL	ACCEPT	REJECT
1	.371	.376	.010	-.004	.002	.0089	X	
2	.371	.379	.013	.004	+.005	.0128	X	

CONVERSION OF COORDINATE MEASUREMENT TO POSITION TOLERANCES - INCH

The chart below will convert coordinate inch measurements to position diameter zone. It is very simply , double the Pythagorean Theorem. The formula is shown below.

	.001	.002	.003	.004	.005	.006	.007	.008	.009	.010	.011
.020	.0400	.0402	.0404	.0408	.0412	.0418	.0424	.0431	.0439	.0447	.0456
.019	.0380	.0382	.0385	.0388	.0393	.0398	.0405	.0412	.0420	.0429	.0439
.018	.0360	.0362	.0365	.0369	.0374	.0379	.0386	.0394	.0403	.0412	.0422
.017	.0340	.0342	.0345	.0349	.0354	.0360	.0368	.0376	.0386	.0394	.0405
.016	.0321	.0322	.0325	.0330	.0335	.0342	.0349	.0358	.0367	.0377	.0388
.015	.0301	.0303	.0306	.0310	.0316	.0323	.0331	.0340	.0350	.0360	.0372
.014	.0281	.0283	.0286	.0291	.0297	.0305	.0313	.0322	.0333	.0344	.0356
.013	.0261	.0263	.0267	.0272	.0278	.0286	.0295	.0305	.0316	.0328	.0340
.012	.0241	.0243	.0247	.0253	.0260	.0268	.0278	.0288	.0300	.0312	.0325
.011	.0221	.0224	.0228	.0234	.0242	.0250	.0261	.0272	.0284	.0297	.0311
.010	.0201	.0204	.0209	.0215	.0224	.0233	.0244	.0256	.0269	.0283	.0297
.009	.0181	.0184	.0190	.0197	.0206	.0216	.0228	.0241	.0254	.0269	.0284
.008	.0161	.0165	.0171	.0179	.0189	.0200	.0213	.0226	.0241	.0256	.0272
.007	.0141	.0146	.0152	.0161	.0172	.0184	.0198	.0213	.0228	.0244	.0261
.006	.0122	.0126	.0134	.0144	.0156	.0170	.0184	.0200	.0216	.0233	.0250
.005	.0102	.0108	.0117	.0128	.0141	.0156	.0172	.0189	.0206	.0224	.0242
.004	.0082	.0089	.0100	.0113	.0128	.0144	.0161	.0179	.0197	.0215	.0234
.003	.0063	.0072	.0085	.0100	.0117	.0134	.0152	.0171	.0190	.0209	.0228
.002	.0045	.0056	.0072	.0089	.0108	.0126	.0146	.0165	.0184	.0204	.0224
.001	.0028	.0045	.0063	.0082	.0102	.0122	.0141	.0161	.0181	.0201	.0221

FORMULA: $Z = 2 \sqrt{X^2 + Y^2}$

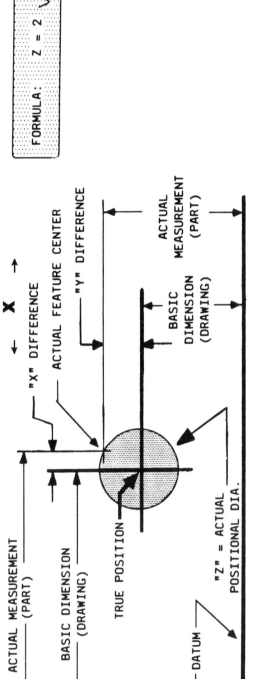

"X" DIFFERENCE

ACTUAL FEATURE CENTER

"Y" DIFFERENCE

ACTUAL MEASUREMENT (PART)

BASIC DIMENSION (DRAWING)

ACTUAL MEASUREMENT (PART)

BASIC DIMENSION (DRAWING)

TRUE POSITION

DATUM

"Z" = ACTUAL POSITIONAL DIA.

CONVERSION OF COORDINATE MEASUREMENT TO POSITION TOLERANCES - METRIC

The chart below will convert coordinate metric measurements to position diameter zone. It is very simply, double the Pythagorean Theorem. The formula is shown below.

POSITION ZONE (Z)

Y \ X	0.02	0.04	0.06	0.08	0.10	0.12	0.14	0.16	0.18	0.20	0.22	0.24	0.26	0.28	0.30	0.32	0.34	0.36	0.38	0.40	0.42	0.44	0.46	0.48	0.50
0.50	1.001	1.003	1.007	1.013	1.020	1.028	1.038	1.050	1.063	1.077	1.093	1.109	1.127	1.146	1.166	1.187	1.209	1.232	1.256	1.281	1.306	1.332	1.359	1.386	1.414
0.48	0.961	0.963	0.967	0.973	0.981	0.990	1.000	1.012	1.025	1.040	1.056	1.073	1.092	1.111	1.132	1.154	1.176	1.200	1.224	1.250	1.276	1.302	1.330	1.358	1.386
0.46	0.921	0.923	0.928	0.934	0.941	0.951	0.962	0.974	0.988	1.003	1.020	1.038	1.057	1.077	1.098	1.121	1.144	1.168	1.193	1.219	1.246	1.273	1.301	1.330	1.359
0.44	0.881	0.884	0.888	0.894	0.902	0.912	0.923	0.936	0.951	0.967	0.984	1.002	1.022	1.043	1.065	1.088	1.112	1.137	1.163	1.189	1.217	1.245	1.273	1.302	1.332
0.42	0.841	0.844	0.849	0.855	0.863	0.874	0.885	0.899	0.914	0.930	0.948	0.967	0.988	1.010	1.032	1.056	1.081	1.106	1.133	1.160	1.188	1.217	1.246	1.276	1.306
0.40	0.801	0.804	0.809	0.816	0.825	0.835	0.848	0.862	0.877	0.894	0.913	0.933	0.954	0.977	1.000	1.024	1.050	1.076	1.103	1.131	1.160	1.189	1.219	1.250	1.281
0.38	0.761	0.764	0.769	0.777	0.786	0.797	0.810	0.825	0.841	0.859	0.878	0.899	0.921	0.944	0.968	0.994	1.020	1.047	1.075	1.103	1.133	1.163	1.193	1.224	1.256
0.36	0.721	0.724	0.730	0.738	0.747	0.759	0.773	0.788	0.805	0.824	0.844	0.865	0.888	0.912	0.937	0.963	0.990	1.018	1.047	1.076	1.106	1.137	1.168	1.200	1.232
0.34	0.681	0.685	0.691	0.699	0.709	0.721	0.735	0.752	0.769	0.789	0.810	0.832	0.856	0.881	0.907	0.934	0.962	0.990	1.020	1.050	1.081	1.112	1.144	1.176	1.209
0.32	0.641	0.645	0.651	0.660	0.671	0.684	0.699	0.716	0.734	0.755	0.777	0.800	0.825	0.850	0.877	0.905	0.934	0.963	0.994	1.024	1.056	1.088	1.121	1.154	1.187
0.30	0.601	0.605	0.612	0.621	0.632	0.646	0.662	0.680	0.700	0.721	0.744	0.768	0.794	0.821	0.849	0.877	0.907	0.937	0.968	1.000	1.032	1.065	1.098	1.132	1.166
0.28	0.561	0.566	0.573	0.582	0.595	0.609	0.626	0.645	0.666	0.688	0.712	0.738	0.764	0.792	0.821	0.850	0.881	0.912	0.944	0.977	1.010	1.043	1.077	1.111	1.146
0.26	0.522	0.526	0.534	0.544	0.557	0.573	0.591	0.611	0.632	0.656	0.681	0.708	0.735	0.764	0.794	0.825	0.856	0.888	0.921	0.954	0.988	1.022	1.057	1.092	1.127
0.24	0.482	0.487	0.495	0.506	0.520	0.537	0.556	0.577	0.600	0.625	0.651	0.679	0.708	0.738	0.768	0.800	0.832	0.865	0.899	0.933	0.967	1.002	1.038	1.073	1.109
0.22	0.442	0.447	0.456	0.468	0.483	0.501	0.522	0.544	0.569	0.595	0.622	0.651	0.681	0.712	0.744	0.777	0.810	0.844	0.878	0.913	0.948	0.984	1.020	1.056	1.093
0.20	0.402	0.408	0.418	0.431	0.447	0.466	0.488	0.512	0.538	0.566	0.595	0.625	0.656	0.688	0.721	0.755	0.789	0.824	0.859	0.894	0.930	0.967	1.003	1.040	1.077
0.18	0.362	0.369	0.379	0.394	0.412	0.433	0.456	0.482	0.509	0.538	0.569	0.600	0.632	0.666	0.700	0.734	0.769	0.805	0.841	0.877	0.914	0.951	0.988	1.025	1.063
0.16	0.322	0.330	0.342	0.358	0.377	0.400	0.425	0.453	0.482	0.512	0.544	0.577	0.611	0.645	0.680	0.716	0.752	0.788	0.825	0.862	0.899	0.936	0.974	1.012	1.050
0.14	0.283	0.291	0.305	0.322	0.344	0.369	0.396	0.425	0.456	0.488	0.522	0.556	0.591	0.626	0.662	0.699	0.735	0.773	0.810	0.848	0.885	0.923	0.962	1.000	1.038
0.12	0.243	0.253	0.268	0.288	0.312	0.339	0.369	0.400	0.433	0.466	0.501	0.537	0.573	0.609	0.646	0.684	0.721	0.759	0.797	0.835	0.874	0.912	0.951	0.990	1.028
0.10	0.204	0.215	0.233	0.256	0.283	0.312	0.344	0.377	0.412	0.447	0.483	0.520	0.557	0.595	0.632	0.671	0.709	0.747	0.786	0.825	0.863	0.902	0.941	0.981	1.020
0.08	0.165	0.179	0.200	0.226	0.256	0.288	0.322	0.358	0.394	0.431	0.468	0.506	0.544	0.582	0.621	0.660	0.699	0.738	0.777	0.816	0.855	0.894	0.934	0.973	1.013
0.06	0.126	0.144	0.170	0.200	0.233	0.268	0.305	0.342	0.379	0.418	0.456	0.495	0.534	0.573	0.612	0.651	0.691	0.730	0.769	0.809	0.849	0.888	0.928	0.967	1.007
0.04	0.089	0.113	0.144	0.179	0.215	0.253	0.291	0.330	0.369	0.408	0.447	0.487	0.526	0.566	0.605	0.645	0.685	0.724	0.764	0.804	0.844	0.884	0.923	0.963	1.003
0.02	0.057	0.089	0.126	0.165	0.204	0.243	0.283	0.322	0.362	0.402	0.442	0.482	0.522	0.561	0.601	0.641	0.681	0.721	0.761	0.801	0.841	0.881	0.921	0.961	1.001

Y → ← Y →

X →

FORMULA:

$$Z = 2 \sqrt{X^2 + Y^2}$$

WORKSHOP EXERCISE 3.1 - PROBLEM #1

The top drawing is a plate with position tolerancing applied. The lower drawing is the produced part. Evaluate the dimensions on the produced part to verify conformance to the position tolerances. Use the chart below to record your calculations.

AS DRAWN

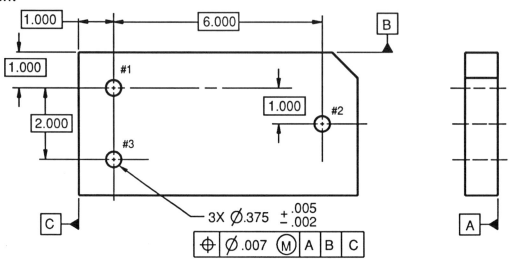

PRODUCED PART

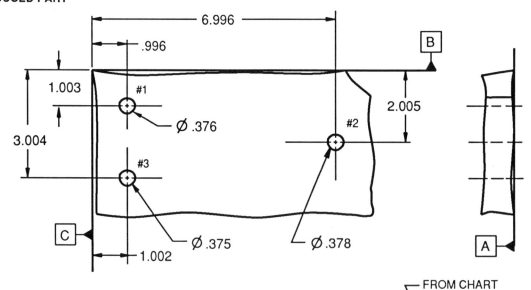

FROM CHART

HOLE NO.	HOLE MMC	HOLE ACTUAL SIZE	POSITION TOLERANCE ALLOWED	"X" DIM.	"Y" DIM.	POSITION TOLERANCE ACTUAL	ACC	REJ
1	.373	.376	.010	-.004	-.003	Ø.010	X	
2	.373	.378	.012	-.004	-.005	Ø.0128		X
3	.373	.375	.009	.002	-.004	Ø.0089	X	

3.8

WORKSHOP EXERCISE 3.1 - PROBLEM #2

The top drawing is a plate with position tolerancing applied. The lower drawing is the produced part. Evaluate the dimensions on the produced part to verify conformance to the position tolerances. Use the chart below to record your calculations.

AS DRAWN

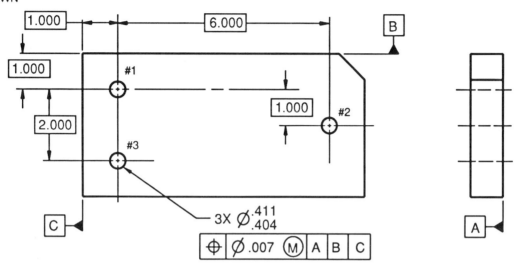

PRODUCED PART

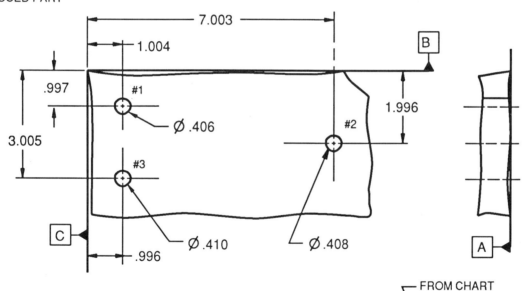

FROM CHART

HOLE NO.	HOLE MMC	HOLE ACTUAL SIZE	POSITION TOLERANCE ALLOWED	"X" DIM.	"Y" DIM.	POSITION TOLERANCE ACTUAL	ACC	REJ
1	.404	.406	.009	.004	+.003	.010		X
2	.404	.408	.011	.003	.004	.010	X	
3	.404	.410	.013	-.004	-.005	.0128	X	

3.9

WORKSHOP EXERCISE 3.1 - PROBLEM #3

The top drawing is a plate with position tolerancing applied. The lower drawing is the produced part. Evaluate the dimensions on the produced part to verify conformance to the position tolerances. Use the chart below to record your calculations.

AS DRAWN

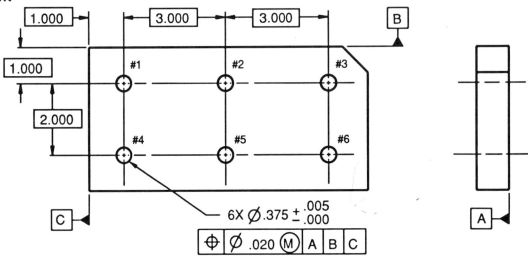

PRODUCED PART

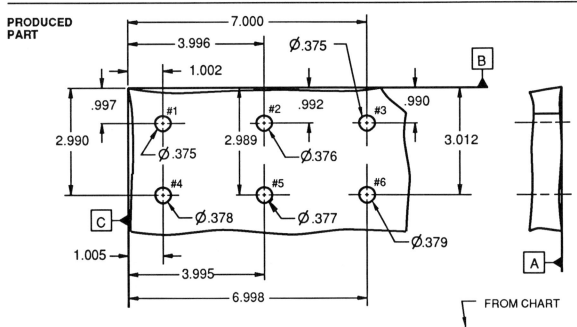

HOLE NO.	HOLE MMC	HOLE ACTUAL SIZE	POSITION TOLERANCE ALLOWED	"X" DIM.	"Y" DIM.	POSITION TOLERANCE ACTUAL	ACC	REJ
1	.375	.375	.020	.002	.003	.0072	X	
2		.376	.021	-.004	.008	.0179	X	
3		.375	.020	.0	.010	(.020)	X	
4		.378	.023	.005	.010	.0224	X	
5		.377	.022	-.005	.011	.0242		X
6		.379	.024	-.002	-.012	.0243		X

3.10

WORKSHOP EXERCISE 3.1 - PROBLEM #4 - LMC

The top drawing is a part with position tolerance applied. The lower drawing is the produced part. Evaluate the dimensions on the produced part to verify conformance to the position tolerances. Use the chart below to record your calculations. Review the previous examples if you have any problems.

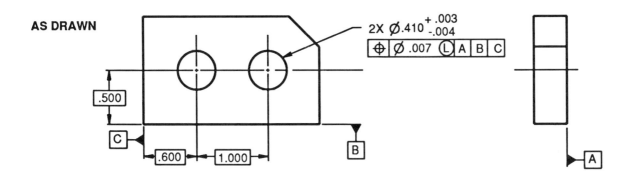

AS DRAWN

2X Ø.410 +.003 -.004

⊕ | Ø .007 Ⓛ | A | B | C

.500

C

.600   1.000

B

A

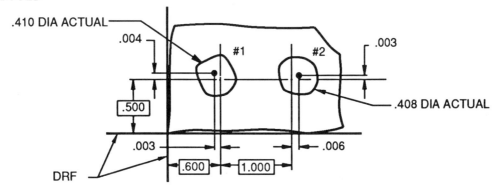

AS PRODUCED

.410 DIA ACTUAL

.004

#1   #2   .003

.500

.408 DIA ACTUAL

.003   .006

DRF

.600   1.000

HOLE NO.	HOLE LMC	HOLE ACTUAL SIZE	POSITION TOLERANCE ALLOWED	"X" DIM	"Y" DIM	POSITION TOLERANCE ACTUAL	ACCEPT	REJECT
1	.413	.410	.010	.003	.004	.010	X	
2	.413	.408	.012	-.006	-.003	.0134		X

PAPER GAGE CONCEPT

The paper gage concept is a good tool for understanding and evaluating geometric tolerancing. Paper gaging concepts are used throughout this workbook to help illustrate many of the geometric concepts. It is suggested that the user complete, at a minimum, the paper gage exercises in unit 3. This hands-on experience will allow the user a greater insight to understanding the concept of virtual size, datum modifiers and hole patterns as datums. These advanced concepts will be explained later in the text. A good knowledge base in paper gaging will go a long way to understanding the geometric system.

The paper gage is also a handy tool for evaluating data that is collected from coordinate measuring machines (CMM's) or surface plate set-ups. Many coordinate measuring machines have software that are capable of evaluating datum modifiers and hole patterns. Completing the manual paper gage concepts will allow the user to understand and evaluate the CMM or surface plate data with confidence. The paper gaging concepts can also be used to evaluate parts instead of using fixed or hard gaging. Paper gaging can also be used to evaluate and adjust the manufacturing process.

The paper gage concept illustrated in this workbook uses a cartesian coordinate and a polar coordinate. In order to complete the exercises, follow the instructions and make a copy of the coordinate systems at the end of this unit.

When plotting points it is important that all users plot consistently and in the same fashion. In this workbook all points are plotted relative to the cartesian coordinate system. In other words, each hole or feature has a basic location. Its address, in theory, is X=0, Y=0. The features should be plotted on the graph relative to their theoretical or basic address in the view in which the holes are specified. If a feature falls to the right of its basic address, it will have a plus X value. If it falls to the left of its basic location, it will have a minus X value. This method will insure consistency between answers.

The paper gage concept explained in this book is a simple example of a very complex calculation. Another method of paper gaging involves inputing the information in a CAD system to evaluate the data using separate tolerance zones on basic locations. This method will allow the features to be rotated as well as shifted in position. The simple method shown in this workbook will be sufficient for all the exercises.

PAPER GAGE EVALUATION OF 3 HOLE PART (PROBLEM #1- EXERCISE 3.1)

The part below was evaluated in the previous exercise 3.1. Hole number 2 was found to out of position. The paper gage exercise on the following page evaluates the problem and suggests possible solutions.

AS DRAWN

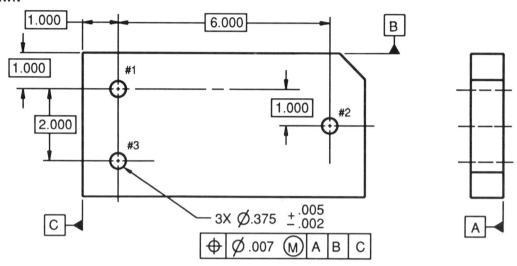

PRODUCED PART

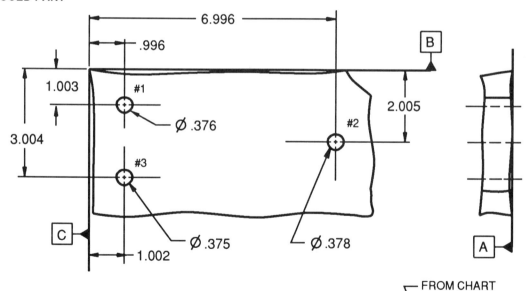

FROM CHART

HOLE NO.	HOLE MMC	HOLE ACTUAL SIZE	POSITION TOLERANCE ALLOWED	"X" DIM.	"Y" DIM.	POSITION TOLERANCE ACTUAL	ACC	REJ
1	.373	.376	.010	- .004	- .003	.010	X	
2	.373	.378	.012	- .004	- .005	.0128		X
3	.373	.375	.009	+ .002	- .004	.0089	X	

3.13

PAPER GAGE EVALUATION OF 3 HOLE PART (PROBLEM 1- EXERCISE 3.1)

The paper gage example below is an evaluation of the three hole part (problem 1) in the workshop exercise 3.1. In the 3.1 exercise, hole #2 was reported out of tolerance. After plotting the holes on the paper gage below we can easily evaluate the problem. The three holes are being manufactured in a close group, but the pattern of the three holes are displaced relative to the datum reference frame. In this example, it seems the process is capable of locating the holes to each other within .007 dia; but there is a problem locating the part relative to the datums.

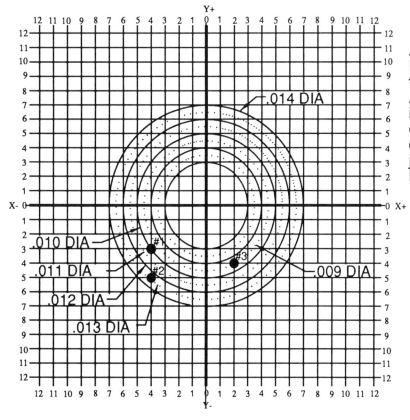

At a first glance it appears that hole #2 must be relocated in the manufacturing process. After a paper gage evaluation, it can be seen that the pattern of 3 holes are good to each other. The group of three are just displaced relative to the DRF.

The example below shows the three holes positioned to each other within a .007 dia zone. In order to center the process, the origin of the holes as a group must be relocated by plus .001 in the X direction and plus .004 in the Y direction.

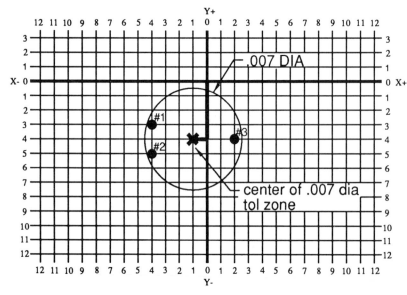

The paper gage concept is a good tool for evaluating and adjusting the manufacturing processes. Of course, a sample of parts should be considered for proper representation.

3.14

WORKSHOP EXERCISE 3.2

In workshop exercise 3.1 you completed verification exercises on three parts using the position formulas and charts. These exercises allowed you to make a determination if the parts were accepted or rejected. These exercises, however, did not give you a clue as to what happened to the process or how to fix the problem.

1. Make a paper copy of the cartesian coordinate system and a clear transparency copy of the polar coordinate system at the end of this unit. Use the paper gage concept to evaluate the three hole part (problem 2) and the six hole part (problem 3) that you evaluated earlier in workshop exercise 3.1. Determine what went wrong in the process with these parts and what can be done in the future to prevent the problem from reoccuring. Discuss and compare your results with others in the classroom.

If you have problems, review the previous example in which the three hole part in problem 1 was illustrated.

PAPER GAGE - CARTESIAN COORDINATE SYSTEM

Make paper copies of this cartesian coordinate system for use in the paper gage exercises in this unit and other exercises later in the text. In using the provided coordinate systems for paper gage calculations, it is important that the circles on the polar coordinate and the squares on the cartesian coordinate system are exactly the same scale. Copies from copiers tend to stretch, and it may be possible that the circles and squares on your copies will not line up exactly. Each .250" division equals .001". One half of a square equals .0005 of an inch. The scale difference on your copies should be close enough to complete the exercises in the workbook. If using this paper gage concept for on-the-job applications, interpolation may be necessary or new graphics can be created. The paper gage calculations can also be completed with a CAD system or appropriate CMM software.

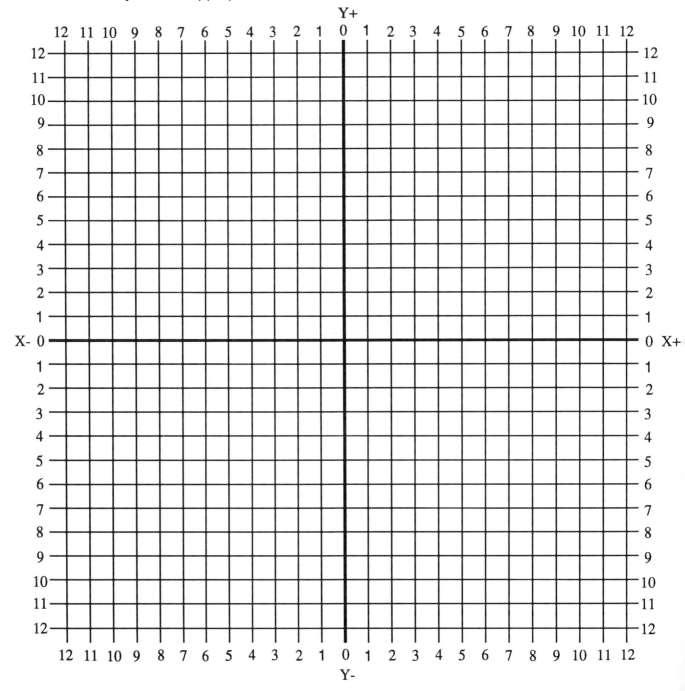

PAPER GAGE - POLAR COORDINATE

Make a transparency copy of this polar coordinate for use in the paper gage exercises in this unit and other exercises later in the workbook.

Scale - .001inch = .250 inch.

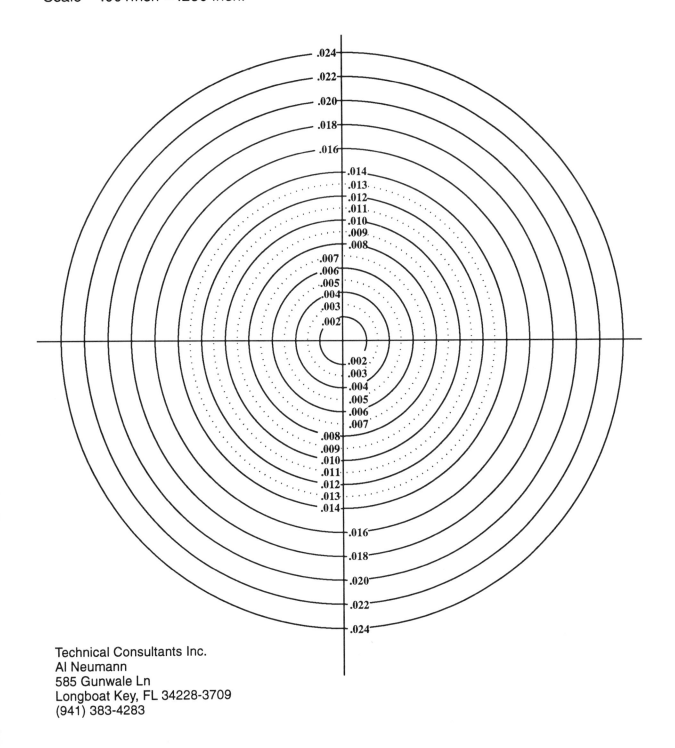

Technical Consultants Inc.
Al Neumann
585 Gunwale Ln
Longboat Key, FL 34228-3709
(941) 383-4283

UNIT 4

LIMITS OF SIZE

Taylor Principle, Rule 1 or Envelope Principle
       Mathematical Definition
Go-No Go Gages for Size
Terms and Definitions for Individual Features of Size
       Actual Size Definitions
Relationship of Individual Features
Perfect Orientation or Position Between Features
The Hub
Three Plans to a Quality Product
       Product Plan
       Manufacturing Plan
       Quality Plan
Virtual Condition
Exercise 4.1

TAYLOR PRINCIPLE - (RULE #1, ENVELOPE PRINCIPLE)

The Taylor Principle is a very important concept that defines the size and form limits for an individual feature of size. (It is called the Taylor Principle because in 1905 William Taylor obtained a patent on the full form go-gage.) The Taylor Principle is also called Rule #1 or limits of size in the Y14.5M, 1994 standard. In the international community the Taylor Principle is also often referred to as the Envelope Principle.

EXTREME VARIATIONS OF SIZE AND FORM ALLOWED ON AN INDIVIDUAL FEATURE OF SIZE BY THE LIMITS OF SIZE

THIS ON THE DRAWING

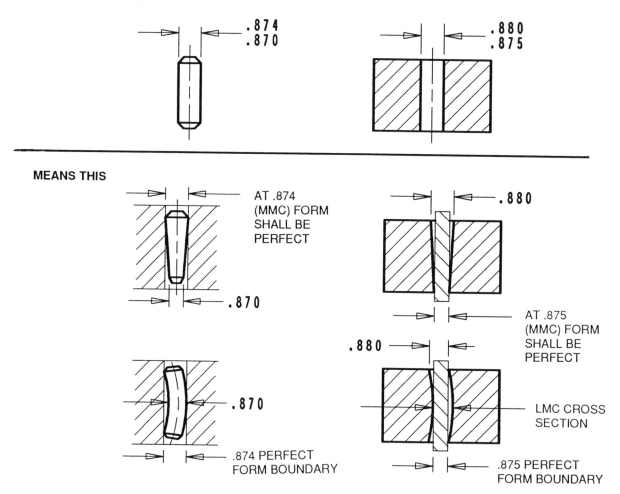

MEANS THIS

The limits of size define the size as well as the form of an individual feature. The feature may be bent, tapered or out of round. The size of the feature may vary within the limits. If the feature is produced at its maximum material condition, the form must be perfect.

Limits of Size
Unless otherwise specified, the limits of size of a feature prescribe the extent within which variations of geometric form, as well as size, are allowed. This control applies solely to individual features of size.

Individual features of size are defined as: one cylindrical or spherical surface (pin, hole, ball etc.), or a set of two opposed elements or opposed parallel surfaces associated with a size dimension (slot, tab etc.).

Variations of size: The actual size of an individual feature at any cross section shall be within the specified limits of size.

Variations of form: The form of an individual feature is controlled by its limits of size to the extent prescribed in paragraphs A, B and C below and in the preceding illustration.

(A) The surface or surfaces of a feature shall not extend beyond a boundary (envelope) of perfect form at MMC. This boundary is the true geometric form represented by the drawing. No variation in form is permitted if the feature is produced at its MMC limit of size.

(B) Where the actual size of a feature has departed from MMC toward LMC, a variation in form is allowed equal to the amount of such departure.

(C) There is no requirement for a boundary of perfect form at LMC. Thus, a feature produced at its LMC limit of size is permitted to vary from true form to the maximum variation allowed by the boundary of perfect form at MMC.

The Taylor Principle basically specifies that the size tolerance of a feature will control the size of that feature as well as the form of that feature. The following example illustrates in 3D the form and size variations allowed on an external feature of size as allowed by the Taylor Principle.

THIS ON THE DRAWING

$\varnothing$ .502 / .498

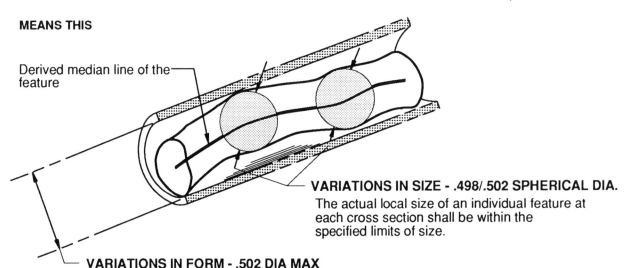

MEANS THIS

Derived median line of the feature

VARIATIONS IN SIZE - .498/.502 SPHERICAL DIA.
The actual local size of an individual feature at each cross section shall be within the specified limits of size.

VARIATIONS IN FORM - .502 DIA MAX
a. The surface or surfaces of a feature shall not extend beyond a boundary (envelope) of perfect form at MMC. This boundary is the true geometric form represented by the drawing. No variation in form is permitted if the feature is produced at its MMC limit of size.

b. Where the actual local size of a feature has departed from MMC toward LMC, a variation in form is allowed equal to the amount of such departure.

c. There is no requirement for a boundary of perfect form at LMC. Thus, a feature produced at its LMC limit of size is permitted to vary from true form to the maximum variation allowed by the boundary of perfect form at MMC.

As you will notice in the above 3D example, the variations in actual local size are verified with spherical cross sections. This concept is a clarification from past issues of the Y14.5 standard. In theory, a continuously expanding and contracting sphere is pulled through the feature. The sphere must never exit the material. This concept is explained in detail in the Y14.5.1M, Mathematical Definition of Dimensioning and Tolerancing principles. The reason for the clarification was to protect design criteria. This definition ensures the part has a minimum diameter cross section area.

In theory, the spherical variations of actual local size can be computed mathematically and verified with a coordinate measuring machine, optical comparator, roundness machine or other similar type equipment. In practical applications on the floor, parts are often verified with a 2 point cross sectional check. This is often accomplished with a micrometer, calipers or functional go and no-go type gages.

The quality engineer should be aware that, in some extreme cases, a two point check might not be sufficient to verify size. The two point measurement of some out of round, lobed or gleichstuck shaped features (parts that are not round but will give equal measurements, such as the impeller on a Wankel engine or reuleaux triangle) may yield different results than if a spherical check were used. In those cases where lobing is possible or of concern, a more stringent spherical check or mapping of the surface might be considered.

USING TWO OPPOSED POINTS TO VERIFY SIZE

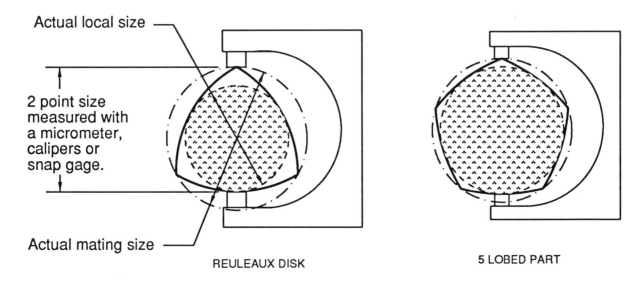

REULEAUX DISK   5 LOBED PART

The graphics above illustrate how a measurement device using two opposed points will not necessarily insure conformance to a size requirement on a round part. Depending on the configuration and the odd lobing on the workpiece, it is possible for the actual local size to be smaller than the two point measurement. At the same time, it is possible for the actual mating size to be larger than the two point measurement. If lobing is a concern a more stringent "mapping" of the surface may be required.

GO AND NO-GO GAGES TO VERIFY SIZE

The two following examples are functional GO and NO-GO type gages used to verify size and form according to the Taylor Principle. These are attribute checks and will not give variable data.

SAMPLE FUNCTIONAL GO and NO-GO GAGES USED TO CHECK SIZE AND FORM ON AN EXTERNAL CYLINDRICAL FEATURE

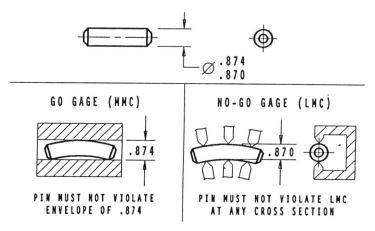

$\emptyset \begin{smallmatrix} .874 \\ .870 \end{smallmatrix}$

GO GAGE (MMC)	NO-GO GAGE (LMC)

.874

.870

PIN MUST NOT VIOLATE
ENVELOPE OF .874

PIN MUST NOT VIOLATE LMC
AT ANY CROSS SECTION

SAMPLE FUNCTIONAL GO, NO-GO GAGES USED TO CHECK SIZE AND FORM ON AN EXTERNAL FEATURE WIDTH

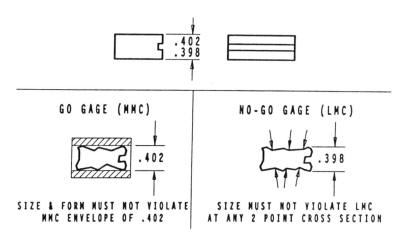

$\begin{smallmatrix} .402 \\ .398 \end{smallmatrix}$

GO GAGE (MMC)	NO-GO GAGE (LMC)

.402

.398

SIZE & FORM MUST NOT VIOLATE
MMC ENVELOPE OF .402

SIZE MUST NOT VIOLATE LMC
AT ANY 2 POINT CROSS SECTION

The GO gage always verifies the maximum material condition. Note that the entire feature length must be fully accepted into the GO gage. If the feature is fully accepted in the gage, both the size and form are controlled within a maximum envelope.

The NO-GO gage always verifies the least material condition. Note that, according to the Taylor Principle, the gage is required to only check cross sectional points across the feature to ensure it does not violate the least material condition boundary.

Note: In reality, gages cannot be made perfect. Gagemaker tolerance must be applied to the gages. Gage tolerances are usually 10% of the product tolerance with another 5% allowable for wear. The gage tolerances are usually arranged in such a manner that a good part may be rejected but bad parts will never be accepted. To avoid complexity with these examples, gage tolerance is not considered. There is additional information on this subject in the standard ANSI B4.4M-1981, Inspection of Workpieces.

TERMS AND DEFINITIONS FOR INDIVIDUAL FEATURES OF SIZE

There are many terms and definitions that are associated with size of an individual feature of size in the ASME Y14.5M, 1994 standard. It is important to understand these terms as they will be used to define other concepts in geometric tolerancing. These terms are also important in engineering conversations, inspection reports and correspondence. These terms are also compatible with those used in ISO standards.

Nominal size - The designation used for purposes of general identification.

Actual size - The general term for the size of a produced feature. This term includes the actual mating size and the actual local sizes.

Actual local size - The value of any individual distance at any cross section of a feature.

Actual mating size - The dimensional value of the actual mating envelope.

Actual mating envelope - This term is defined according to the type of feature as follows:
a. For an external feature. A similar perfect feature counterpart of smallest size that can be circumscribed about the feature so that it just contacts the surface at its highest points. For example, a smallest cylinder of perfect form or two parallel planes of perfect form at minimum separation that just contact(s) the highest points of the surface(s).
For features controlled by orientation or positional tolerances, the actual mating envelope is oriented relative to the appropriate datum(s), for example, perpendicular to a primary datum plane.
b. For an internal feature. A similar perfect feature counterpart of largest size that can be inscribed within the feature so that it just contacts the surface at the highest points. For example, a largest cylinder of perfect form or two parallel planes of perfect form at maximum separation that just contact(s) the highest points of the surface(s).
For features controlled by orientation or positional tolerances, the actual mating envelope is oriented to the appropriate datum(s).

The limits of size (Taylor Principle) specify the size as well as the form of an individual feature. In discussing the actual size of an individual feature it is important to recognize that the size of an individual feature will have two values. One is for the actual mating size and another is the actual local size. The following graphic illustrates common terms and definitions used with actual size.

ACTUAL SIZE DEFINITION

EXTERNAL FEATURE

AS DRAWN

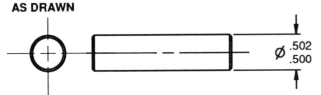

$\varnothing\ {.502 \atop .500}$

ACTUAL SIZE

The general term for the size of a produced feature. The term includes:

One value for the Actual Mating Size
One value for the Actual Local Size

PRODUCED PART

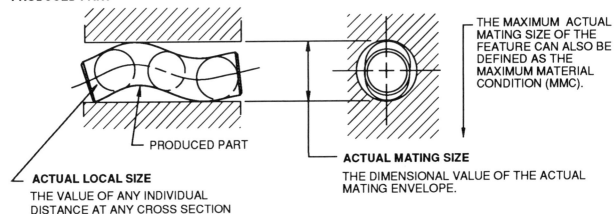

THE MAXIMUM ACTUAL MATING SIZE OF THE FEATURE CAN ALSO BE DEFINED AS THE MAXIMUM MATERIAL CONDITION (MMC).

PRODUCED PART

ACTUAL MATING SIZE

THE DIMENSIONAL VALUE OF THE ACTUAL MATING ENVELOPE.

ACTUAL LOCAL SIZE

THE VALUE OF ANY INDIVIDUAL DISTANCE AT ANY CROSS SECTION OF A FEATURE.

INTERNAL FEATURE

AS DRAWN

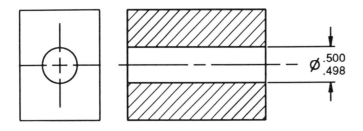

$\varnothing\ {.500 \atop .498}$

ACTUAL SIZE

The general term for the size of a produced feature. The term includes:

One value for the Actual Mating Size
One value for the Actual Local Size

PRODUCED PART

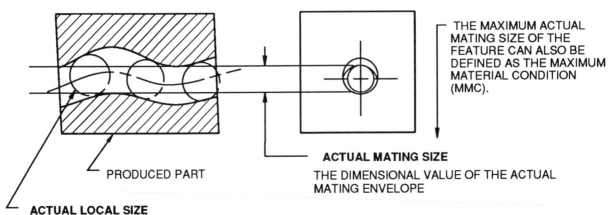

THE MAXIMUM ACTUAL MATING SIZE OF THE FEATURE CAN ALSO BE DEFINED AS THE MAXIMUM MATERIAL CONDITION (MMC).

PRODUCED PART

ACTUAL MATING SIZE

THE DIMENSIONAL VALUE OF THE ACTUAL MATING ENVELOPE

ACTUAL LOCAL SIZE

THE VALUE OF ANY INDIVIDUAL DISTANCE AT ANY CROSS SECTION OF A FEATURE.

In theory, the Taylor Principle is automatically invoked when we specify the use of ASME Y14.5M, 1994. Practically, in industry it is found that in many cases a relaxation or a more stringent enforcement of the Taylor Principle is left to the discretion of inspection or quality function. A micrometer check to verify size is an example of relaxation. A micrometer check will not verify the perfect form at MMC requirement.

In some cases the actual local size of a feature is important but the form may be allowed to exceed the boundary of perfect form at MMC. If this is required, a note such as PERFECT FORM AT MMC NOT REQD is specified to the feature, exempting the pertinent size dimension from the variations of form requirements. The form requirements for Taylor Principle can also be adjusted or rearranged by using straightness of an axis/median plane control or the average diameter concept, both of which are described in the form control section of this text.

The control of geometric form prescribed by the limits of size does not apply to the following:

A. Stock such as bars, sheets, tubing, structural shapes, and other items produced to established industry or government standards that prescribe limits for straightness, flatness, and other geometric characteristics. Unless geometric tolerances are specified on the drawing of a part made from these items, standards for these items govern the surfaces that remain in the "as-furnished" condition on the finished part.

B. Parts subject to free state variation in the un-restrained condition.

Special note: The Taylor Principle is widely accepted and recognized by the United States and the International Standards Organization (ISO). The ISO standards also allow the Principle of Independency. In the Principle of Independency, size is verified by two point measurements. Form is not included in size. Depending on the particular standards invoked, either case may apply.

If company drawings are exchanged with other countries, more information should be sought in the ISO standards. ISO 8015 and ISO 1938 both cover individual features of size. There is a complete listing of ISO standards related to dimensioning and tolerancing earlier in the text. In some cases, the circle E symbol is used in ISO to designate features that must conform to the Envelope Principle. The Envelope Principle is another name used in ISO for the Taylor Principle.

RELATIONSHIP OF INDIVIDUAL FEATURES

Parts are made up of individual features. There is no relationship between individual features implied in the ASME Y14.5M, 1994 standard. The Taylor Principle (Rule#1) is a very important rule that covers the size and form of individual features of size. It does not cover the interrelationship of features of size or non-features of size. Features (size or non-size) shown perpendicular, coaxial or symmetrical to each other must be controlled for location or orientation to avoid incomplete drawing requirements. Once you fully understand there is no relationship between individual features, the need for the datum reference frame and the geometric characteristics will become clear.

To further clarify the above statement that there is no relationship between individual features, consider the example below. A shaft is shown with many diameters. All these diameters are individual features of size. All diameters are shown on the same center and are probably required to be on the same center, but the drawing does not state how much tolerance one feature can be off location or orientation to the other.

RULE #1 DOES NOT CONTROL LOCATION OR ORIENTATION BETWEEN INDIVIDUAL FEATURES

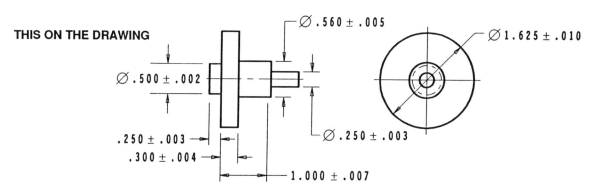

THIS ON THE DRAWING

$\varnothing .560 \pm .005$

$\varnothing 1.625 \pm .010$

$\varnothing .500 \pm .002$

$.250 \pm .003$

$.300 \pm .004$

$\varnothing .250 \pm .003$

$1.000 \pm .007$

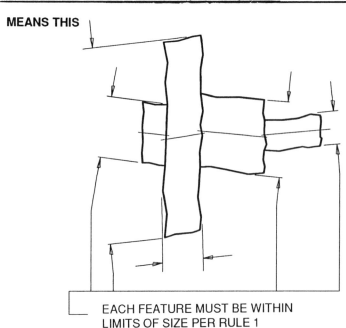

MEANS THIS

THE LIMITS OF SIZE DO NOT CONTROL THE ORIENTATION OR LOCATIONAL RELATIONSHIP BETWEEN FEATURES.

EACH FEATURE MUST BE WITHIN LIMITS OF SIZE PER RULE 1

Rule 1 only covers the size and form requirements of each individual feature of size and not the relationship between the features. If there is a need to control the locational relationship of these features, we usually use datums and a position, runout or profile tolerance. An example of such a control is shown below. If nothing is called out, it probably means that it is not very important and generally accepted manufacturing practices apply. Certainly, it is not a clear specification. In some cases the relationship may be specified in a general note or a general specification that is referenced on the drawing The following drawing of the shaft shows the coaxial requirements very clearly specified with position tolerance.

POSITION BETWEEN   INDIVIDUAL FEATURES MUST BE SPECIFIED TO AVOID INCOMPLETE DRAWING REQUIREMENTS

THIS ON THE DRAWING

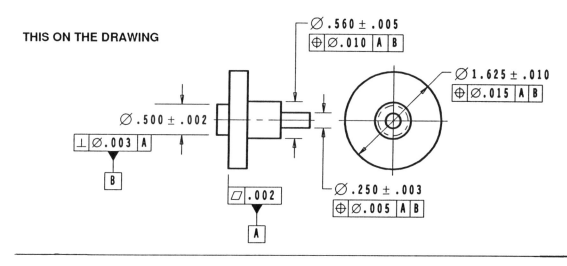

MEANS THIS

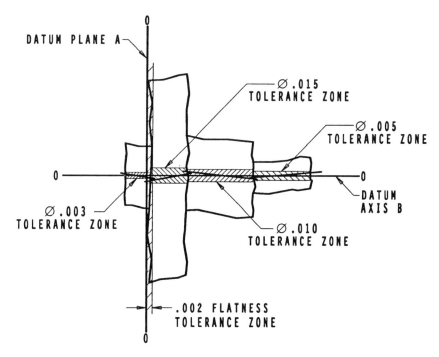

A further clarification to show the non-interrelationship between features can be seen below. The plus/minus tolerance is only a 2 dimensional control that is being used to control a 3 dimensional part. All the dimensions have angle tolerance. The machined block below has many individual features of size and non-size features. The size and form of each individual feature of size is covered under Rule 1, but the interrelationship between features is not covered. The corners are shown square and are implied to be 90 degrees. The tolerance on these corners, unless otherwise specified, is usually stated in the general tolerance note for angles. The angle tolerance only controls the angular relationship between the features. This illustration does not imply that plus/minus can not be used, but be aware that at times it can be unclear and very complicated.

RULE 1 DOES NOT CONTROL ORIENTATION OR LOCATION BETWEEN INDIVIDUAL FEATURES

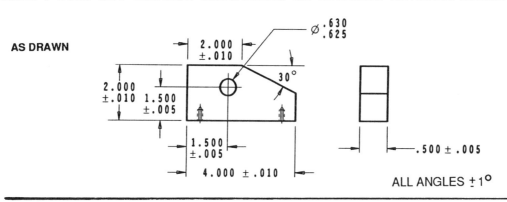

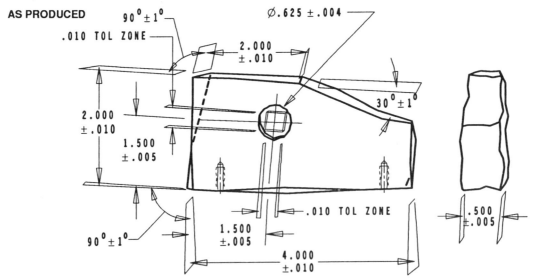

The limits of size do not control the orientation or locational relationship between features. Many of the plus/minus dimensions on the drawing above are location and not size because the features do not have opposing elements.

A 90 degree angle applies where centerlines and lines depicting features are shown on a drawing at right angles and no angle is specified. The tolerance on these angles can usually be found in the title block.

4.12

The drawing below illustrates the machined block again with geometric tolerancing applied. In order to clearly and mathematically define the part in 3D, a datum reference frame is established and a position and profile tolerance are specified. The plus/minus tolerances locating the surfaces are replaced with the profile tolerance. The interrelationships between the features are controlled. This illustration is not to suggest that all parts must have profile tolerance, but notice that the part definition is very clear and concise. The example below illustrates one method of how this part may be toleranced with profile and position tolerances. There are more application examples of profile, orientation and position tolerances later in the text.

PROFILE, POSITION AND ORIENTATION TOLERANCES ARE USED TO CONTROL RELATIONSHIPS BETWEEN INDIVIDUAL FEATURES.

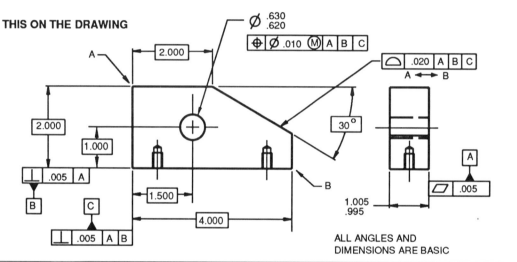

THIS ON THE DRAWING

ALL ANGLES AND
DIMENSIONS ARE BASIC

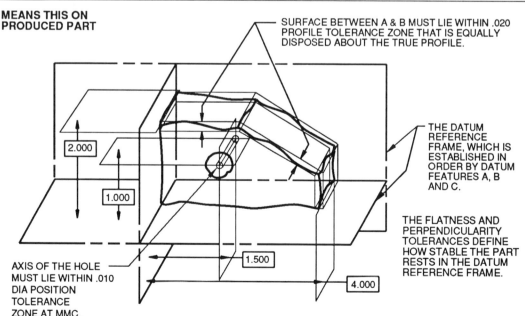

MEANS THIS ON PRODUCED PART

SURFACE BETWEEN A & B MUST LIE WITHIN .020 PROFILE TOLERANCE ZONE THAT IS EQUALLY DISPOSED ABOUT THE TRUE PROFILE.

THE DATUM REFERENCE FRAME, WHICH IS ESTABLISHED IN ORDER BY DATUM FEATURES A, B AND C.

THE FLATNESS AND PERPENDICULARITY TOLERANCES DEFINE HOW STABLE THE PART RESTS IN THE DATUM REFERENCE FRAME.

AXIS OF THE HOLE MUST LIE WITHIN .010 DIA POSITION TOLERANCE ZONE AT MMC.

4.13

Another clarification of the Taylor Principle (Rule#1) is the simple sheet metal part shown below. Notice that it has a plus/minus tolerance giving the height of the part. This plus/minus dimension does not fall under the provisions of Rule# 1. Rule #1 only covers features of size, and this is not a feature of size. The definition of a feature of size requires the two parallel planes to be opposing. The feature below is not opposing planes.

THE LIMITS OF SIZE DOES NOT CONTROL UN-OPPOSED FEATURES

THIS ON THE DRAWING

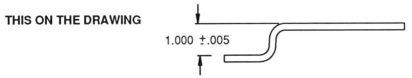

1.000 ±.005

PLUS/MINUS TOLERANCES TO UN-OPPOSED SURFACES DO NOT PROVIDE A CLEAR DEFINITION OF THE PRODUCT. PLUS/MINUS DIMENSIONS ARE NOT ASSOCIATED WITH DATUMS. THE DATUMS ARE IMPLIED. EVEN IF DATUMS ARE SPECIFED, THEY DO NOT IMPLY THE ORIGIN OF MEASUREMENTS.

INTERPRETATION IS UNCLEAR MAY MEAN EITHER OF THESE

PRODUCED PARTS

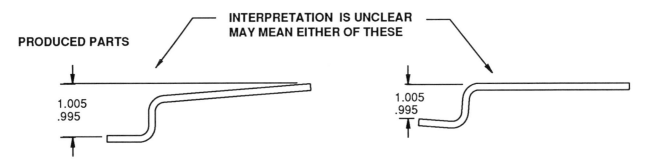

1.005
.995

1.005
.995

NOTE: THERE IS A PROVISION IN THE ASME Y14.5M, 1994 STANDARD THAT ALLOWS PLUS/MINUS TOLERANCING TO BE ASSOCIATED WITH DATUMS. THE DATUMS ARE IDENTIFIED AND ORDER OF PRECEDENCE IS SPECIFIED IN A NOTE ON THE DRAWING. THE NOTED DIMENSIONS DEFINE ONLY THE MAXIMUM MATERIAL CONDITION ENVELOPE RELATED TO THE DATUM REFERENCE FRAME DEFINED BY THE DATUMS. THIS METHOD CAN BE COMPLICATED AND UNCLEAR IF PARTS ARE CHAIN DIMENSIONED.

The part would not meet the actual local size or the actual mating requirement of the Taylor Principle. Since the planes do not oppose each other, either one surface or the other would have to serve as the origin of measurements. If the part were mounted on the short surface and the long surface were checked, it would yield a different result than if the long surface were used as the origin. The plus/minus dimension on the sheet metal part is not size but rather the location of one surface to another. The design intent is not clear. This is not to imply that plus/minus tolerance can not be used to locate features, but be aware of possible interpretations when using this type of tolerancing.

In order to ensure design intent, one surface may be located to the other through the use of a dimension origin symbol. The dimension origin symbol is only a simple 2D control and may work with this example but may not be sufficient for complex examples such as the previous example of the machined block. Notice that there are also un-opposed features on the machined block.

The two examples shown below illustrate a sheet metal part defined according to how it functions and how it mates in the assembly. In the upper example the sheet metal part mounts on the long surface and the short surface must clear the button. In the lower example the part mounts on the short surface and the long surface must clear the button. In each case the appropriate surface is defined as a datum, and then the other surface is located with a profile specification. The design intent is very clearly stated. In this case, the profile tolerance replaces the plus/minus tolerance. (Note: If plus/minus tolerances are used with datums, this does not imply the dimensions originate from the datums as there is no order of precedence specified.)

SPECIFIED DATUMS AND GEOMETRIC TOLERANCE CLEARLY DEFINE FUNCTIONAL REQUIREMENTS.

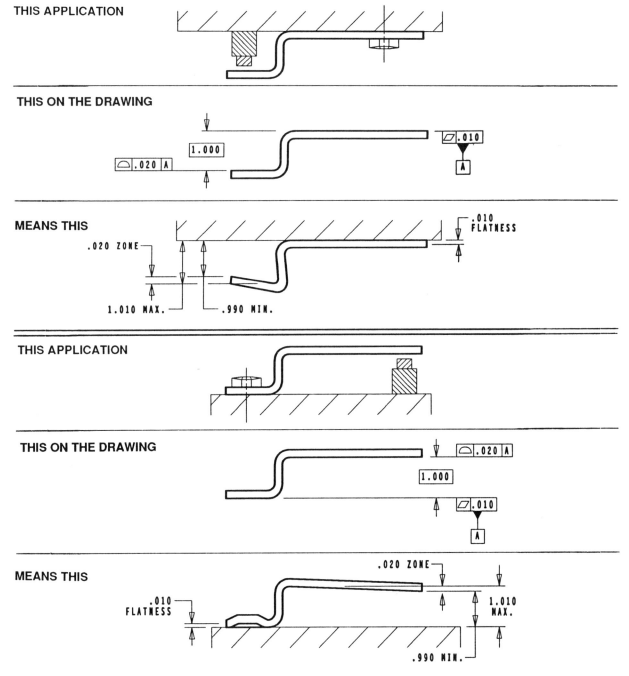

4.15

PERFECT ORIENTATION OR POSITION BETWEEN FEATURES

There is no interrelationship implied between features. In some cases though, it may be necessary to control a perfect orientation or position boundary at MMC between features. This is often required when controlling the relationship between datum features. There are four methods shown below that can be applied. These concepts are used and illustrated later in the text.

a. Specify a zero tolerance of orientation at MMC including a datum reference to control angularity, perpendicularity or parallelism of a feature. See page 9.8, 11.16 and 12.7.

b.   Specify a zero positional tolerance at MMC including a datum reference to control coaxial or symmetrical features. See page 9.8.

c. Indicate this control by a general note such as PERFECT ORIENTATION (or COAXIALITY or SYMMETRY) AT MMC REQUIRED FOR RELATED FEATURES. See page 13.9.

d. Relate dimensions to a datum reference frame by a local or general note indicating datum precedence.

The four methods shown above are all ways to control perfect orientation and/or position relationships at MMC between features. In many cases there is not a need for perfect relationships between all the features. The usual method is to establish a datum reference, relate the datum features and then relate the features as necessary with the use of orientation, profile, runout and position tolerances. These concepts are explained in detail later in the text.

GEOMETRIC TOLERANCING APPLICATION - HUB

The drawing below is a sample application of geometric tolerancing. The datum features are all related and qualified. There are also applications of form, orientation, position and profile tolerances.

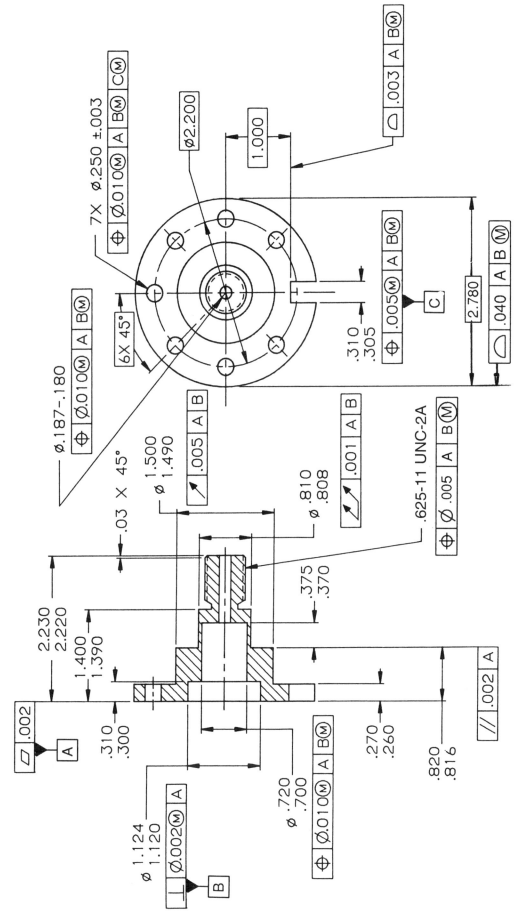

INTERPRET THIS DRAWING
PER ASME Y14.5M, 1994

THREE PLANS TO A QUALITY PRODUCT

In order to build quality parts we must have a plan. In fact, we need three plans. Making quality parts requires the input from the three main players; design, manufacturing and quality. The building of quality parts requires three documents or plans.

The three documents or plans are:

1. **The Design Product Drawing - Defines the functional requirements of the product.**

2. **The Manufacturing Process Plan - Defines how to make the product.**

3. **The Quality Dimensional Measurement Plan - Defines how to verify the product.**

These three documents can vary in complexity. In some cases, these steps can be very detailed written documents. In other cases, these steps can be very informal verbal instructions. How formal these documents are depends on the complexity of the part, the amount of parts to be produced, and the particular organization.

Design Product Drawing. This document defines the product. This product drawing is usually very formal and contains very detailed information about the functional requirements of the product. It is the most valuable document that we have. Without this document we do not have a product. This document is usually controlled by design engineering and kept locked in a vault where no one can change this drawing without the approval of design engineering.

The selected datums and specified requirements on this drawing define the functional requirements of the part. It explains what the part is supposed to look like before it goes out the door or is assembled with mating parts. The datums defined on the product drawing are not for manufacturing. This drawing does not tell how to build the part. For example, a hole call out lists the size of a hole but does not list whether it is drilled, reamed or punched unless that information is critical to the function of the part. Datums are selected on the basis of part function. It defines how the part mounts or how it is set up for functional requirements.

Manufacturing Process Plan. This document tells us how to make the product. Every organization has manufacturing process plans. They may be formal or informal. They may be called by another name, such as routing or operation sheets. This document is sometimes very formal, as in the manufacturing of a cylinder block. Or, it can be very informal, as in the manufacturing of a prototype part.

This document tells how to build the product and is usually drawn up by a manufacturing or process engineer. The process engineer designs the process to meet the requirements of the product drawing. This process sheet can be very detailed, such as listing the processes, procedures, sequence of operations and the equipment necessary to build the product. In other cases, the process plan may be as simple as a CNC tape and the drawing of a fixture that holds the part.

These process plans are flexible and can be changed at will by the manufacturing engineer. In fact, a single part may be made by two different processes. In this case, the part has two process plans to reflect the operations of the two processes.

Dimensional Measurement Plan. This document establishes the methods of measurement. It tells how to inspect or verify the part. Just as with the manufacturing process plan, this dimensional measurement plan can be a very formal written document or it can be very informal, and left up to the individual inspectors discretion.

This inspection plan is usually drawn up by a quality engineer and lists step by step which characteristics to check, how often they should be checked and the method and type of equipment that should be used. The plan formulated by the quality engineer is usually developed after a careful evaluation of the functional requirements specified on the product drawing. Once the quality engineer understands the functional requirements, he then examines the manufacturing process sheet to determine how the part is made. The first article inspection may be a complete verification of all the characteristics. Once we know something about the process variation and what could go wrong, it helps define what characteristics must be checked and how often. It also helps determine the inspection tools, procedures and methods to be used.

Different inspection methods or devices yield different results during the verification of the products. This is evident by the different results obtained from a CMM vs. open set-up surface plate techniques. This is called methods divergence. The dimensional measurement plan should define the acceptable measurement method and the associated uncertainty of results.

As you can see, it is very important to have some plan or organization to build interchangeable parts. This is especially important as we buy and sell parts all over the world and strive to achieve world class quality. The three plan system allows our parts to be followed and tracked from beginning to end no matter who works with the product. This makes it easy to identify and solve any problems or differences we may have along the way. In order to have quality parts, you must have quality throughout the design, manufacturing and verification process.

VIRTUAL CONDITION

Depending upon its function, a feature may be controlled by tolerances such as size, form, orientation and location. Consideration must be given to the collective effects of these factors in determining the clearances between mating parts and in establishing gage feature sizes. The collective effect of these factors is termed virtual condition.

Virtual condition is a constant boundary generated by the collective effects of a size feature's specified MMC or LMC material condition and the geometric tolerance for that material condition.

The pin in the illustration below has two virtual sizes. The .257 diameter virtual size is a result of the perpendicularity tolerance relative to datum A. The .262 diameter virtual size is a result of the position tolerance relative to datums A, B and C. The virtual sizes on this part could also be called outer boundary. It can also be viewed as a 3D solid.

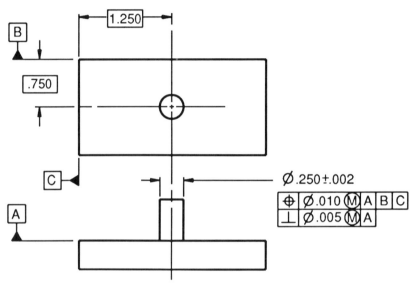

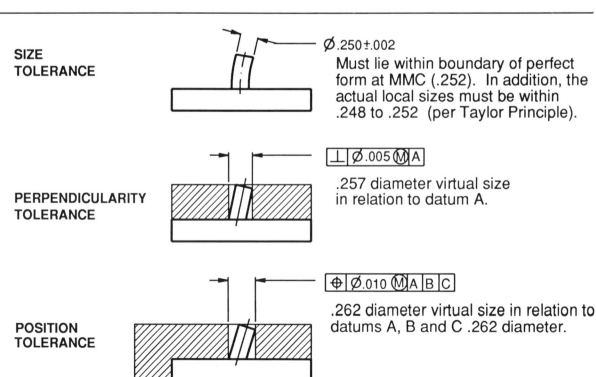

| SIZE TOLERANCE | $\emptyset$.250±.002 Must lie within boundary of perfect form at MMC (.252). In addition, the actual local sizes must be within .248 to .252 (per Taylor Principle). |

| PERPENDICULARITY TOLERANCE | ⊥ $\emptyset$.005 Ⓜ A |
.257 diameter virtual size in relation to datum A. |

| POSITION TOLERANCE | ⊕ $\emptyset$.010 Ⓜ A B C |
.262 diameter virtual size in relation to datums A, B and C .262 diameter. |

WORKSHOP EXERCISE 4.1

1. The principle used in the ASME Y14.5M-1994 standard that includes form within the limits of size is often referred to by three different names. Name all three.

_____

2. What is the name of the ISO principle that does not include form within the limits of size?

_____

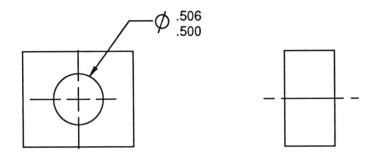

3. On the part above what is the smallest the hole can get?   _____

4. What is the largest the hole can get?   _____

5. How much can the form on the hole vary?   _____

6. The limits of size principle (Taylor Principle) in the ASME Y14.5M-1994 standard specifies the size, as well as the form for individual features. The actual size of a feature will therefore have two values, actual mating size and actual local size. Name and label these two values on the produced part below.

AS DRAWN

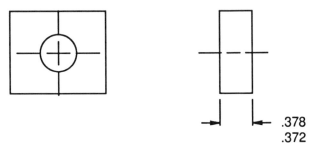

PRODUCED PART

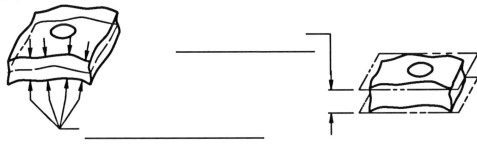

7. On the shaft below, with only the limits of size defined, what coaxiality tolerance is implied among the three diameters?

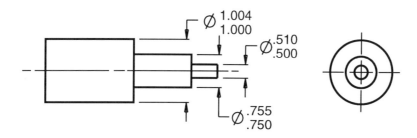

8. On the pin below, with only the limits of size defined, what perpendicularity tolerance is implied between the pin and the bottom surface?

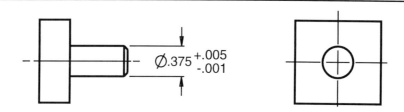

9. On the machined block below, circle only the size dimensions.

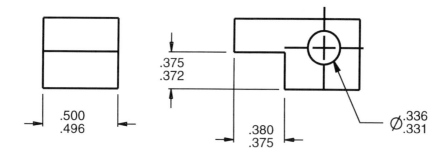

10. On the machined block above, you circled the size dimensions. Explain the difference between the dimensions you selected and the dimensions you did not select.

_____

_____

_____

_____

11. On the drawing below, calculate the virtual sizes for the identified features.
Remember, when calculating for virtual with MMC, holes get smaller and pins larger.

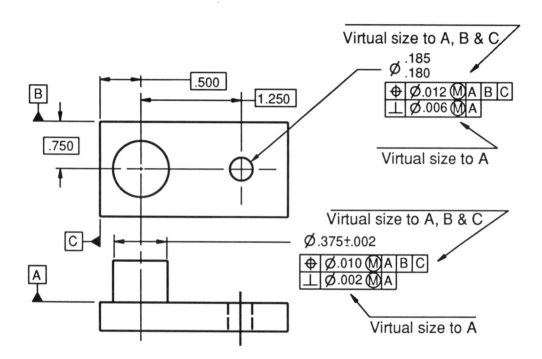

UNIT 5

DATUM REFERENCE FRAME THEORY

The Datum Reference Frame
     Datum Reference Frame Simulators
Positioning Parts in a Datum Reference Frame
Implied Datums Are Not Clear
Specified Datums Are Clear
Datum Features With and With Out Size
     Datum Feature Symbols Past Practice
     Datum Features Without Size
     Datum Features With Size
Datum Simulators
Partial Datum Features
Establishing a DRF and Qualifying Datums - Plane Surfaces
Establishing a DRF
     Point of Origin
Establishing a DRF and Qualifying Datums -Secondary Axis
Establishing a DRF and Qualifying Datum Features - Primary Axis
Workshop Exercise 5.1

DATUM REFERENCE FRAME (DRF)

The datum reference frame is, by far, the most important concept in the geometric tolerancing system. The datum reference frame is the skeleton of the system. It is the "frame of reference" to which all of the requirements are attached. It is often claimed that position or profile tolerancing is difficult to understand. Profile and position are very simple. They are very simply zones of tolerance. Position and profile are usually associated with datums, and people inherently do not understand datums. This misunderstanding of datums and is what makes the characteristics difficult for them. Let us talk about the datum reference frame. (DRF)

THE DATUM REFERENCE FRAME (DRF)

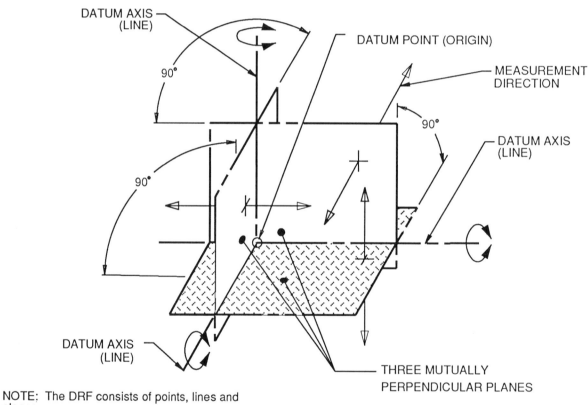

NOTE: The DRF consists of points, lines and planes.

Engineering, manufacturing, and inspection all share a common three plane concept. These three mutually perpendicular planes are perfect and exactly 90 degrees to each other. In geometric tolerancing we call this the datum reference frame. In mathematics we call this the cartesian coordinate system. The cartesian coordinate system was invented by a French mathematician, Rene Descartes (1596-1650). The two concepts are essentially the same.

In design engineering the three mutually perfect planes can be found mathematically in the CAD system or practically in the surface of the drafting table, the straight edge, and a right triangle. The three mutually perpendicular planes in inspection can be found mathematically in the coordinate measuring machines (CMM's) or practically defined as a surface plate and two angle plates. In manufacturing, these three perfect planes can be found in computer numerical control (CNC) and practically defined as the bed on a machine, a fence, and a stop.

Datum reference frame simulators

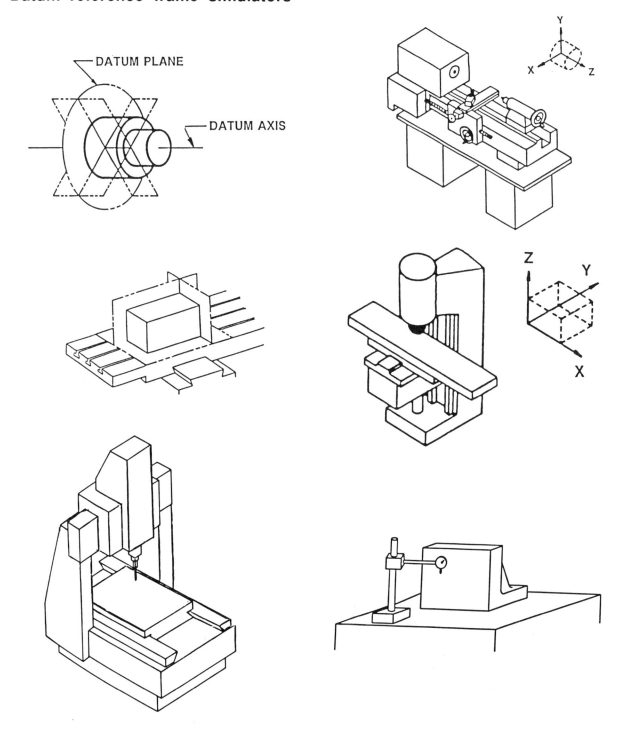

Engineering, manufacturing and inspection each consider their respective planes perfect for purposes of calculation and reference. In reality, we know that their planes are not perfect, as they are all simulated by processing equipment. The planes are usually considered perfect because they are usually created 10 times better than the parts we work with.

In geometric tolerancing we can relate engineering, manufacturing, and inspection together by using this datum reference frame. The three planes are called the primary plane, secondary plane and tertiary plane. The datum reference frame is made up from a series of individual components. The individual components are planes, axes and points.

Datums and the datum reference frame indicate the origin of a dimensional relationship to a toleranced feature or features on a part. When a feature serves as a datum feature, its true geometric counterpart actually establishes the datum. Since measurements cannot be made from a true geometric counterpart, which is theoretical, simulated datums are assumed to exist and be simulated with our manufacturing, processing and inspection equipment such as the bed on a machine, a collet or chuck, gage pin, a surface plate, angle plate etc. Measurements then originate from the simulated planes or axes that the manufacturing or inspection equipment simulates and not the features themselves.

DATUM PLANE, SIMULATED DATUM PLANE, DATUM FEATURE, DATUM SIMULATOR AND ASSOCIATED MEASUREMENT UNCERTAINTY

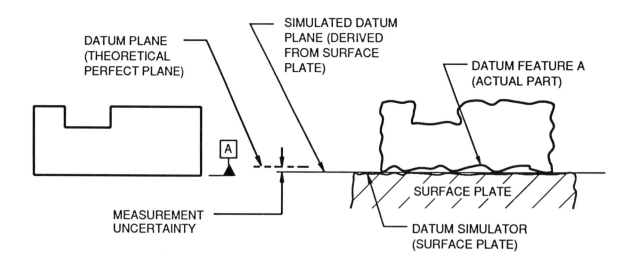

The datum plane is the theoretically , mathematically perfect plane. The simulated datum plane is derived from the datum feature contacting the high points on the datum simulator or surface plate. All measurements originate from the simulated datum plane and not the datum feature or part surface. The difference between the theoretically perfect datum plane and the simulated datum plane is the measurement uncertainty found in our inspection equipment.

5.4

POSITIONING PARTS IN A DATUM REFERENCE FRAME

Understanding that the datum reference frame is perfect, the next concept we must understand is that the parts (often called work pieces) are not perfect. They are never perfect, and we can never build a perfect part. In order to engineer, manufacture, or inspect this part, we must somehow "load" this imperfect part onto the datum reference frame to communicate to each other.

An illustration for orientating this imperfect part into the datum reference frame relative to the primary, secondary, and tertiary planes is shown in the figure below. Each unsupported object or part has six degrees of freedom. The part must be fixed in relation to the datum framework in order to stop this freedom. The part is related to the primary plane by contacting a minimum of three points. It is related to the secondary datum by contacting a minimum of two points. It is then related to the tertiary datum by contacting a minimum of one point.

All measurements are now made from the planes and not the part. This orientation of the part onto the datum reference frame ensures a common base of communication and measurement between engineering, manufacturing, and inspection.

DATUM REFERENCE FRAME

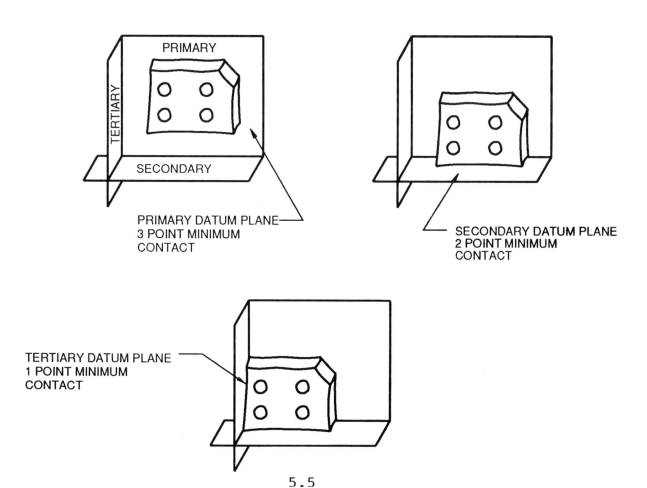

5.5

IMPLIED DATUMS - PROBLEMS

The order of precedence in the selection and establishment of datums is very important. In the figure below there is a part with four holes. The four holes are located from the edges with basic dimensions. The datums are not called out in the feature control frame. Rather, they are what we call implied datums. We call them implied datums because the dimensions originate from the bottom edge and the left hand edge. Thus, we imply that these edges are the datums.

Part with implied datums

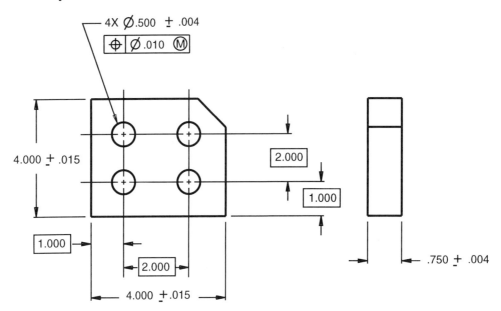

The produced part shown below (exaggerated for clarity) is not perfect. It will never be perfect; we cannot build perfect parts. None of the edges are perfectly square. It is a good part though. It is well within the acceptable limits for size and squareness.

The problem we now have with these implied datums is that we do not know the order in which the datums are to be used. We know that the datum reference frame is perfect, but the parts are not perfect. The part is shown with square corners, but we are allowed tolerance on the size. And, the 90 degree corners also have tolerance. Even if we put a perpendicularity on the corners of .0001, the part will still, in theory, "rock" back and forth in the datum reference frame.

Produced part

Implied datums - In what order do we load an imperfect part into a perfect reference frame?

Is the bottom edge our secondary datum with two points contact, or is the left hand edge our secondary datum? Is the large surface the primary datum? Which datum is the tertiary? It is not clear. There are a number of different combinations that this part can be oriented to the datum reference frame.

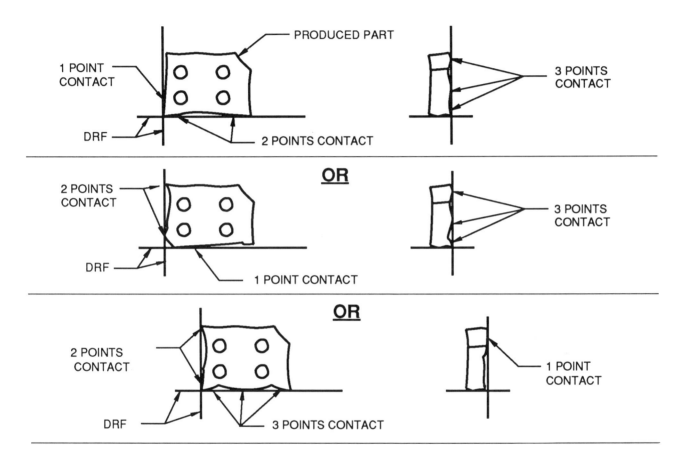

Engineering, manufacturing, and inspection may all have different interpretations as to the order that this part is loaded into the datum reference frame. This will result in different interpretations as to where the holes must lie on the part. All personnel should do it exactly the same way. The four hole part explained above is a very simple example. As the parts become more complex, with some datum features being holes, slots, surfaces and some being shorter or longer in length, this phenomenon of order becomes even more important. If the datums are not specified, it is possible to misinterpret design intent and reject good parts and accept bad parts.

Engineering, manufacturing and inspection can not work without a datum reference frame. Creating a datum reference frame is mandatory in order to achieve interchangeable parts. Our modern computerized methods of engineering, manufacturing and inspection require a datum reference frame.

SPECIFIED DATUMS CLEARLY DEFINE DESIGN INTENT

The part shown in the figure below has four holes located in relation to datums. The requirements for the holes are shown in the feature control frame along with the specified datums. The datums identify how the part is located in the datum reference frame.

In the first datum compartment of the feature control frame, you will always find the primary datum (in this case, 3 highest points of contact on face). In the second datum compartment, you will always find the secondary datum (in this case, 2 highest points of contact on bottom edge). In the third datum compartment, you will always find the tertiary datum (in this case, 1 highest point of contact on left hand edge). The alphabetical order in the compartments is not important. It is the order that the datums are placed in the compartments that is important. (first, second and third). All measurements for the holes will originate from the established datum reference frame, and not the part.

Part shown with specified datums

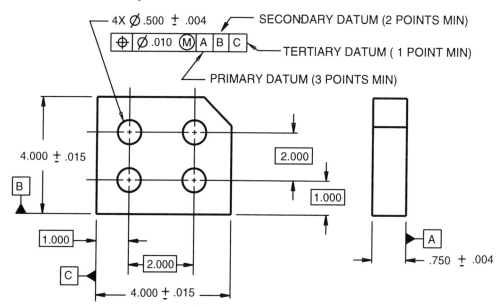

The produced part shown below (exaggerated for clarity) is not perfect. It will never be perfect; we cannot build perfect parts. None of the edges are perfectly square. It is a good part though. It is well within the acceptable limits for size and squareness.

Produced part

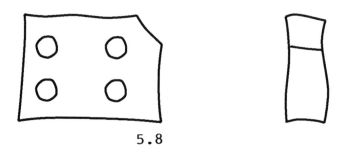

Datums in the feature control frame specify the order the part is loaded in the DRF.

THE ORDER OF THE DATUMS IN THE FEATURE CONTROL FRAME SPECIFY THE ORDER IN WHICH TO LOAD THE IMPERFECT PART IN THE PERFECT DATUM REFERENCE FRAME. WE THEN MEASURE FROM THE DRF AND NOT THE PART.

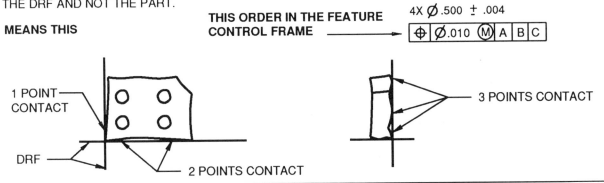

IF THE ORDER OF THE DATUMS IN THE FEATURE CONTROL FRAME ARE CHANGED, IT WILL CHANGE THE ORDER IN WHICH THE PART IS LOADED IN THE DATUM REFERENCE FRAME.

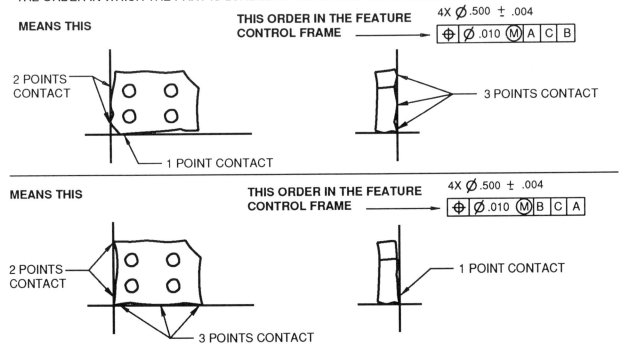

The order of the datums in the feature control frame specify the order in which to "load" this imperfect part onto our perfect datum reference frame. We measure from the planes and not the part. If the order of the datums in the feature control frame are changed, it will change the order in which the part is loaded in the datum reference frame.

From our discussion, we can see that it is very important for the product designer to specify the correct order of datum precedence in the feature control frame. This will ensure design intent. The selection of the datums depends on part function. The designer must visualize how the part will function with its mating part. When this part is located into an assembly, what is its primary, secondary, and tertiary planes?

How are the three planes defined, and what is their order of importance? These questions must be answered and the order of the datums then placed into the feature control frame. There is no one answer for these questions. As you know, all parts are different, and each one has a new twist.

The part with four holes that we used earlier to specify datums is a very simple example. As the parts become more complex, so will our selection of datums. In cylindrical parts the tertiary datum (sometimes called an auxiliary datum) may be used to stop rotation about the axis. What we are trying to accomplish with our datum reference frame is to restrict the six degrees of freedom from movement a part may have.

In the following sections there are a number of more complex type parts on which datums are specified. These examples will act as an aid to the proper selection and order of datums.

Datums restrict the six degrees of freedom

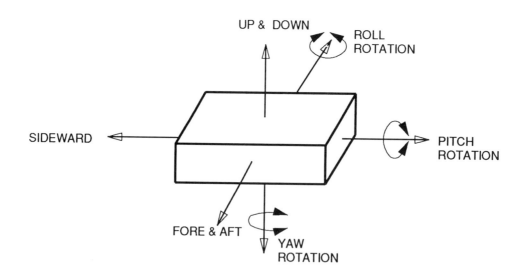

DATUM FEATURES

Parts are composed of features. The features can have size, such as holes, slots, tabs, pins etc. The features can also be without size, such as plane surfaces. Any feature can be a datum feature. Therefore, datum features can be datum features with size or datum features without size.

Datum features without size

If the datum feature is a plane surface, the datum feature symbol is directed to the surface. It may be directed to the surface by means of a leader line with an arrow or attached to an extension line of that surface. See examples on following page.

Datum features with size

If the datum feature is a feature of size, the datum feature symbol is attached or associated with the size dimension. Depending on the shape of the feature, this defines the axis, median plane or point of the feature as the datum. The datum feature symbol is not attached to a center line on a drawing because in many cases a number of features on the drawing may share this same theoretical centerline and drawing intent would not be clear. By associating the datum feature symbol with a particular size tolerance we are defining a specific definable axis, median plane or point of a feature as the datum. See examples on following pages.

The correct placement of the datum feature symbol is very important. The procedure for placement differs for size and non-size features. Take particular care in examining the figure below for datum features without size and datum features with size. Datum features K and M are plane surfaces or datum features without size. Datums features A and B are datum features with size or centerplanes. The difference is how the datum feature symbol is attached. Datums B and A have size and the datum feature symbol is aligned with the dimension line. Datums M and K have no size and the datum feature is applied to the plane surface and is not aligned with the dimension line.

PLACEMENT OF DATUM FEATURE SYMBOL IS VERY IMPORTANT

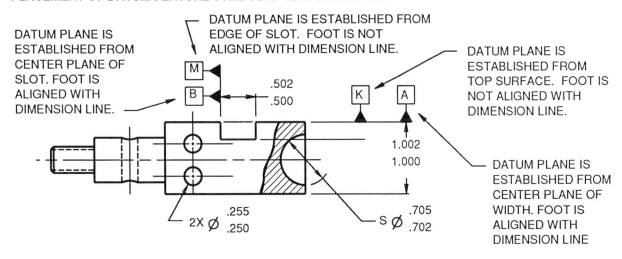

DATUM PLANE IS ESTABLISHED FROM CENTER PLANE OF SLOT. FOOT IS ALIGNED WITH DIMENSION LINE.

DATUM PLANE IS ESTABLISHED FROM EDGE OF SLOT. FOOT IS NOT ALIGNED WITH DIMENSION LINE.

DATUM PLANE IS ESTABLISHED FROM TOP SURFACE. FOOT IS NOT ALIGNED WITH DIMENSION LINE.

DATUM PLANE IS ESTABLISHED FROM CENTER PLANE OF WIDTH. FOOT IS ALIGNED WITH DIMENSION LINE

.502
.500

1.002
1.000

2X Ø .255 / .250

S Ø .705 / .702

DATUM FEATURE SYMBOLS - FORMER PRACTICE

The datum feature symbols shown below are a former practice per the ANSI Y14.5-1982 standard. If the datum feature is a plane surface, the datum feature symbol is attached to the surface or an extension line of the surface. If the datum feature is a feature of size, the datum feature symbol is placed under or associated with the size tolerance.

DATUM FEATURES WITHOUT SIZE

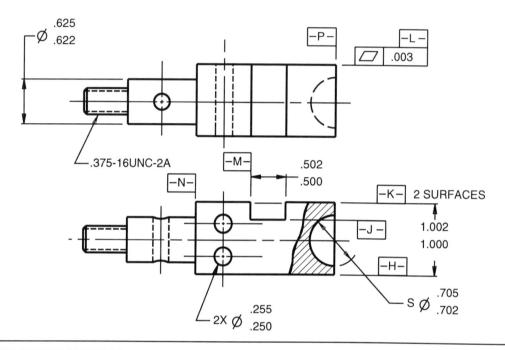

DATUM FEATURES WITH SIZE

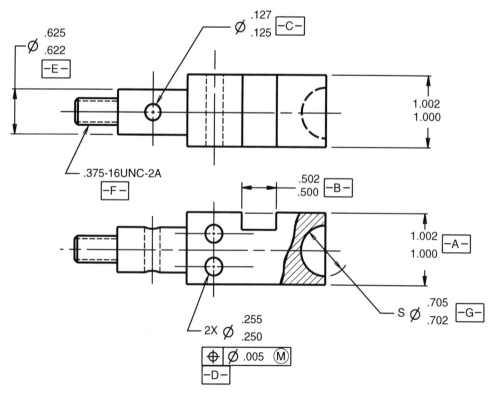

5.12

DATUM FEATURES WITHOUT SIZE

If the datum feature is a plane surface, the datum feature symbol is attached to the surface or an extension line of the surface as shown below. This will establish a datum plane.

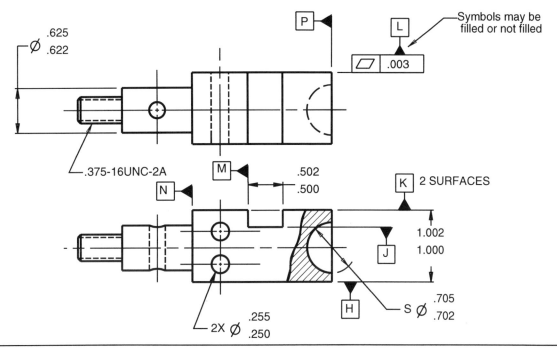

Symbols may be filled or not filled

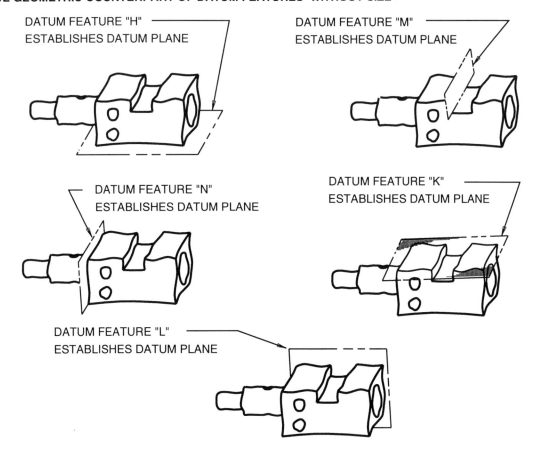

TRUE GEOMETRIC COUNTERPART OF DATUM FEATURES WITHOUT SIZE

DATUM FEATURE "H" ESTABLISHES DATUM PLANE

DATUM FEATURE "M" ESTABLISHES DATUM PLANE

DATUM FEATURE "N" ESTABLISHES DATUM PLANE

DATUM FEATURE "K" ESTABLISHES DATUM PLANE

DATUM FEATURE "L" ESTABLISHES DATUM PLANE

DATUM FEATURES WITH SIZE

If the datum feature is a feature of size, the datum feature symbol is attached to or associated with the size dimension. By attaching or associating the datum feature symbol with a particular size dimension, it defines a specific point, axis or median plane that is derived from that feature as the datum.

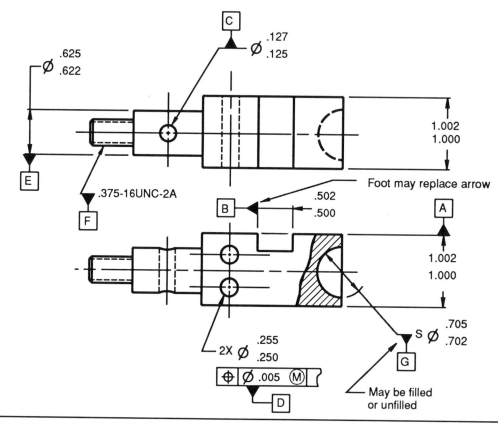

TRUE GEOMETRIC COUNTERPART OF DATUM FEATURES WITH SIZE

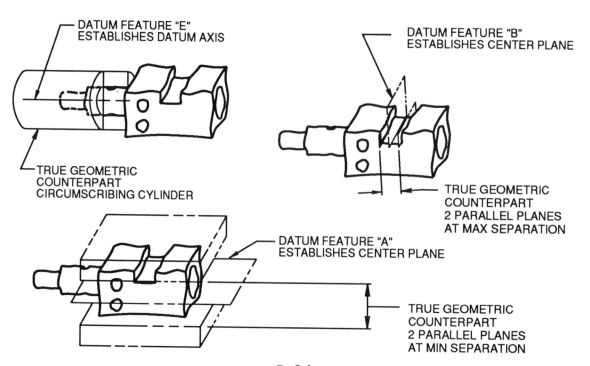

DATUM FEATURES WITH SIZE

TRUE GEOMETRIC COUNTERPART OF DATUM FEATURES WITH SIZE

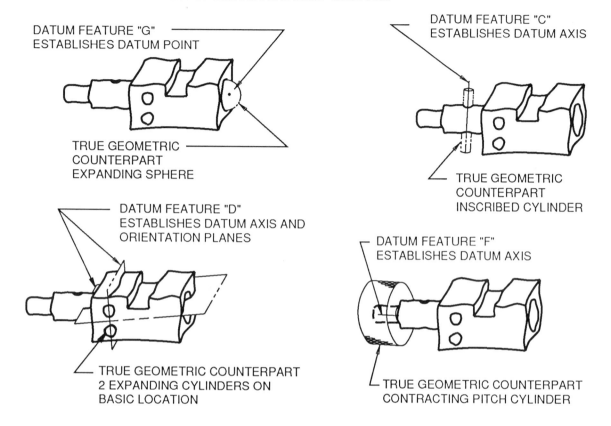

True geometric counterpart

The true geometric counterpart is the theoretical perfect boundary (virtual condition or actual mating envelope) or best-fit tangent plane of a specified datum feature. In the theoretical world we understand that parts and also the part features are imperfect. If a feature is referenced on a drawing as a datum or toleranced feature, a product designer will replicate the feature with a mathematically true geometric counterpart of itself and then dimension to or from that exact counterpart.

Notice that each true geometric counterpart of a feature creates a point, plane, line or axis. These components are used in an order in the feature control frame to establish a complete datum reference frame. As measurements cannot be made from a true geometric counterpart that is theoretical, a datum is simulated by datum simulators. The true geometric counterpart is used only to mathematically represent and understand the concept.

DATUM SIMULATORS

A datum simulator is a surface of adequately precise form (such as a surface plate, a gage surface or a mandrel) contacting the datum feature(s) and used to establish the simulated datum(s). In order to get from the theoretical world of true geometric counterparts to the practical world, we use datum simulators. We understand that these datum simulators are not perfect; but they are of such quality that their axes, centerplanes and surfaces are used for measurements and verifications. The true geometric counterparts can also be simulated mathematically with the use of a coordinate measuring machine. Shown below are examples of datum simulators.

DATUM SIMULATORS

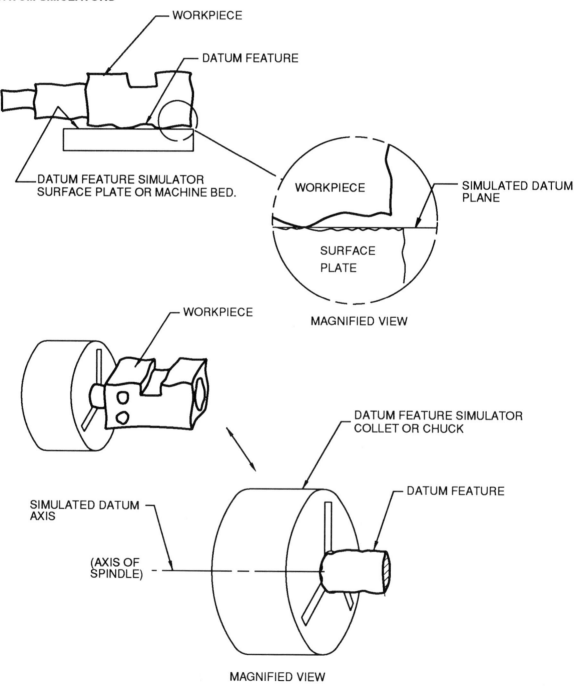

DATUM SIMULATORS

DATUM SIMULATORS

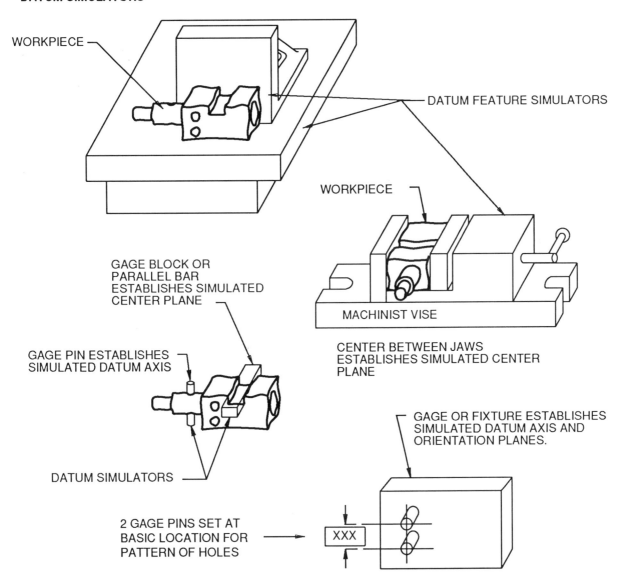

WORKPIECE

DATUM FEATURE SIMULATORS

WORKPIECE

GAGE BLOCK OR
PARALLEL BAR
ESTABLISHES SIMULATED
CENTER PLANE

MACHINIST VISE

GAGE PIN ESTABLISHES
SIMULATED DATUM AXIS

CENTER BETWEEN JAWS
ESTABLISHES SIMULATED CENTER
PLANE

GAGE OR FIXTURE ESTABLISHES
SIMULATED DATUM AXIS AND
ORIENTATION PLANES.

DATUM SIMULATORS

2 GAGE PINS SET AT
BASIC LOCATION FOR
PATTERN OF HOLES

XXX

The above figures are just a small sample of possible datum simulators. As the parts get more complex, so do the simulators. A thread pitch dia can be simulated with a thread gage. The pattern of holes shown above can also be simulated with paper gage techniques or "best fit", "auto fit" software programs on a coordinate measuring machine. In some cases special gages or fixtures will be built to simulate the datum features. This is especially evident in the aircraft or automotive industries where contoured surfaces are specified as datum features.

PARTIAL DATUM FEATURES

In some cases only a particular area of a feature will serve as a datum feature. This may be specified by the use of a thick chain line drawn parallel to the surface profile and dimensioned as to required length and width. If the area needs clarification it may be cross hatched. Partial datums may also be specified by means of a note or datum target. See examples below.

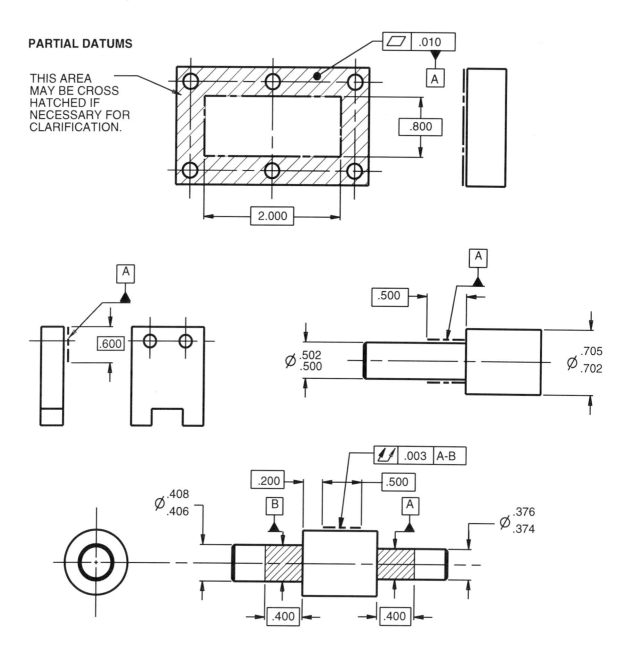

Note: Where a datum is established by two datum features (two slots, two diameters etc.), both datum reference letters are entered in a single datum compartment in the feature control frame and separated by a dash.

ESTABLISHING A DRF AND QUALIFYING DATUM FEATURES - PLANE SURFACES

THIS ON THE DRAWING

ESTABLISHMENT OF THIS DRF.

$\oplus$ | $\emptyset$.010 Ⓜ | H | L | P

The qualification of the datum features with perpendicularity and flatness control the unstability of the imperfect part in the perfect DRF and, ultimately, in the functional assembly.

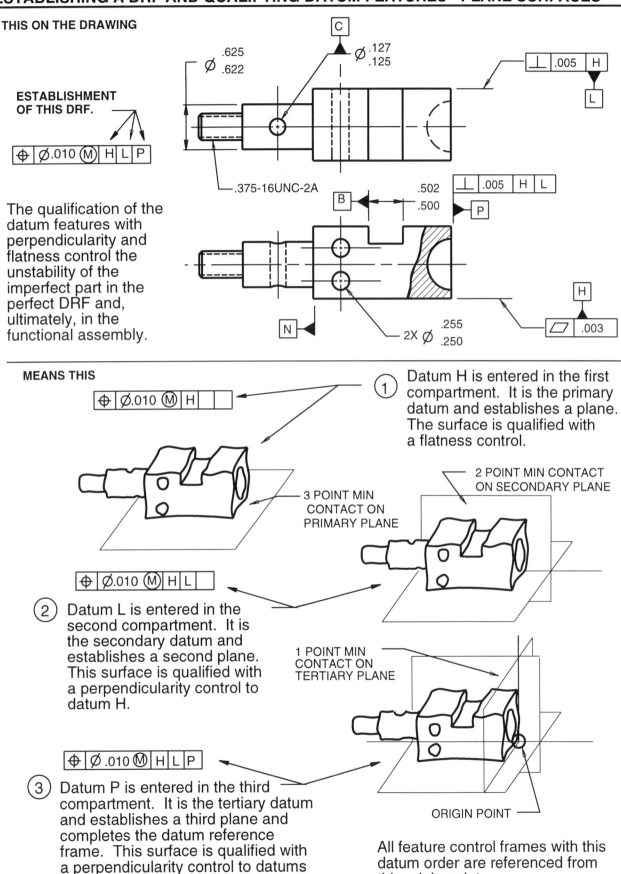

MEANS THIS

$\oplus$ | $\emptyset$.010 Ⓜ | H | |

① Datum H is entered in the first compartment. It is the primary datum and establishes a plane. The surface is qualified with a flatness control.

3 POINT MIN CONTACT ON PRIMARY PLANE

2 POINT MIN CONTACT ON SECONDARY PLANE

$\oplus$ | $\emptyset$.010 Ⓜ | H | L |

② Datum L is entered in the second compartment. It is the secondary datum and establishes a second plane. This surface is qualified with a perpendicularity control to datum H.

1 POINT MIN CONTACT ON TERTIARY PLANE

$\oplus$ | $\emptyset$ .010 Ⓜ | H | L | P

③ Datum P is entered in the third compartment. It is the tertiary datum and establishes a third plane and completes the datum reference frame. This surface is qualified with a perpendicularity control to datums H and L.

ORIGIN POINT

All feature control frames with this datum order are referenced from this origin point.

ESTABLISHING A DRF AND THE ORIGIN POINT

The origin point is the point at which all three planes in the datum reference frame intersect. It is the point of origin of all measurements. All feature control frames that reference the same datums in the same order of precedence, with the same modifying symbols will originate from this point, regardless of where the basic dimensions originate from. The location of the origin point will change as the order of the datums are changed in the feature control frame. The three examples below reference the part in the previous example. Notice that as the tertiary datum changes in the feature control frame so does the location of the origin point on the part.

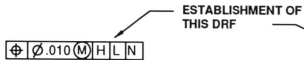

ESTABLISHMENT OF THIS DRF

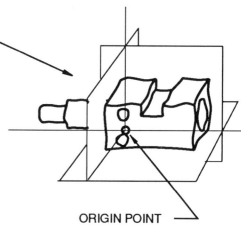

If the tertiary datum in the feature control frame is changed from P to N, the tertiary datum plane is established by the rear surface. The point of origin will move to the intersection of the three planes. To qualify the datum feature, an orientation tolerance (perpendicularity) to datums H and L would be applied.

ORIGIN POINT

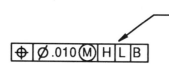

ESTABLISHMENT OF THIS DRF

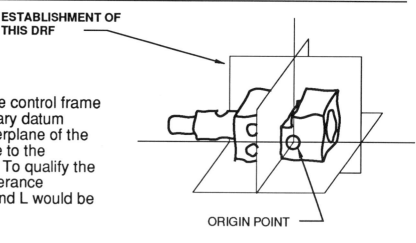

If the tertiary datum in the feature control frame is changed from N to B, the tertiary datum plane is established by the centerplane of the slot. The point of origin will move to the intersection of the three planes. To qualify the datum feature, an orientation tolerance (perpendicularity) to datums H and L would be applied.

ORIGIN POINT

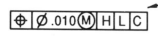

ESTABLISHMENT OF THIS DRF

ORIGIN POINT

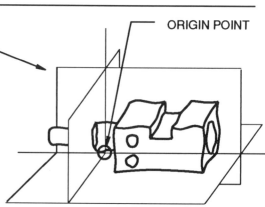

If the tertiary datum in the feature control frame is changed from B to C, the tertiary datum plane is established at the axis of the hole. The point of origin will move to the intersection of the three planes. To qualify the datum feature, a location (position) tolerance to datums H and L would be applied.

THIS ON THE DRAWING

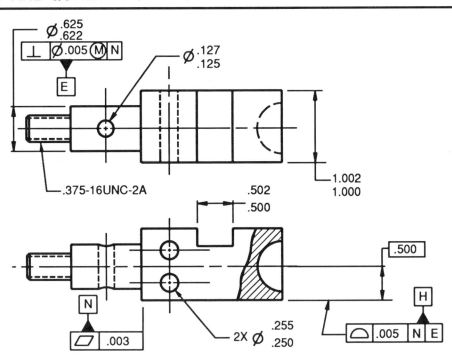

Ø .625 .622

⊥ | Ø .005 Ⓜ N

E

Ø .127 .125

.375-16UNC-2A

.502 .500

1.002 1.000

ESTABLISHMENT OF THIS DRF.

⊕ | Ø .010 Ⓜ | N | E | H

The qualification of the datum features with perpendicularity and profile control the unstability of the imperfect part in the perfect DRF and, ultimately, in the functional assembly.

N

▱ | .003

2X Ø .255 .250

.500

H

⌓ | .005 | N | E

MEANS THIS

⊕ | Ø .010 Ⓜ | N | |

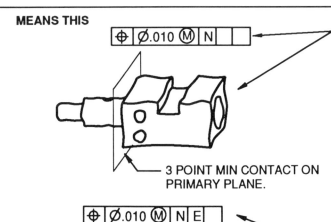

3 POINT MIN CONTACT ON PRIMARY PLANE.

① Datum N is entered in the first compartment. It is the primary datum and establishes a plane. The surface is qualified with a flatness control.

DATUM AXIS

3 POINTS MIN CONTACT FOR SECONDARY AXIS.

⊕ | Ø .010 Ⓜ | N | E |

② Datum E is entered in the second compartment. It is the secondary datum and establishes an axis. The surface is qualified with a perpendicularity control.

⊕ | Ø .010 Ⓜ | N | E | H

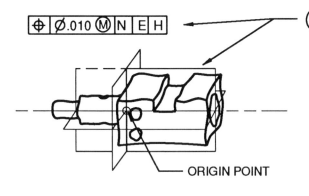

ORIGIN POINT

③ Datum H is entered in the third compartment. It is the tertiary datum. The feature surface establishes an orientation plane. The feature is qualified with a profile tolerance as it is located from the datum E axis. It stops rotation. This completes the DRF.

All feature control frames with this order are referenced from this origin point.

THIS ON THE DRAWING

**ESTABLISHMENT
OF THIS DRF.**

$\boxed{\oplus \; \boxed{\varnothing .010 \, Ⓜ} \; E \; N \; S}$

The qualification of the datum features with perpendicularity and flatness control the unstability of the imperfect part in the perfect DRF and, ulitmately, in the functional assembly.

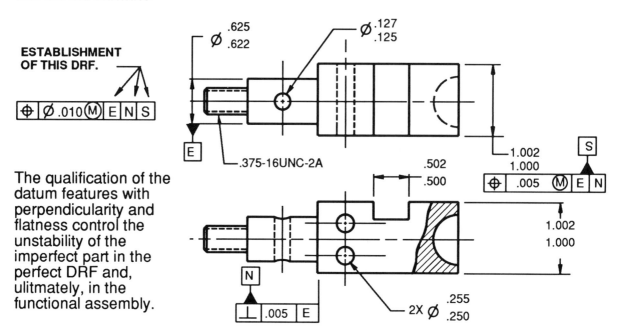

MEANS THIS

$\boxed{\oplus \; \boxed{\varnothing .010 \, Ⓜ} \; E \; \; \;}$

① Datum E is entered in the first compartment. It is the primary datum and establishes an axis. If necessary the feature may be qualified with a form (circularity or cylindricity) control.

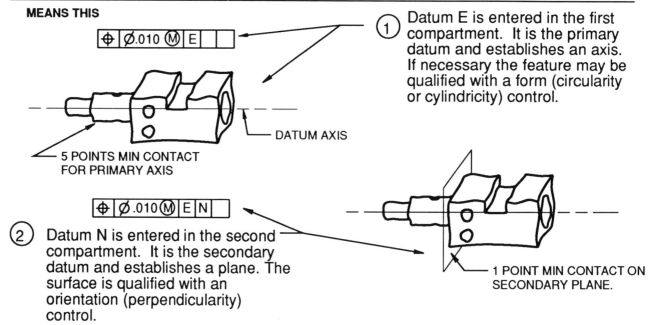

— DATUM AXIS

— 5 POINTS MIN CONTACT FOR PRIMARY AXIS

$\boxed{\oplus \; \boxed{\varnothing .010 \, Ⓜ} \; E \; N \; \;}$

1 POINT MIN CONTACT ON SECONDARY PLANE.

② Datum N is entered in the second compartment. It is the secondary datum and establishes a plane. The surface is qualified with an orientation (perpendicularity) control.

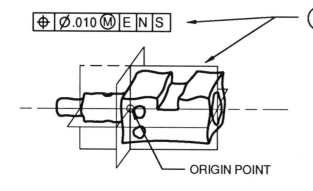

$\boxed{\oplus \; \boxed{\varnothing .010 \, Ⓜ} \; E \; N \; S}$

③ Datum S is entered in the third compartment. It is the tertiary datum. The median plane of the feature establishes an orientation plane. The feature is qualified with a location (position) tolerance as it is located from the datum E axis. This completes the datum reference frame.

All feature control frames with this order are referenced from this origin.

— ORIGIN POINT

1. The datum reference frame (DRF) consists of points, lines (axes) and planes. Label these components in the illustration below.

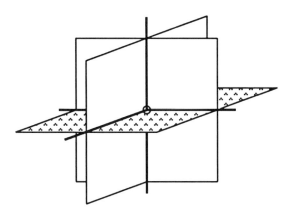

2. The datum reference frame is three mutually perpendicular planes. In engineering, it represents the surface of a drafting table, the straight edge and the 90 degree triangle or the coordinate system in CAD. What equipment does it represent in manufacturing?

_____

_____

_____

3. What equipment does the datum reference frame represent in inspection?

_____

_____

4. What does the datum reference frame represent in mathematics? Hint: It is named after a French mathematician.

_____

5. The two symbols below are called _____

6. Next to the above symbols, write the name and date of the American National Standard in which they are used.

7. In the drawing below, label the following terms;
    Datum feature
    Datum simulator
    Simulated datum

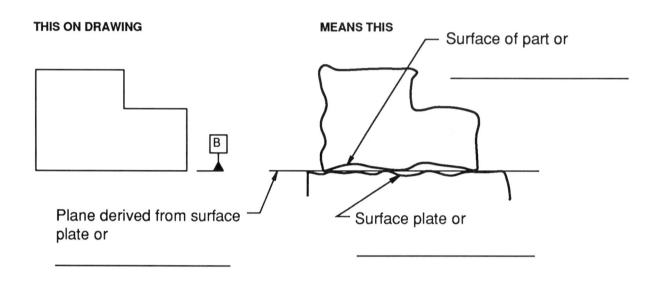

THIS ON DRAWING　　　　　　　**MEANS THIS**

Surface of part or

_____

B

Plane derived from surface
plate or

Surface plate or

_____

_____

8. Name a datum simulator which will establish a simulated datum line or axis on an internal feature such as a hole.

_____

9. Name a datum simulator which will establish a simulated datum line or axis on an external feature such as a shaft.

_____

10. Name a datum simulator which will establish a simulated datum plane on a slot.

_____

11. Name a datum simulator which will establish a simulated datum plane on a plane surface.

_____

12. The primary datum requires a minimum of how many points of contact?

_____

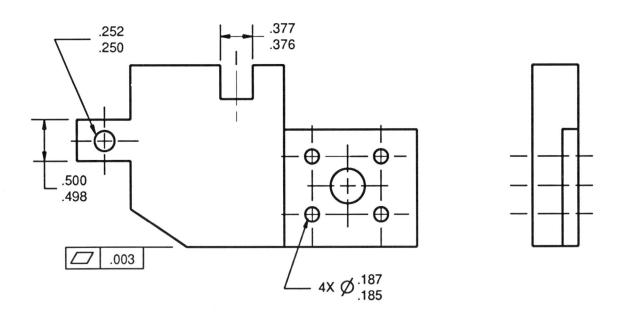

On the drawing above, apply datum feature symbols as directed below.

13. Establish the right hand face in the side view as datum feature A.

14. Establish the bottom surface in the plan view as datum B.

15. Establish the top of the tab as datum feature C.

16. Establish the tab as datum feature D. (Median plane of the feature)

17. Establish the bottom surface of the slot as datum feature E.

18. Establish the slot as datum feature F. (Median plane of the feature)

19. Establish a partial datum G on the right hand surface of the part in the plan view. The length is 1.750 from the bottom, up.

20. Establish the 4 hole pattern as datum feature H. (Axis and orientation planes of the features.

21. Identify the .250 diameter hole as datum feature J. (Axis of the feature)

22. Datum features can be placed in two categories - datum features that have size and datum features that have no size. Place an "X" next to all the datums that have size.

THIS ON THE DRAWING

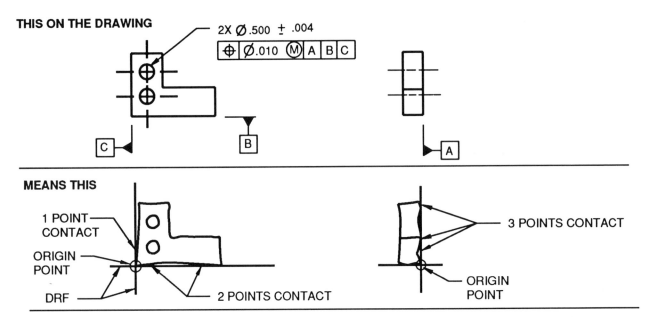

2X Ø .500 ± .004

⊕ | Ø.010 Ⓜ | A | B | C

MEANS THIS

1 POINT CONTACT

ORIGIN POINT

DRF

2 POINTS CONTACT

3 POINTS CONTACT

ORIGIN POINT

In the figure above, an imperfect part is shown loaded on the datum reference frame according to the order designated in the feature control frame. In the following examples, sketch an imperfect part loaded on the datum reference frame according to the applicable feature control frame. Also, specify the points of contact on each plane and identify the origin point.

23. This order in the feature control frame. ⟶

2X Ø .500 ± .004

⊕ | Ø .010 Ⓜ | A | C | B

MEANS THIS

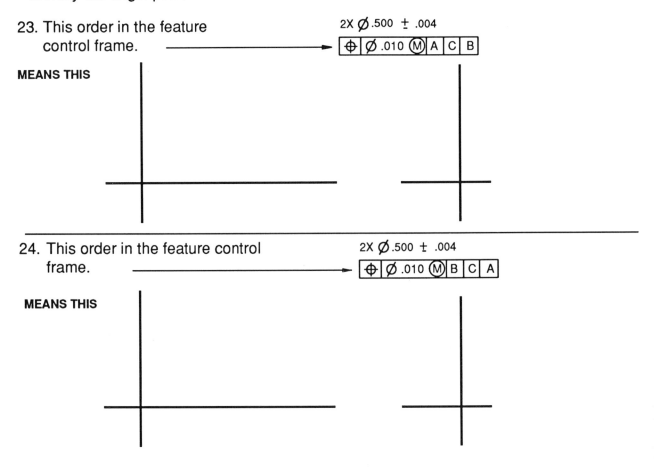

24. This order in the feature control frame. ⟶

2X Ø .500 ± .004

⊕ | Ø .010 Ⓜ | B | C | A

MEANS THIS

UNIT 6

DATUM MODIFIERS

EFFECT OF DATUM MODIFIERS - PLATE

A group of features may be controlled relative to a datum feature at MMC as shown below. Datum feature B at MMC establishes the location of the axis of the datum reference frame (DRF) for the location of all the features. As datum feature B departs from MMC, its axis may be displaced relative to the datum B at MMC axis in a diameter zone equal to the difference between the virtual size and the actual virtual size.

THIS ON THE DRAWING

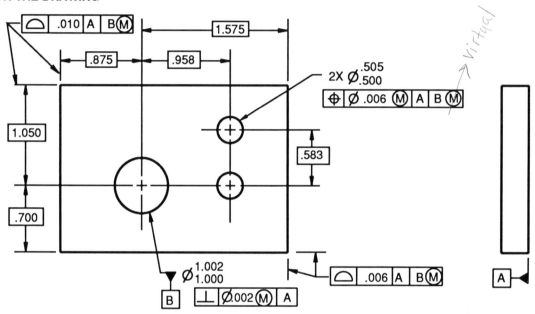

MEANS THIS

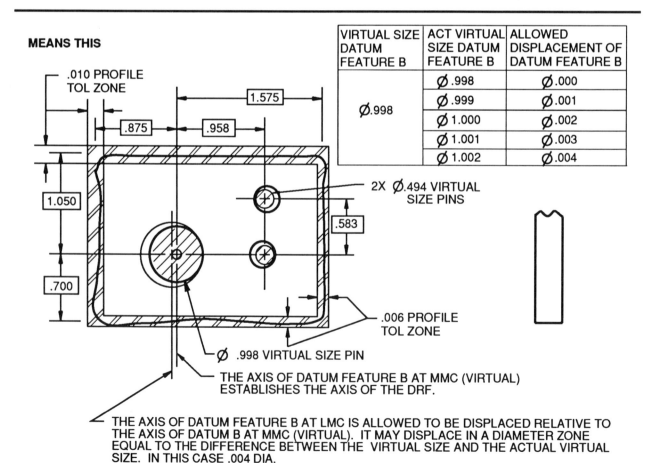

VIRTUAL SIZE DATUM FEATURE B	ACT VIRTUAL SIZE DATUM FEATURE B	ALLOWED DISPLACEMENT OF DATUM FEATURE B
Ø.998	Ø.998	Ø.000
	Ø.999	Ø.001
	Ø1.000	Ø.002
	Ø1.001	Ø.003
	Ø1.002	Ø.004

THE AXIS OF DATUM FEATURE B AT MMC (VIRTUAL) ESTABLISHES THE AXIS OF THE DRF.

THE AXIS OF DATUM FEATURE B AT LMC IS ALLOWED TO BE DISPLACED RELATIVE TO THE AXIS OF DATUM B AT MMC (VIRTUAL). IT MAY DISPLACE IN A DIAMETER ZONE EQUAL TO THE DIFFERENCE BETWEEN THE VIRTUAL SIZE AND THE ACTUAL VIRTUAL SIZE. IN THIS CASE .004 DIA.

6.2

The effect of a datum modifier is accommodated automatically in a functional gage as shown below. The datum feature and the related features must fall within the appropriate tolerance zones. As the datum feature and related features depart from their stated material condition, they may displace and shift as long as they clear the virtual size pins and remain within the profile tolerance zones. Notice that the datum modifier has no effect on the relationship between the related features. Unless otherwise stated there is an implied simultaneous requirement between all the features and the datum features.

FUNCTIONAL GAGE ILLUSTRATING EFFECT OF DATUM MODIFIERS

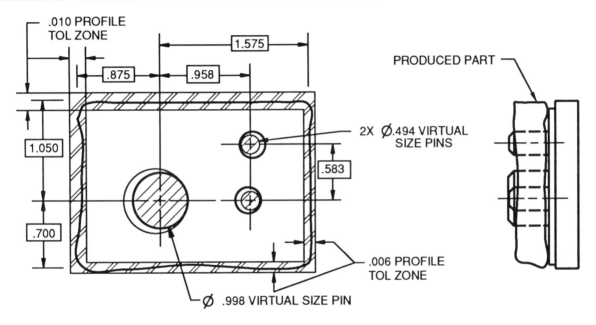

If the produced part were to be evaluated using a CMM or open set-up techniques, the shift of the datum feature may also have to be taken into account. If inspection "zeroed in" on the datum feature and all the related features checked good, then the part would pass inspection. If the part checked bad, the zero or origin could be reset in a diameter zone equal to the departure of the feature from its virtual size to its actual virtual size.

It is often explained that the features shift as a group. In actuality, it can be seen that the actual datum features, as they depart from virtual, do the shifting or displacement relative to the DRF at virtual. The datum shift can be calculated with the appropriate software on a CMM or by the paper gage concept. It can also be accommodated with functional gaging techniques.

Datum Features at Virtual Condition

A virtual condition exists for a datum feature of size where its axis or centerplane is controlled by a geometric tolerance. In such cases, the datum feature applies at its virtual condition even though it is referenced at MMC or LMC. Where a virtual condition equal to the maximum material condition is required, a zero tolerance at MMC or LMC is specified.

SECONDARY DATUM - PAPER GAGE EVALUATION

The example below is used to illustrate the use of the paper gage concept to gain the additional tolerance from datum B as it departs from virtual. (The extra datum departure tolerance cannot be just added to the position tolerance for the holes.) The produced part below has been set up by "zeroing" on the datums A and B and "balancing" the pattern of two holes according to the specifications below. Data has been collected from inspection and the results of the data are illustrated in the chart below.

The position of hole #1 checks bad and the position of hole # 2 checks good. After evaluating datum feature B it is found that datum B departed from virtual by .004. This allows a shift tolerance of .004 to the two holes as a group. A paper gage evaluation has been completed on the following page and the part is acceptable.

THIS ON THE DRAWING

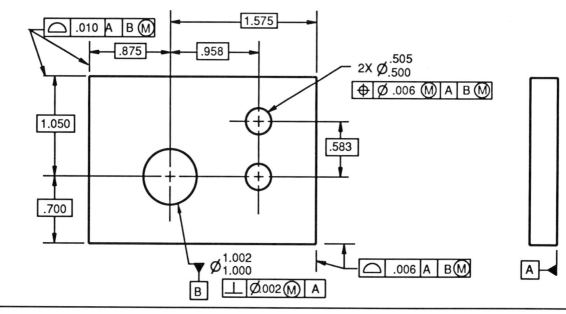

PRODUCED PART

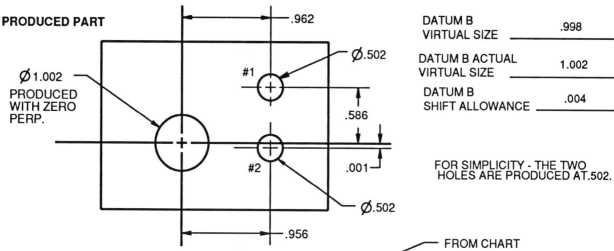

Ø 1.002
PRODUCED
WITH ZERO
PERP.

DATUM B
VIRTUAL SIZE _____ .998

DATUM B ACTUAL
VIRTUAL SIZE _____ 1.002

DATUM B
SHIFT ALLOWANCE _____ .004

FOR SIMPLICITY - THE TWO
HOLES ARE PRODUCED AT.502.

HOLE NO.	HOLE MMC	HOLE ACTUAL SIZE	POSITION TOLERANCE ALLOWED	"X" DIM.	"Y" DIM.	POSITION TOLERANCE ACTUAL	ACC or REJ	DATUM EVALUATION	ACC or REJ
1	.500	.502	.008	+.004	+.003	.010	R	DATUM B DEPART FROM VIRTUAL	A
2	.500	.502	.008	-.002	-.001	.0045	A	ALLOWS SHIFT AND ACCEPTS 2 HOLES	A

6.4

SECONDARY DATUM - PAPER GAGE EVALUATION

The paper gage solution below is for the flat plate shown on the previous page. Notice that the position for all the features check good. The pilot hole departed from virtual by .004 diameter. This allows a shift of the pattern of two holes as a group by .004 diameter.

Notice below that the polar coordinate can shift relative to the cartesian coordinate as long as the axis of datum feature B (center of the cartesian system) stays within a .004 dia polar coordinate circle. At the same time the two holes must also remain in their alloted position tolerance which is a .008 dia circle.

The shifted center of the polar coordinate relative to the center of the cartesian coordinate is the location of the axis of the virtual size of datum feature B. In this case, the origin has shifted by +.001 in the X and +.001 in the Y direction. To prove the results the part could be reset in inspection to the new origin and the holes checked again.

Basically, since all the features (datum feature B and the two holes) have all departed from MMC, all they have to do is fall within their respective tolerance zones. Remember though, because of the simultaneous requirement, the profle tolerance must also be acceptable at the new datum origin location.

THE PAPER GAGE EVALUATION BELOW SHOWS THE POSITION FOR THE HOLES TO BE ACCEPTABLE.

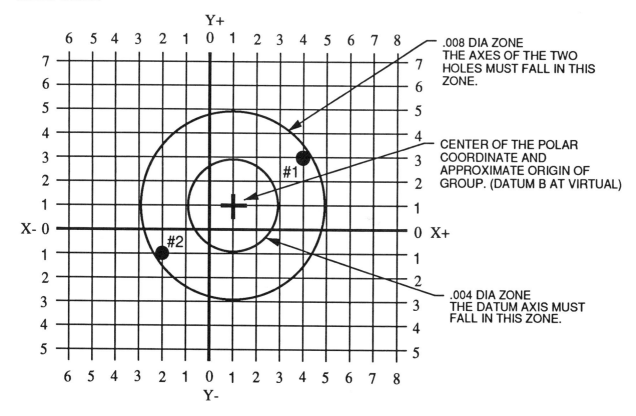

ROUND PART WITH SLOT

The part below mounts on the right hand face, the pilot and the key. These features are identified as datum features and are related to each other by the perpendicularity and position tolerances. The holes and outside contour have been related to the datum features by the position and profile tolerances. The .060 profile tolerance on the outside contour controls size, form, location and orientation.

THIS ON THE DRAWING

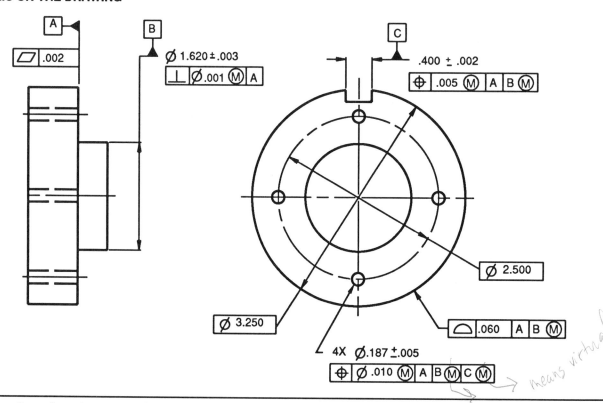

The functional gage or boundaries shown below reflect the geometric controls above. Since the features and datums are all applied at MMC, all the features may shift and/or displace as long as they meet the requirements set forth below.

FUNCTIONAL GAGE OR BOUNDARIES

.393 VIRTUAL KEY
(MMC - POSITION)

1.624 VIRTUAL SIZE
(MMC + POSITION)

.060 MACHINED
STEP
(PROFILE CHECK)

4X ⌀.172 VIRTUAL PINS
(MMC - POSITION)

SAMPLE INSPECTION SET-UP FOR ROUND PART WITH SLOT

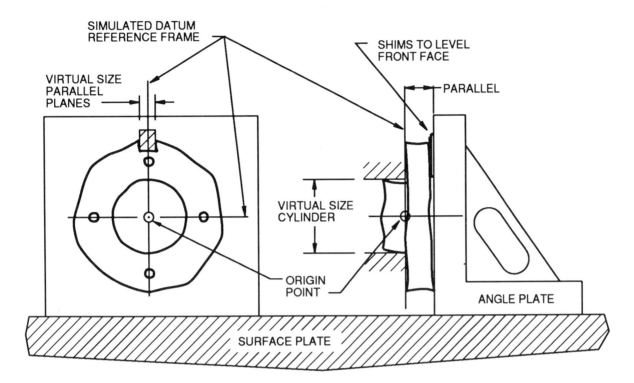

The illustration above is a sample open set-up inspection procedure for the verification of the round part in the previous example. Notice that the front face, datum feature A, is leveled or balanced with shims to make it parallel with the angle plate. The virtual size of the pilot is centered and the part is anti-rotated with the slot to set up the datum reference frame. All measurements are made from the origin point.

This part can also be verified with a coordinate measuring machine (CMM). The front face is leveled, the origin is set at the pilot and the part is oriented by the slot.

Since the MMC modifier is referenced on the features and datum features, additional tolerance is available as the datum features and features depart from virtual size.

The geometric tolerancing defines the product without defining the measurement procedure. All measurement procedures have associated uncertainty or risk. Some procedures will have more risk than others. This is called methods divergence. This part can also be verified wtih calipers to provide a rough check. The type of verification procedure is defined in the measurement plan. The measurement plan is designed by the quality engineer with knowledge of the manufacturing process.

TERTIARY DATUM - PAPER GAGE EVALUATION

This example is intended to illustrate the use of the paper gage concept to gain the additional positional tolerance from datum C as it departs from virtual. (The extra datum departure tolerance cannot be just added to the position tolerance.) The produced part below has been set up by "zeroing" on the the datums A, B and C according to the specification on the previous page. Data has been collected from the workpiece and the results of the data are illustrated in the chart below. The paper gage evaluation of the tertiary datum is shown on the next page.

THIS ON THE DRAWING

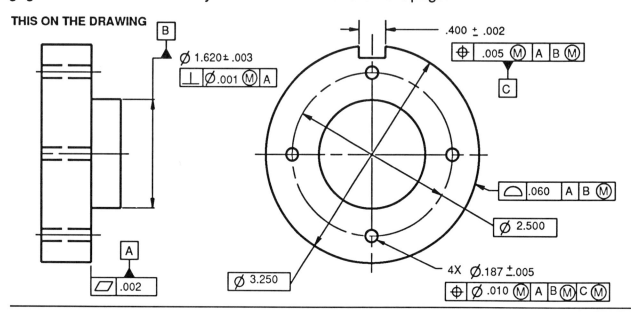

FOR SIMPLICITY - ALL HOLES ARE PRODUCED AT MMC (.182)

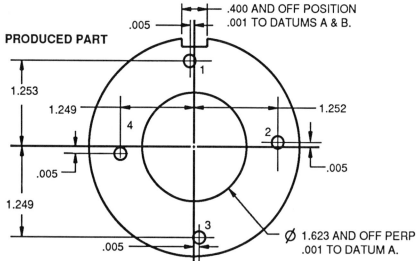

PRODUCED PART

DATUM B VIRTUAL SIZE	1.624
DATUM B ACTUAL VIRTUAL SIZE	1.624
DATUM B SHIFT ALLOWANCE	NONE
DATUM C VIRTUAL SIZE	.393
DATUM C ACTUAL VIRTUAL SIZE	.399
DATUM C SHIFT ALLOWANCE	.006

FROM CHART

HOLE NO.	HOLE MMC	HOLE ACTUAL SIZE	POSITION TOLERANCE ALLOWED	"X" DIM.	"Y" DIM.	POSITION TOLERANCE ACTUAL	ACC or REJ	DATUM EVALUATION	ACC or REJ
1	.182	.182	.010	-.005	+.003	.0117	R	DATUM C DEPARTURE	A
2	.182	.182	.010	+.002	+.005	.0108	R	FROM VIRTUAL ALLOWS	A
3	.182	.182	.010	+.005	+.001	.0102	R	ROTATION AND ACCEPTS 4	A
4	.182	.182	.010	+.001	-.005	.0102	R	HOLES	A

6.8

TERTIARY DATUM - PAPER GAGE EVALUATION

The position of the four holes are, at first, rejected without considering the datum modifiers. After evaluating the datums it is found that datum feature B, the pilot, did not depart from virtual and, therefore, offers no additional tolerance. Datum feature C, the slot, however, did depart from virtual. Additional position tolerance is available from the slot. The slot is the tertiary datum, and it controls rotation. Since the slot departed from virtual, additional rotational tolerance is available for the four holes as a group.

The paper gage solution below illustrates the four holes all check good and are all within their alloted .010 position. The slot departed from virtual by .006. This allows a rotation of the pattern of holes by approximately .005 total or .0025 in either direction. The holes are estimated to have only .005 rotation as they are closer to the center than the slot and does not receive full benefit of the .006 departure of datum feature C.

Notice each hole moved on the graph in the same direction as it would have moved if the part were rotated to the right. Hole #1 rotates to the right. Hole #2 rotates down. Hole #3 rotates to the left. Hole #4 rotates up. After the rotation, all of the hole locations are acceptable. To prove the results obtained with the paper gage, the part could be reset in the inspection procedure with the new datum feature C alignment.

The calculations below are an approximate evaluation of the hole pattern. Other factors, such as orientation, form and number of data points taken from the features, will also affect the outcome. This example is intended to illustrate the general concepts. Data may also be evaluated by appropriate CMM software or on a CAD system.

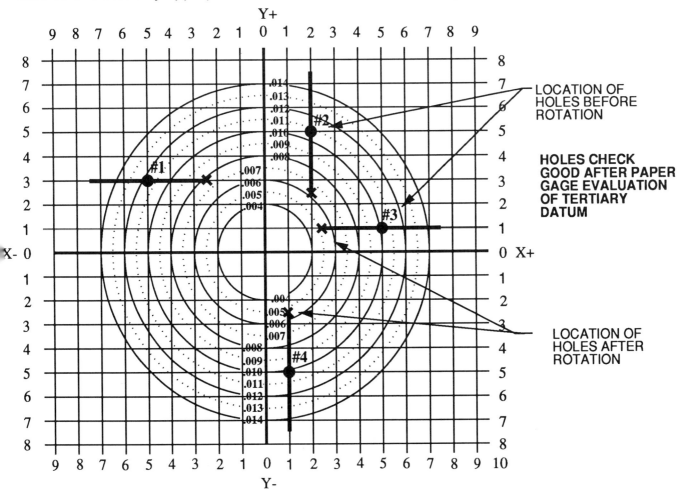

SIMULTANEOUS VS. SEPARATE REQUIREMENT

The simultaneous requirement concept applies to both position and profile specifications. It does not apply to the lower segment in a composite tolerance.

Multiple patterns of features, located by basic dimensions from common datum features of size, are considered a single composite pattern if their respective feature control frames contain the same datums in the same order of precedence with the same material condition modifiers.

If such an interrelationship is not required between patterns of features, the notation SEP REQT is placed under each applicable feature control frame. This allows each pattern of features to shift and/or rotate independently about the established datum reference frame.

SIMULTANEOUS REQUIREMENT

In the part below all the features are related with basic dimensions using the same datums, in the same order of precedence, with the same modifiers. This constitutes a single pattern and all the features must shift or rotate together as a group as datum feature B departs from MMC. In this case, since all the features are considered a single pattern, another datum to control orientation is not necessary.

SIMULTANEOUS REQUIREMENT IS IMPLIED

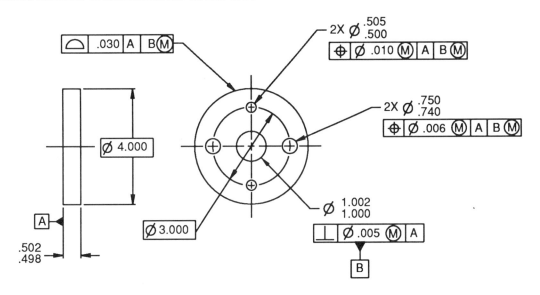

FUNCTIONAL GAGE OR 3D BOUNDARIES ILLUSTRATING VIRTUAL SIZES FOR SIMULTANEOUS REQUIREMENT.

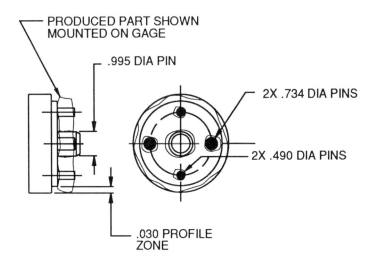

The sample functional gage shown above will verify all the features at once. The gage is for illustration only, as other methods could be used as well. In open set-up procedures, care should be taken when datum shift or feature rotation is factored into the acceptance criteria. Remember, datum shift and rotation must be factored to all the features as a group, rather than allowing each pattern of features to shift independently as the datum feature departs from MMC.

SEPARATE REQUIREMENT

In the part below all the features are related with basic dimensions using the same datums, in the same order of precedence, with the same modifiers. The notation SEP REQT is placed under all the feature control frames. This constitutes three separate patterns and all the features may rotate, and/or shift independently as datum feature B departs from MMC. If orientation between the features is required, it must be specified.

SEPARATE REQUIREMENT MUST BE SPECIFIED

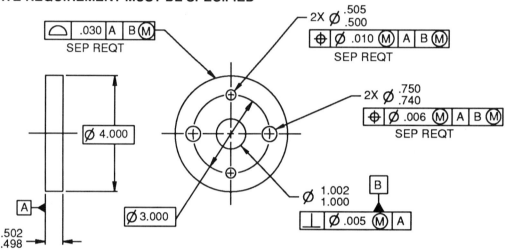

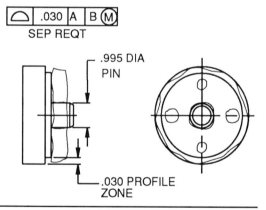

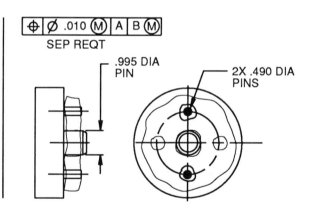

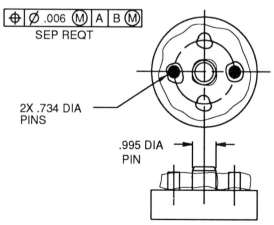

The three functional gages to verify the separate requirements are shown for illustration only. The requirements may be verified with other procedures as well. The produced parts are shown on the gages.

Notice that the separate requirement specification allows three independent verifications of the features. This allows the features to rotate and/or shift to each other. Each pattern, however, is related to datum feature A and datum feature B at MMC.

WORKSHOP EXERCISE 6.1 - PROBLEM 1

The top drawing is a plate with a datum of size and position tolerancing on the two holes. The lower drawing is the produced part. Evaluate the dimensions on the produced part to verify conformance to the tolerances. For this problem ignore the profile tolerances. Use the chart below to record your calculations.

Use the paper gage materials from unit 3 to calculate the shift tolerance available from datum feature B. As an alternative to completing the entire paper gage evaluation, make sure you can at least calculate the available shift tolerance. Record your numbers below.

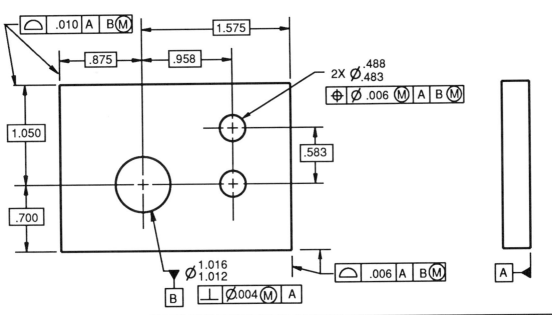

PRODUCED PART

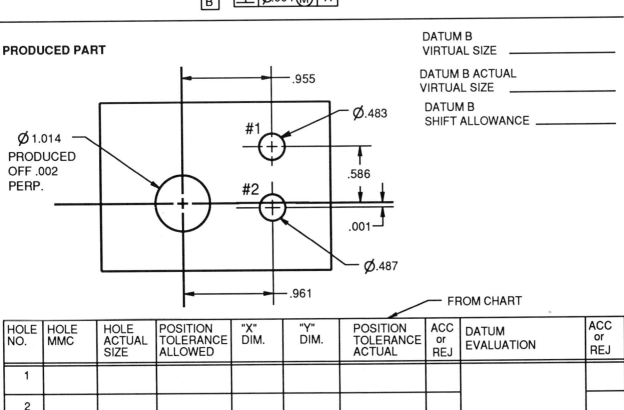

DATUM B
VIRTUAL SIZE _____

DATUM B ACTUAL
VIRTUAL SIZE _____

DATUM B
SHIFT ALLOWANCE _____

FROM CHART

HOLE NO.	HOLE MMC	HOLE ACTUAL SIZE	POSITION TOLERANCE ALLOWED	"X" DIM.	"Y" DIM.	POSITION TOLERANCE ACTUAL	ACC or REJ	DATUM EVALUATION	ACC or REJ
1									
2									

6.13

WORKSHOP EXERCISE 6.1 - PROBLEM 2

The top drawing has two datums of size and position tolerancing on the 4 holes. The lower drawing is the produced part. Evaluate the dimensions on the produced part to verify conformance to the tolerances. For this problem ignore the profile tolerance. Use the chart below to record your calculations. Use the paper gage materials from unit 3 to calculate shift or rotation tolerance. As an alternative to completing the entire paper gage evaluation, make sure you can at least calculate the availble shift or rotation tolerance from the datums. Record your numbers below.

THIS ON THE DRAWING

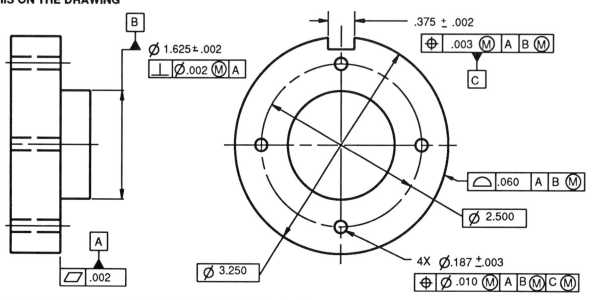

PRODUCED PART

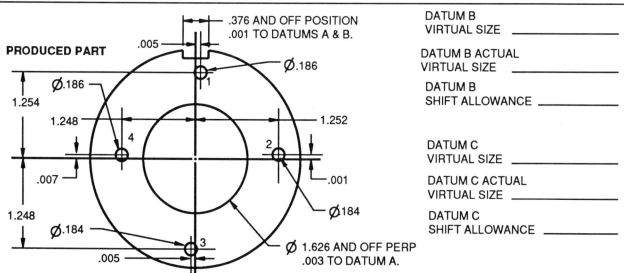

DATUM B
VIRTUAL SIZE _____

DATUM B ACTUAL
VIRTUAL SIZE _____

DATUM B
SHIFT ALLOWANCE _____

DATUM C
VIRTUAL SIZE _____

DATUM C ACTUAL
VIRTUAL SIZE _____

DATUM C
SHIFT ALLOWANCE _____

FROM CHART

HOLE NO.	HOLE MMC	HOLE ACTUAL SIZE	POSITION TOLERANCE ALLOWED	"X" DIM.	"Y" DIM.	POSITION TOLERANCE ACTUAL	ACC or REJ	DATUM EVALUATION	ACC or REJ
1									
2									
3									
4									

UNIT 7

DATUM REFERENCE FRAME SELECTION AND QUALIFICATION EXERCISES

DATUM SELECTION

The selection of datums should be based on the function of the part - how it mates, how it assembles or how it is installed. If datums are selected based on function, it will provide the designer with an organized, methodical, realistic method to calculate tolerances. This, in turn, will provide maximum manufacturing tolerances based on function.

Parts are comprised of features. Some of these features are leaders, and some features are followers. The designer should evaluate the assembly and determine global datum reference frames on the assembly as well as individual reference frames on the parts. He should determine which features are the leaders and which features are the followers. The leaders are selected as datum features. The features are selected in order of precedence and qualified to the DRF. The features are then located to the established DRF. A sequence or outline to aid the designer in applying geometric tolerancing can be found below.

1. **ESTABLISH THE DATUM REFERENCE FRAME (DRF) - Establish the leaders**
   Select datum features as necessary in order of precedence.
   Primary
   Secondary
   Tertiary

2. **QUALIFY THE DATUM FEATURES TO THE DRF - Qualify the leaders**
   Apply form, orientation, profile and location tolerances to the datum features as necessary.

3. **RELATE THE FEATURES TO THE DRF - Relate the followers**
   Apply form, orientation, profile and location tolerances to the remaining features as necessary.

The designer should be careful when establishing datum reference frames. A common mistake is for a designer to establish too many DRFs on simple parts. Parts as complex as a gear box housing or automotive cylinder block may have many DRFs. Parts as simple as a shaft or cover will have only one DRF. In fact, most parts will have only one DRF. Modern manufacturing methods usually try to manufacture parts in one set-up.

Parts incorrectly drawn with every feature control frame having different datum precedence or different datums constitutes multiple DRFs. Inspection must reorient parts to the different DRFs. This has the possibility of inspection rejecting functionally good parts. The designer should try to keep the part simple and, if possible, establish only one reference frame on a part relating everything to this DRF. If it is necessary to establish multiple reference frames, make sure the DRFs are related to each other as shown in the multiple DRF section later in the text.

WORKSHOP EXERCISES - INTRODUCTION

The datum reference frame exercises in this unit are designed to help the user in selecting and applying a datum reference frame to parts. These exercises will also require the user to qualify the datum features relative to the datum reference frame. Problem #1 on the following page is a sample exercise that has been completed. The remaining exercises should be completed in a similar manner.

In the middle of the page, on the right hand side of every exercise is an assembly drawing showing how the part fits in the functional assembly. It is your task to establish a datum reference frame on the engineering drawing based on function. Apply feature control frames to insure the datum features are qualified with any necessary form, orientation, position or profile tolerances.

Afterwards, **VERY CLEARLY** on the bottom produced part, show by hand sketch the datum reference frame, datum origin point and applicable tolerance zones. To keep the example simple, make all the tolerance zones .005 unless it is necessary to insure a fit with virtual sizes on the mating part in problem #3 and beyond.

On the sample problem #1, tolerances were also applied to other features as well, such as the profile tolerance and the other four position tolerances. It is not necessary for you to apply tolerances to the features. Since the function of these other features are not defined, it would be impossible to specify tolerances for the remaining features. The tolerances on the sample problem #1 were added to expand the concept that, once the designer establishes the DRF, he then can locate the remaining features to the DRF. Position is used to locate features of size and profile is used to locate surfaces.

The exercises in this chapter are designed to select DRF's and qualify the datum features. As you work on problem #3 and beyond, datum modifiers must also be selected. Again to keep the problems focused and simple, apply the MMC modifier as necessary to all the datum features.

CLASS EXERCISE 7.1 - PROBLEM 1

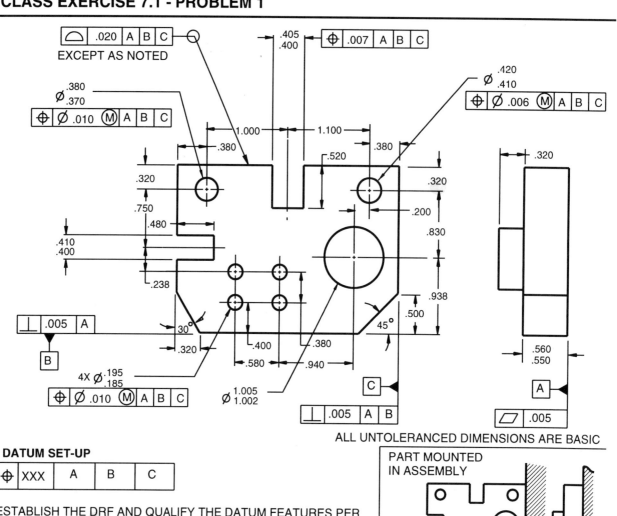

⌒ | .020 | A | B | C
EXCEPT AS NOTED

⊕ | .007 | A | B | C

⌀ .380 / .370
⊕ | ⌀ | .010 | Ⓜ | A | B | C

⌀ .420 / .410
⊕ | ⌀ | .006 | Ⓜ | A | B | C

1.000   1.100

.380   .520   .380

.320   .320

.750   .200

.480   .830

.410 / .400   .938

.238   .500

⊥ | .005 | A

B

30°   45°

.320   .400   .380

4X ⌀ .195 / .185
⊕ | ⌀ | .010 | Ⓜ | A | B | C

.580   .940

⌀ 1.005 / 1.002

C

⊥ | .005 | A | B

.320

.560 / .550

A

▱ | .005

ALL UNTOLERANCED DIMENSIONS ARE BASIC

DATUM SET-UP

⊕ | XXX | A | B | C

ESTABLISH THE DRF AND QUALIFY THE DATUM FEATURES PER THE ASSEMBLY REQUIREMENTS. THE PART MOUNTS ON THE BACK SURFACE, BOTTOM EDGE AND RIGHT HAND SURFACE.

PART MOUNTED IN ASSEMBLY

.005 FLATNESS TOL ZONE

PRODUCED PART

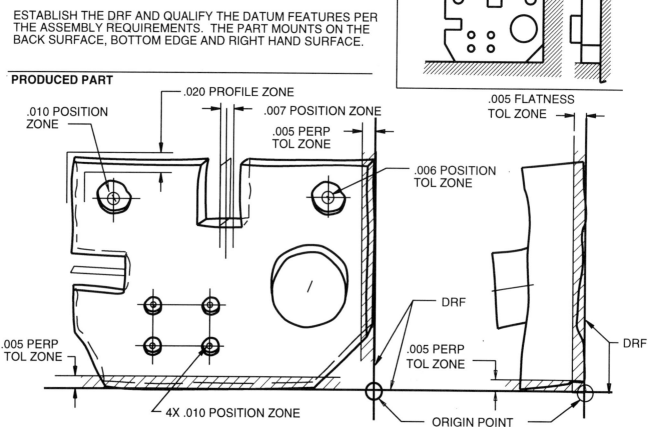

.020 PROFILE ZONE

.007 POSITION ZONE

.010 POSITION ZONE

.005 PERP TOL ZONE

.006 POSITION TOL ZONE

DRF

.005 PERP TOL ZONE

.005 PERP TOL ZONE

DRF

4X .010 POSITION ZONE

ORIGIN POINT

7.4

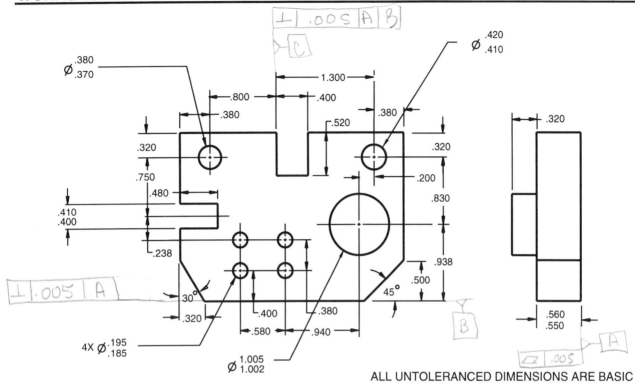

⊥ |.005| A | B |
C

Ø .380 / .370

Ø .420 / .410

1.300

.800   .400

.380

.520

.380

.320

.320

.750

.480

.200

.830

.410 / .400

.238

.938

.500

30°   45°

.320

.400   .380

.580   .940

4X Ø .195 / .185

Ø 1.005 / 1.002

⊥ |.005| A

B

.320

.560 / .550

⊘ |.005

A

ALL UNTOLERANCED DIMENSIONS ARE BASIC

DATUM SET-UP

⊕	XXX	A	B	C

ESTABLISH THE DRF AND QUALIFY THE DATUM FEATURES PER THE ASSEMBLY REQUIREMENTS. THE PART MOUNTS ON THE BACK SURFACE, BOTTOM EDGE AND LEFT HAND SURFACE OF THE SLOT.

PART MOUNTED IN ASSEMBLY

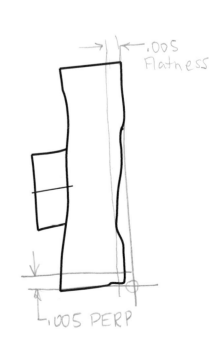

PRODUCED PART

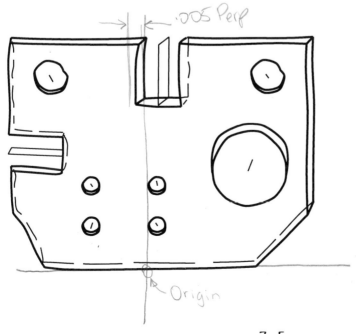

.005 Perp

Origin

.005 Flatness

.005 PERP

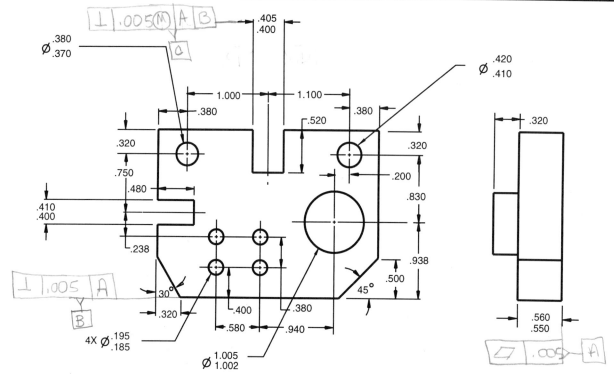

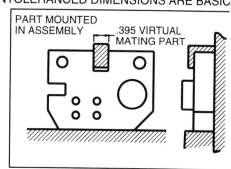

ALL UNTOLERANCED DIMENSIONS ARE BASIC

DATUM SET-UP

⊕	XXX	A	B	C

ESTABLISH THE DRF AND QUALIFY THE DATUM FEATURES PER THE ASSEMBLY REQUIREMENTS. THE PART MOUNTS ON THE BACK SURFACE, BOTTOM EDGE AND SLOT.

PART MOUNTED IN ASSEMBLY   .395 VIRTUAL MATING PART

PRODUCED PART

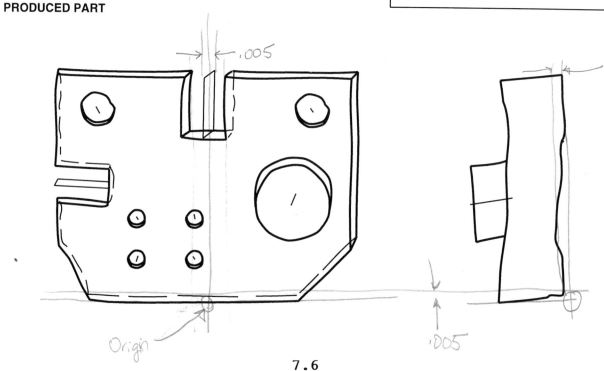

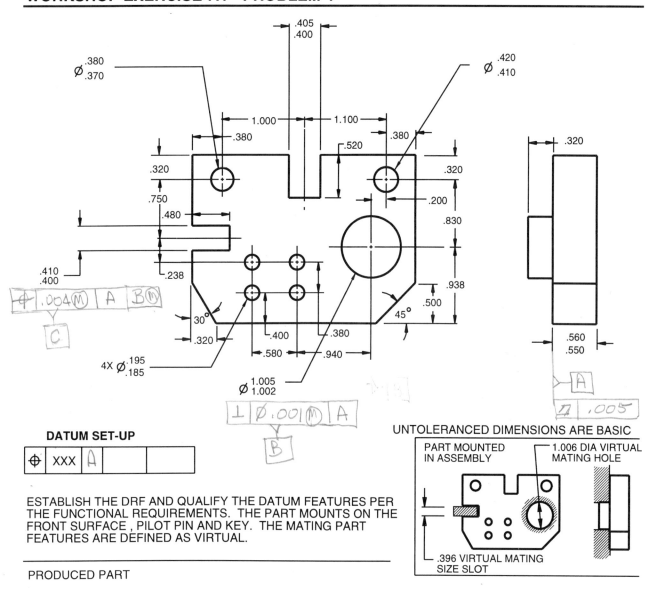

DATUM SET-UP

⊕	XXX	A		

ESTABLISH THE DRF AND QUALIFY THE DATUM FEATURES PER THE FUNCTIONAL REQUIREMENTS. THE PART MOUNTS ON THE FRONT SURFACE , PILOT PIN AND KEY. THE MATING PART FEATURES ARE DEFINED AS VIRTUAL.

UNTOLERANCED DIMENSIONS ARE BASIC

PART MOUNTED IN ASSEMBLY — 1.006 DIA VIRTUAL MATING HOLE

.396 VIRTUAL MATING SIZE SLOT

PRODUCED PART

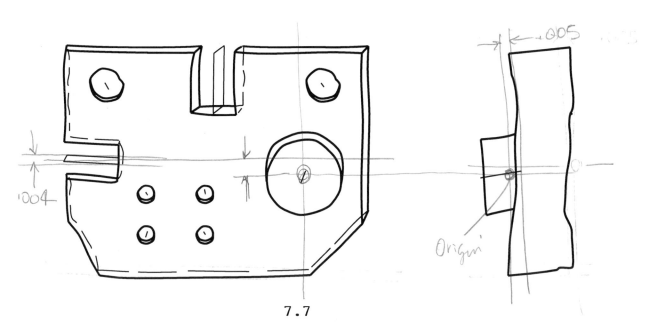

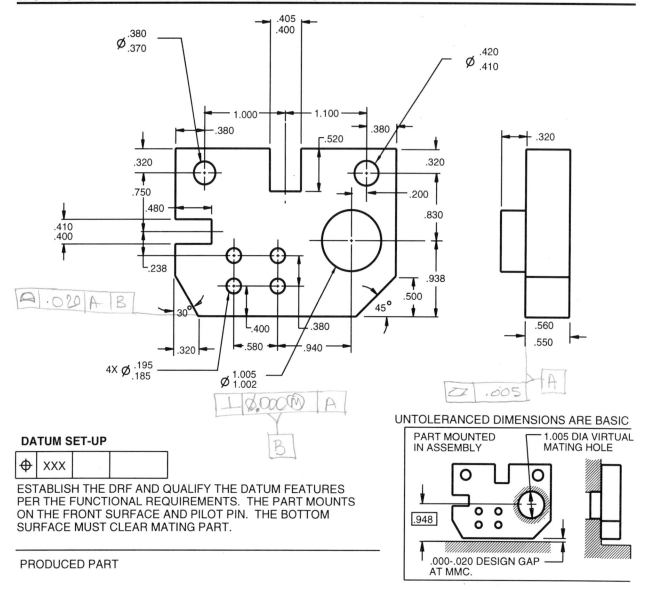

UNTOLERANCED DIMENSIONS ARE BASIC

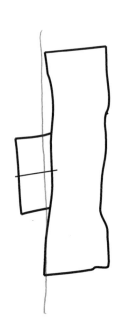

DATUM SET-UP

⊕	XXX		

ESTABLISH THE DRF AND QUALIFY THE DATUM FEATURES PER THE FUNCTIONAL REQUIREMENTS. THE PART MOUNTS ON THE FRONT SURFACE AND PILOT PIN. THE BOTTOM SURFACE MUST CLEAR MATING PART.

PART MOUNTED IN ASSEMBLY — 1.005 DIA VIRTUAL MATING HOLE

.948

.000-.020 DESIGN GAP AT MMC.

PRODUCED PART

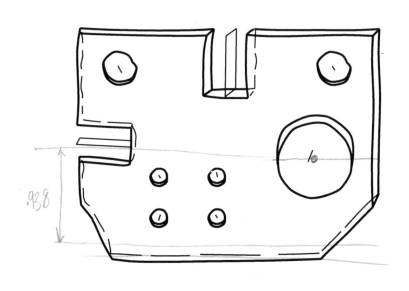

.938

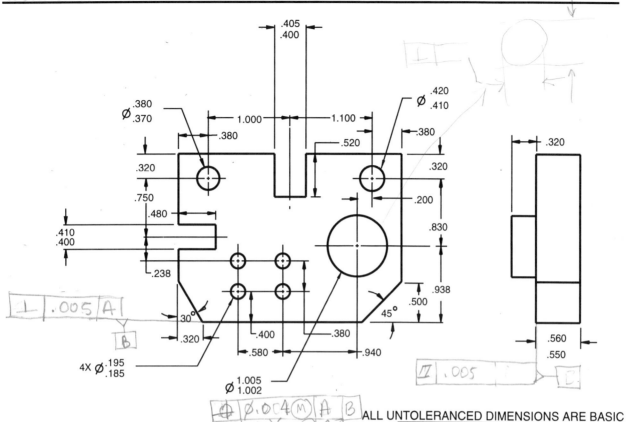

.405
.400

Ø .380
.370

1.000

1.100

Ø .420
.410

.380

.520

.380

.320

.320

.750

.200

.480

.830

.410
.400

.238

.938

.500

30°

45°

.320

.400

.380

.940

.580

4X Ø .195
.185

Ø 1.005
1.002

.320

.560

.550

⌀ Ø.004 Ⓜ A B
Ⅰ Ø.002 Ⓜ A

Ⅰ .005 A

B

Ⅱ .005

C

ALL UNTOLERANCED DIMENSIONS ARE BASIC

DATUM SET-UP

⊕	XXX			

ESTABLISH THE DRF AND QUALIFY THE DATUM FEATURES PER
THE ASSEMBLY REQUIREMENTS. THE PART MOUNTS ON
THE FRONT SURFACE, BOTTOM EDGE AND THE PILOT PIN. THE
PILOT PIN STOPS THE MOVEMENT OF THE PART FROM SIDE TO
SIDE. UP AND DOWN THE PIN MAY MOVE .040 TOTAL AT MMC.

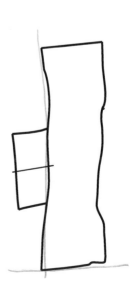

PART MOUNTED
IN ASSEMBLY

1.007 DIA VIRTUAL
MATING SLOT

PRODUCED PART

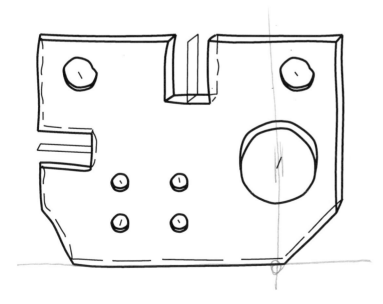

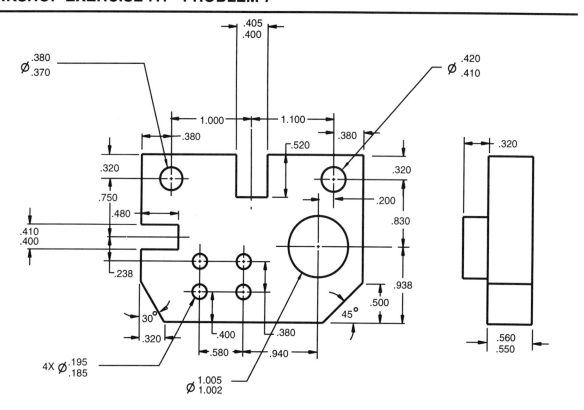

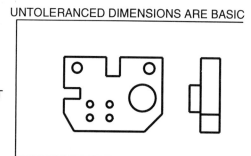

DATUM SET-UP

⊕	XXX			

THIS IS AN EXERCISE WHERE YOU WILL SKETCH THE ASSEMBLY
FUNCTION IN THE RIGHT SIDE PICTURE. THEN TRADE YOUR
EXERCISE WITH A CLASSMATE. SEE IF EACH OTHER CAN SELECT
THE DATUMS AND ILLUSTRATE THE TOLERANCE ZONES PER THE
SELECTED FUNCTION.

PRODUCED PART

UNTOLERANCED DIMENSIONS ARE BASIC

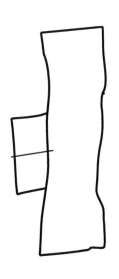

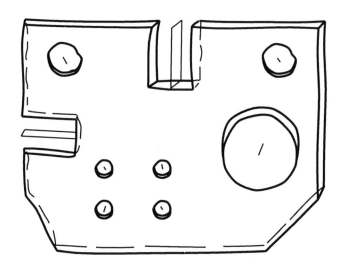

UNIT 8

DATUM REFERENCE FRAME
APPLICATIONS I

Plate with Pilot Hole and Surface Orientation
      Verification
Inclined Datum Features
      Verification
Contoured Surface as a Datum
Datum Targets
Datum Targets Application
      Hood Panel Outer - Targets
      Hood Panel Outer - Profile
Workshop Exercise 8.1

PILOT HOLE WITH SURFACE ORIENTATION

On the part below, the large pilot hole has been located from a datum reference frame established by the A,B and C datum features. To refine the orientation of the pilot hole, a more restrictive perpendicularity tolerance is applied. The four holes are located around the pilot hole with the orientation of the four hole pattern being controlled by the bottom surface. Notice the position call-out for the four holes. The datums in the feature control frame reference datum features A, D at MMC and B. A relaxation in the orientation of the four hole pattern can be accomplished by methods shown later in the position section of this text.

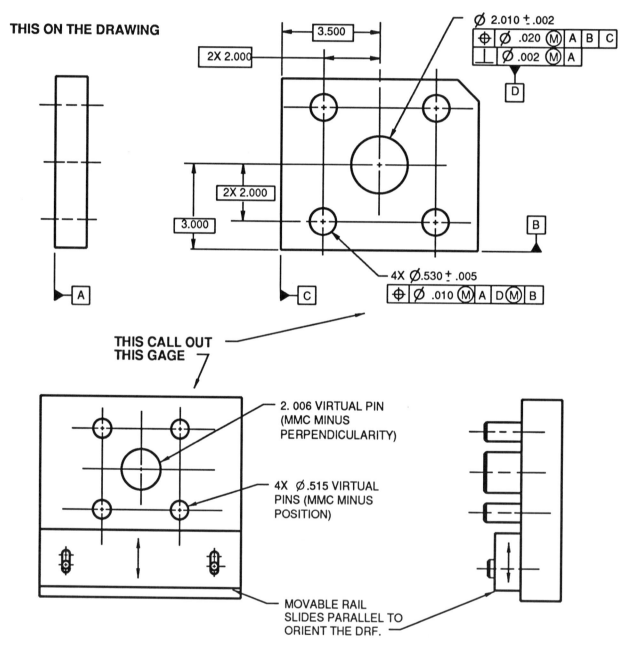

The simple functional gage above illustrates the position call out for the four holes. The part will mount on the back face and locate on the large pin. The sliding rail moves up and down while contacting the bottom surface of the part at a minimum of two points. The rail ensures proper orientation of the pattern of holes with the bottom surface. A gage is not necessary to verify the position call out. It is shown only to allow visualization of the concept. A surface plate set-up is shown on the following page.

SAMPLE INSPECTION SET-UP FOR PATTERN OF FOUR HOLES.

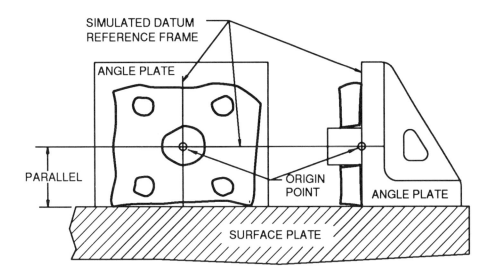

The illustration above is a sample open set-up inspection procedure for the pattern of four holes in the previous example. The datum face A is mounted against the angle plate to establish the primary plane. The large pilot hole, daturm feature D, is the intersection of the two planes in the datum reference frame and is the origin of measurements. The bottom surface, datum B, is used to orient the datum reference frame and must have a minimum of two points of contact. If datum B were not referenced in the feature control frame, the pattern of holes would have an implied orientation to the part.

This part can also be verified with a coordinate measuring machine (CMM). Datum feature A is leveled and establishes the primary plane. A cylinder alignment is set on the large hole establishing an axis. The orientation is supplied by leveling the bottom surface. The origin of measurements is at the intersection of the cylinder and the contacting plane of the back face.

Since the MMC modifier is applied to the features as well as the datum feature, additional tolerance is available as the features and datum feature D departs from virtual size. This may be calculated by the paper gage method, trial and error (re-setting the origin), or with appropriate CMM software.

INCLINED DATUM FEATURES

Often, a datum reference frame may need to be established from datum features that are not mutually perpendicular to each other. All the angles between the datum features must be indicated as basic. True contacting planes are orientated and established at these angles relative to a datum reference frame (cartesian system). All measurements originate from the datum reference frame. See following examples. Note that the angularity control specified on the angled surface could also have been correctly applied as a perpendicularity control. The angled surface is perpendicular to the primary datum, with the secondary datum providing orientation.

THIS ON THE DRAWING

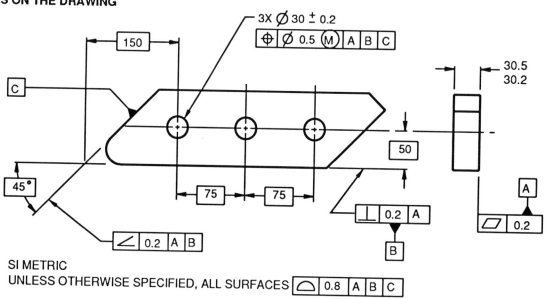

SI METRIC
UNLESS OTHERWISE SPECIFIED, ALL SURFACES ⌓ 0.8 A B C

THIS APPLICATION

Part mounts on back face, bottom edge and inclined edge. The three bolts keep it in place.

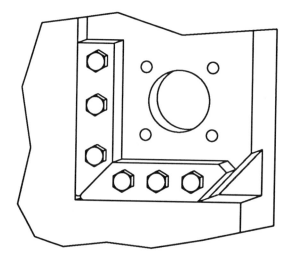

MEANS THIS

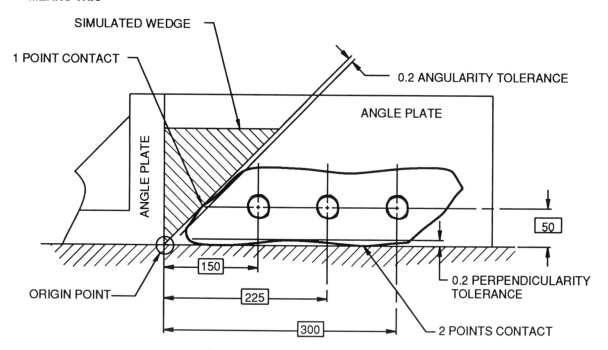

Part may be verified using functional gage, coordinate measuring machine or set up on a sine plate and rotated at the basic angle.

MEANS THIS

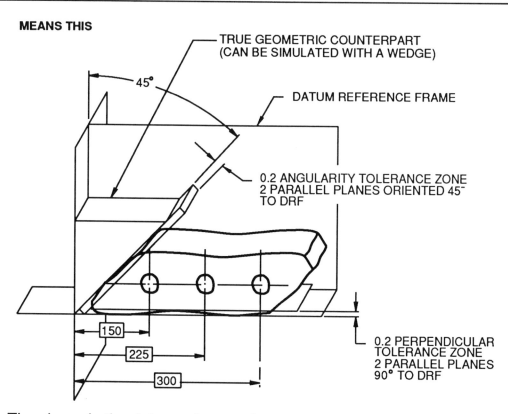

The planes in the datum reference frame are always at 90 degrees to each other. The datum features may be at any angle. A true contacting plane is oriented at the basic angle of the feature. All measurements originate from the datum reference frame.

CONTOURED SURFACE AS A DATUM - AIRCRAFT ACCESS DOOR

The assembly shown is an aircraft access door on a wing. It mounts on the bottom free form contoured surface and is fastened with 12 screws. The surface contour and hole locations are defined with math data relative to the ship coordinate system (water lines, butt lines and station lines).

The access door drawing illustrates geometric tolerancing applied to reflect the functional requirements of the part. The mounting contour is defined with CAD math data and is implied as all basic dimensions. A partial area on the contour is selected as the primary datum feature and is defined on the drawing.

The true geometric counterpart of the datum feature is the CAD math data. The math data is relative to a coordinate system, and this coordinate system is the datum reference frame. The partial datum is controlled with a unilateral profile tolerance. This profile tolerance defines the gap allowed when the datum feature is compared to the true geometric counterpart. (It acts similar to a flatness on a plane surface.)

In theory, since the math data has X, Y and Z coordinates, the part is immobilized. In a practical situation, the contour definition alone is not sufficient to immobilize the part. It is for this reason the 12 holes were selected as a secondary datum. The holes are positioned to the DRF. The remaining features are defined relative to the DRF established by the contoured surface and the pattern of holes.

As an alternative, datum targets can be specified on the contour instead of identifying the contour itself as a datum feature. This alternative method is often used when the part is assembled with other components using fixture points in an assembly fixture.

The access door gage reflects a possible verification method for the access door. The datum features are simulated by a fixture or laminate model that is produced relative to the math data. The contoured surface in the gage is the datum simulator and is relative to a coordinate system which acts as the DRF. The gage is shown as an illustration only to aid the reader in understanding the boundaries in which the part may lie.

This part can also be verified with a coordinate measuring machine with surface profiling capabilities. The surface and the holes are "best fit" and then the remaining features are checked for conformance. Although the part shown is an aircraft application, the concept can be applied to other contoured parts as well.

CONTOURED SURFACE AS A DATUM - AIRCRAFT ACCESS DOOR

The drawing below is an aircraft access door. The primary datum is the contoured surface. The true geometric counterpart is the CAD math data. The datum can be sumulated with a fixture or a CMM with surface profile capabilities.

THIS APPLICATION

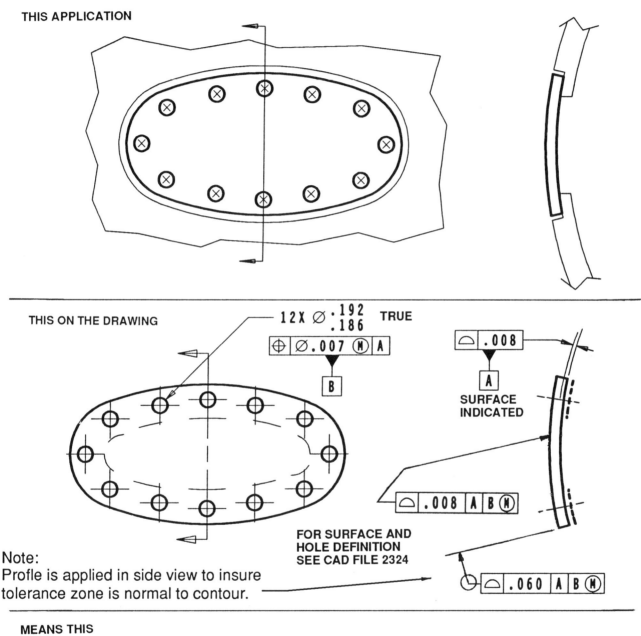

THIS ON THE DRAWING

$12X \oslash \begin{smallmatrix} .192 \\ .186 \end{smallmatrix}$ **TRUE**

⊕ | ∅.007 Ⓜ | A

B

△ | .008

A

SURFACE INDICATED

△ | .008 | A | B Ⓜ

FOR SURFACE AND HOLE DEFINITION SEE CAD FILE 2324

△ | .060 | A | B Ⓜ

Note:
Profile is applied in side view to insure tolerance zone is normal to contour.

MEANS THIS

SAMPLE GAGE

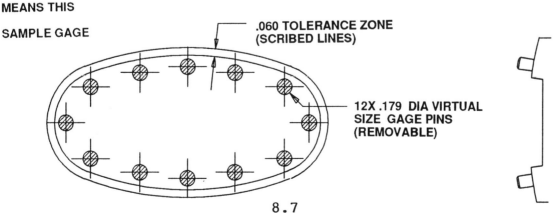

.060 TOLERANCE ZONE (SCRIBED LINES)

12X .179 DIA VIRTUAL SIZE GAGE PINS (REMOVABLE)

DATUM TARGETS

There are many cases where, in order to set up a datum reference frame, we might prefer to pick specific points of contact on a surface rather than an entire surface. This occurs because of manufacturing concerns or surface irregularities that make it impractical. Examples of this application might be castings, forgings, sheet metal parts, plastic parts and weldments.

Datum targets may be used to set up the datum reference frame. In the past these points had a variety of different names such as: set up points, principle locating points, tooling points, fixture points etc. All of these names have been discarded and we now call these points datum targets.

The accompanying figure is a casting drawing. The contour and hole location dimensions are left off for clarity. The dimensions are all basic and originate from the datum reference frame.

The datum targets can be points, lines or planes. If the target is an area contact, the area is indicated by section lines inside a phantom outline with size dimensions. If the definition of the target area is clear, the section lines may be omitted. If the target shape is a circle or square, the size and shape may be designated in the upper half of the datum target symbol. In the side view this target area is designated as an "X".

DATUM TARGETS USED ON A CASTING

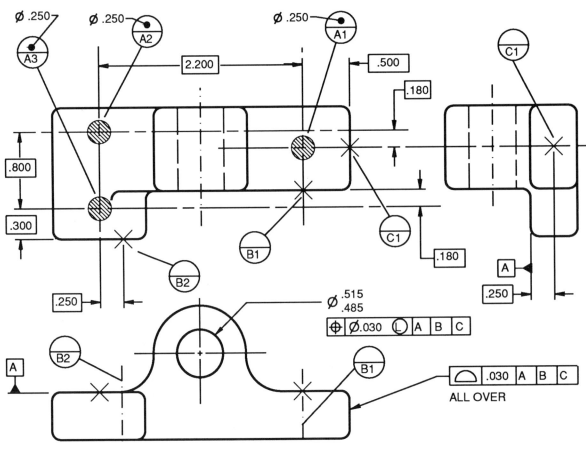

**TOOLING FIXTURE USED TO SET UP DATUM
REFERENCE FRAME**

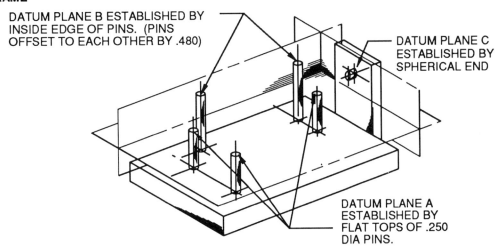

DATUM PLANE B ESTABLISHED BY
INSIDE EDGE OF PINS. (PINS
OFFSET TO EACH OTHER BY .480)

DATUM PLANE C
ESTABLISHED BY
SPHERICAL END

DATUM PLANE A
ESTABLISHED BY
FLAT TOPS OF .250
DIA PINS.

If the target is a line contact, it is indicated by an "X" on the edge in the plan view. In the front view it is indicated with a phantom line. If the target is a point contact, the point is indicated by an "X" on the edge in the side view. It is also identified with an "X" in the front view. Datum targets are usually located with basic dimensions. If the targets are located with basic dimensions, tooling or gaging tolerances are assumed to apply. The targets can also be located with toleranced dimensions.

Notice that in this particular part, the datum reference frame is set up by the 3, 2, 1 concept. The primary datum has three (.250 diameter) area contacts identified as A plane, target 1, target 2, target 3. The secondary datum has two line contacts identified as B plane, target 1, target 2. The tertiary datum has one point of contact identified as C plane, target 1.

The pattern maker needs these targets to build and inspect the part. The machinist will use these same target points to set up the part for the initial machining operations. The datum target selection is usually a joint effort between design, manufacturing, inspection and casting personnel.

**CAST PART IN THE TOOLING
FIXTURE**

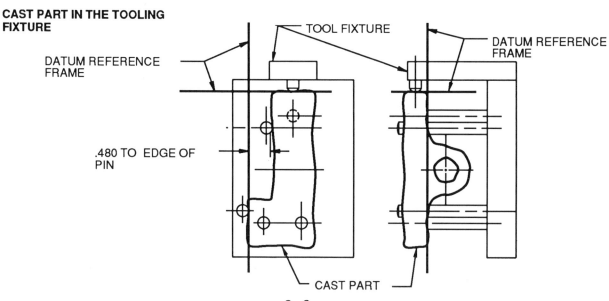

TOOL FIXTURE

DATUM REFERENCE
FRAME

DATUM REFERENCE
FRAME

.480 TO EDGE OF
PIN

CAST PART

DATUM TARGETS APPLICATION - HOOD PANEL OUTER

The panel below is flexible. The datum target selection was based on how the part loads in the assembly fixture with the inner reinforcement. See profile drawing for tolerancing.

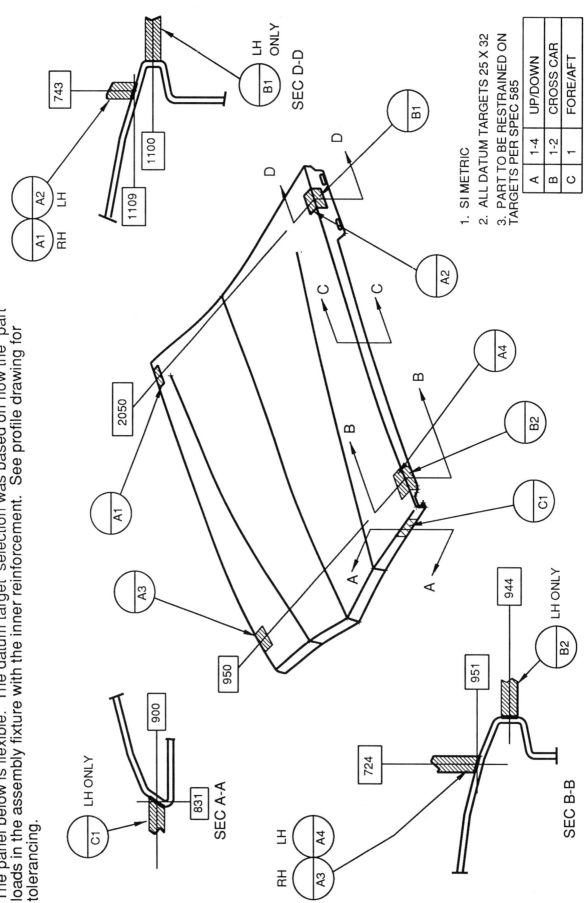

SEC D-D

LH ONLY

SEC A-A

SEC B-B

1. SI METRIC
2. ALL DATUM TARGETS 25 X 32
3. PART TO BE RESTRAINED ON TARGETS PER SPEC 585

		UP/DOWN	CROSS CAR	FORE/AFT
A	1-4			
B	1-2			
C	1			

PROFILE APPLICATION - HOOD PANEL OUTER

The flexible panel below is mathematically defined in a CAD data base. The data base file is considered basic. This drawing defines the tolerancing, and the datum target drawing defines the datum set-up. The two drawings can be combined. There are no other drawings needed. The profile controls are all in the restrained condition.

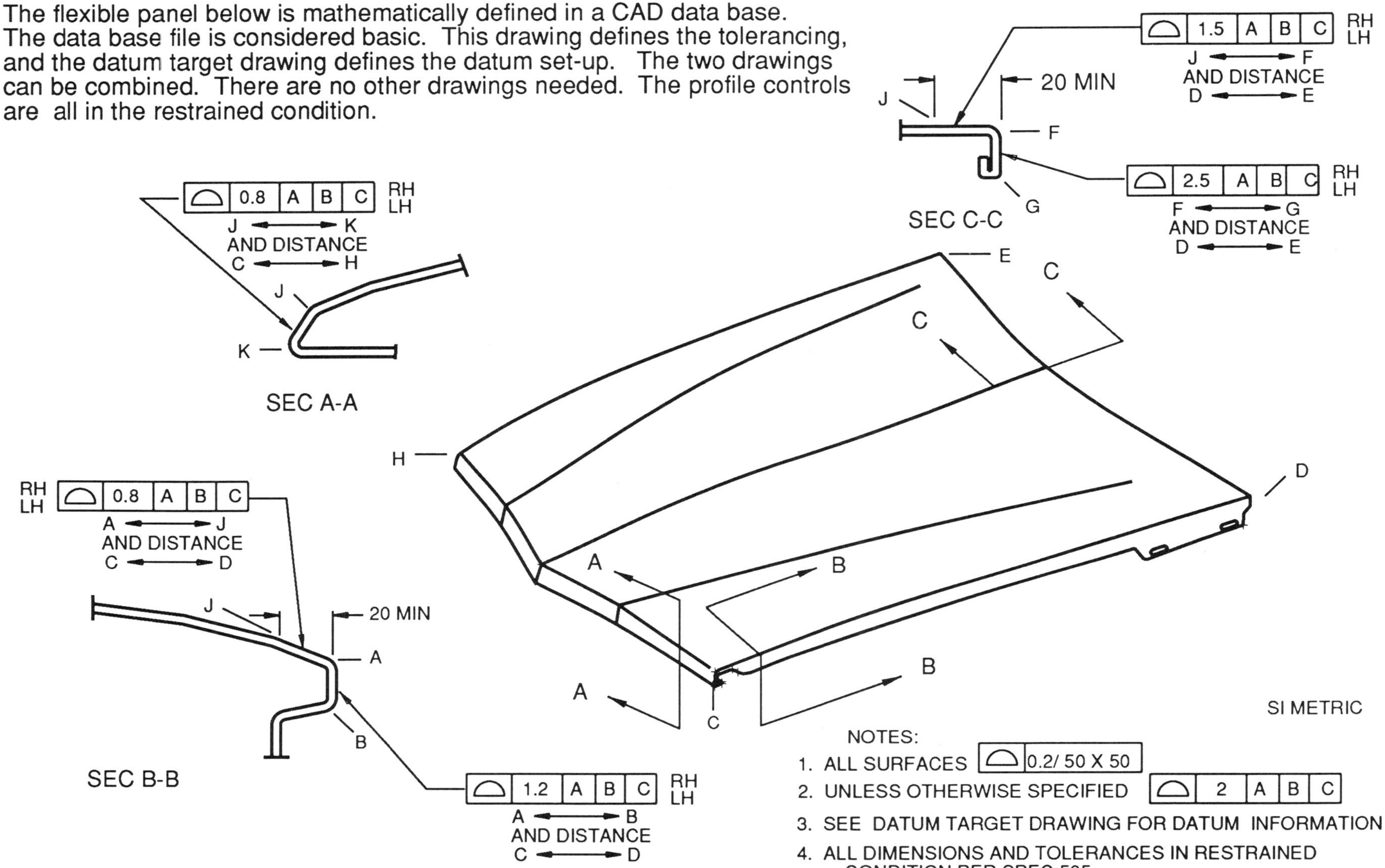

SEC C-C

SEC A-A

SEC B-B

SI METRIC

NOTES:
1. ALL SURFACES ⌓ 0.2/ 50 X 50
2. UNLESS OTHERWISE SPECIFIED ⌓ 2 A B C
3. SEE DATUM TARGET DRAWING FOR DATUM INFORMATION
4. ALL DIMENSIONS AND TOLERANCES IN RESTRAINED CONDITION PER SPEC 585.
5. SEE CAD FILE 2324 FOR ALL DIMENSIONS AND MATH DATA.

8.11

WORKSHOP EXERCISE 8.1

The drawing below is a cast part. In a meeting between engineering, manufacturing and the casting vendor, it was decided that datum targets must be selected. On the drawing below draw the datum targets in proper format as you would expect to see them on an engineering drawing. Consult the previous examples if you need help.

1. Select three pads on the top two surfaces of the part in the plan view. Select two pads on the front face of the part in the front view. Select a point contact on the right face of the part. Make all the pads .250 dia. Apply basic dimensions as necessary and simply estimate the distances. Keep all targets above the parting line.

2. Apply a profile tolerance of .020 on the part above the parting line and a profile tolerance of .030 on the part below the parting line.

THIS ON THE DRAWING

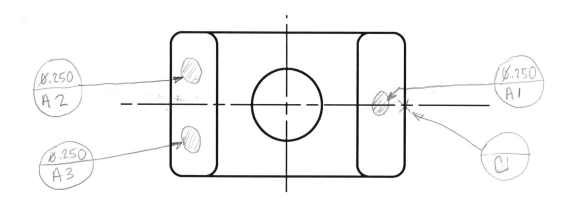

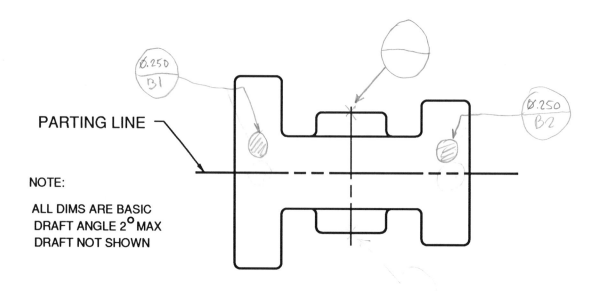

PARTING LINE

NOTE:

ALL DIMS ARE BASIC
DRAFT ANGLE 2° MAX
DRAFT NOT SHOWN

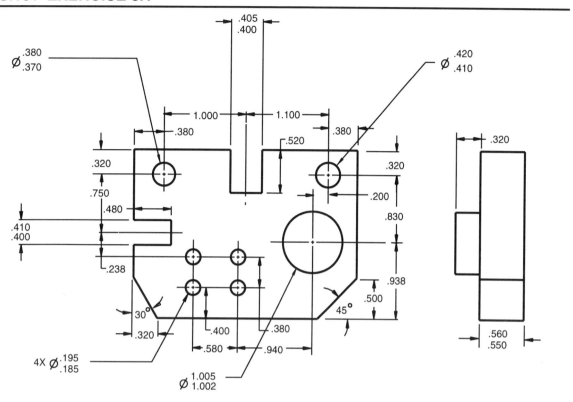

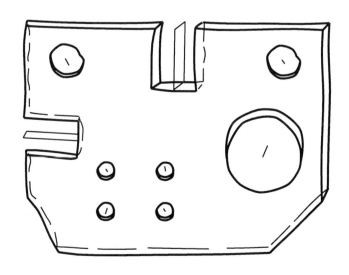

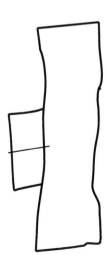

DATUM SET-UP

⊕	XXX			

UNTOLERANCED DIMENSIONS ARE BASIC

3. This is an exercise where you will establish and qualify a DRF in the product drawing above. Select the DRF and relate the features in any manner you wish. Then trade your exercise with a classmate. See if each other can correctly communicate information with geometric tolerancing. Find the DRF, origin point and illustrate the tolerance zones on the produced part below.

SHOW YOUR WORK VERY CLEARLY

PRODUCED PART

UNIT 9

DATUM REFERENCE FRAME
APPLICATIONS II

Hole and Slot
Holes as Datums
Hole Patterns as Datums
        Paper Gage Evaluation
Coaxial Holes as Datums -Seat Latch
Interrelationship of Multiple DRF's
Workshop Exercise 9.1

HOLE AND SLOT

The scalloped part below mounts on the right hand face, the .250 diameter hole and the slot. These features are established as datums. A virtual size relationship between the datum features is established by the perpendicularity and position tolerances. The remaining features on the part are then related to the datums through the profile and position tolerances. The importance of the relationship between the features and the datums is evidenced by the size of the tolerance zone.

THIS ON THE DRAWING

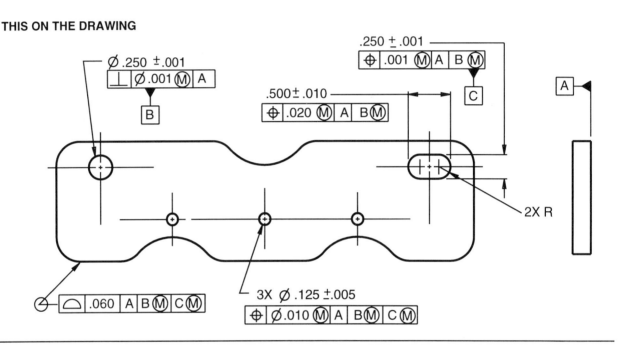

The functional gage or 3D solid shown below reflects the geometric controls applied above. Since all the features and the datum features are referenced at MMC, both the features and the datum features may shift or be displaced from their virtual size location by the amount they depart from virtual. In other words, all the holes may move around as long as they clear the virtual boundaries. At the same time, simultaneously, all the profiled surfaces may also move as long as they fall within the boundaries described.

FUNCTIONAL GAGE OR BOUNDARIES

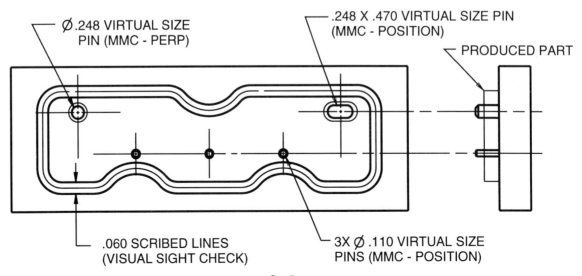

9.2

HOLES AS DATUMS

The two parts below mount in an assembly with bolts in the holes. The two parts are geometrically toleranced in a different manner, but both have an identical final interpretation. See the functional gage or 3D solid below. The top part is certainly more complicated and has more symbology and yet obtains the same results. The top part gives the impression that the left hole is more important than the right hole. In fact, both holes will locate the part equally in the assembly. If the MMC modifier is referenced on the datum features in the feature control frames, the datum precedence of datum features B and C becomes irrelevant as both datums are established at their virtual size.

The point is: KEEP IT SIMPLE.

If the part bolts and is located on the two holes equally, then merely identify both holes as a datum. The verification inspectors still might choose to set-up on one hole or both. It is their choice. The object is to balance or shift the pattern using the MMC modifiers to bring the part to conformance. If the datum features are referenced at RFS, then datum precedence is certainly important. Also see the discussion on a pattern of holes later in the text.

INDIVIDUAL HOLES AS DATUMS

SAME FUNCTIONAL GAGE OR 3D SOLID FOR BOTH PARTS

The two parts are both geometrically toleranced but the top part is more complicated. The final end result is the same. The bottom method of calling both holes a datum is preferred. KEEP IT SIMPLE!

HOLE PATTERN AS A DATUM

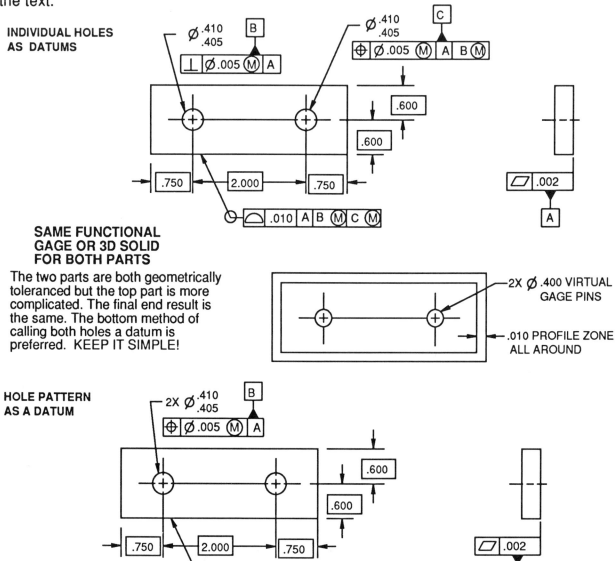

PATTERN OF HOLES ESTABLISHING A DATUM REFERENCE FRAME (DRF)

When function dictates, multiple features of size, such as a pattern of holes at MMC, may be used to establish a datum reference frame. The holes are located to each other and plane A with a position tolerance. The axes of the holes must fall within these tolerance cylinders. These cylinders are basically located to each other and in effect establish the second and third planes in the DRF. Since datum B is referenced in the feature control frame at MMC, additional shift or displacement may be available as the datum features depart from MMC (virtual).

APPLICATION ⟶

The light switch cover mounts on four holes and must clear the switch. The outside edges on the cover are not important.

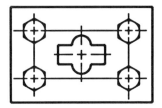

THIS ON THE DRAWING

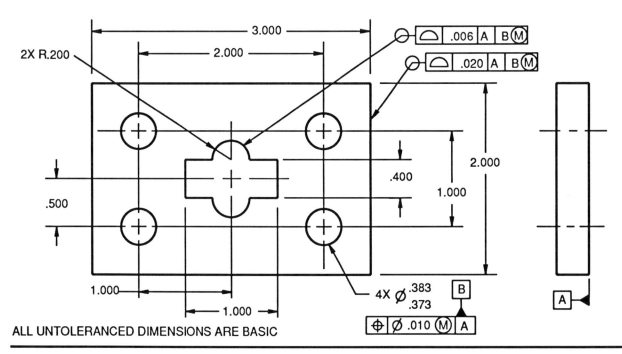

ALL UNTOLERANCED DIMENSIONS ARE BASIC

FOUR .010 POSITION TOLERANCE ZONE CYLINDERS WITHIN WHICH THE AXES OF THE FOUR HOLES MUST LIE. THE TOLERANCE CYLINDERS ARE BASICALLY LOCATED TO EACH OTHER AND IN EFFECT ESTABLISH THE SECOND AND THIRD PLANE IN THE DRF.

MEANS THIS

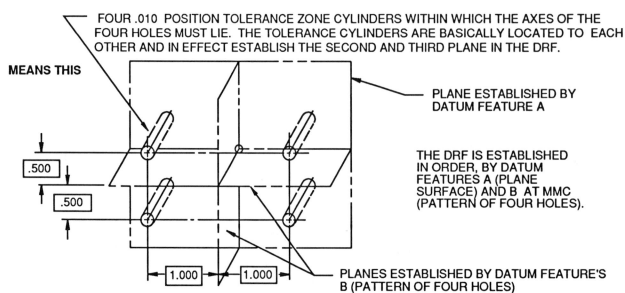

PLANE ESTABLISHED BY DATUM FEATURE A

THE DRF IS ESTABLISHED IN ORDER, BY DATUM FEATURES A (PLANE SURFACE) AND B AT MMC (PATTERN OF FOUR HOLES).

PLANES ESTABLISHED BY DATUM FEATURE'S B (PATTERN OF FOUR HOLES)

9.4

A pattern of holes used as a datum is a very common application. Many parts mount on a group of holes. There is no need to identify two holes of the pattern as datums. The four holes restrict the movement of the part in the X and Y direction and rotationally. Function dictates that the entire "best fit" pattern establishes the DRF.

This will provide function as well as obtaining maximum manufacturing tolerance since a pattern of holes is "best fit" during the manufacturing process. Holes are usually processed in one operation, and good parts may show bad if inspection aligns on only one or two holes. In our part the edges are unimportant to function. The part mounts on the holes and the contour, and edges are located from the pattern. Basic dimensions on the drawing may originate from any of the holes since all features are connected with basic dimensions. Mathematically, the basic dimensions originate from the axis of the tolerance cylinders and not the individual actual produced hole.

To establish a DRF from a pattern of holes on a CMM or open set-up inspection, the pattern of holes must be "averaged, balanced or best fit" to insure conformance to the position requirement. If the position requirement is met, then a DRF has been established; and, without reorienting the part, the remaining features are checked relative to this set-up. Once the DRF origin has been established, it may be relocated since all the features are related to each other with basic dimensions.

If inspection chooses only two holes to set-up the part and the part checks good, it is good. If the part checks bad, it may be "balanced or best fit" similar to a functional gage to insure acceptance. This balancing method of "eking out the last bit of tolerance" may be accomplished by CMM's with appropriate software or by the paper gage method.

The figure below represents a functional gage used to verify features related to the pattern of holes. The four virtual size pins verify the position tolerance as well as establish the DRF from which the other features are located.

FUNCTIONAL GAGE OR 3D SOLID TO CHECK FEATURES RELATED TO PATTERN OF HOLES

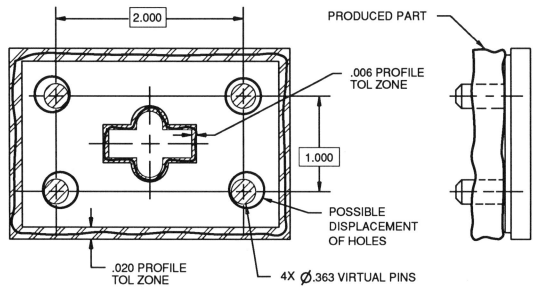

9.5

PATTERN OF HOLES AND DATUM ORIGIN - PAPER GAGE EVALUATION

This is a simple illustration on how to use the paper gage concept to evaluate a pattern of holes and then use this pattern as a datum feature. The product drawing and the associated discussion of the produced part shown below is shown on the previous pages. All of the holes are shown produced at MMC to keep the example simple. The check of the profile specification must also be completed, but it is not part of this exercise.

Caution: this is a very simple explanation of a very complex problem. There are many other factors which will contribute to a complete and thorough evaluation of the parts. The following paper gage explanation will give an approximate answer. Parts should be reset to the calculated origin and checked again to insure conformance.

Notice the four holes were balanced or alligned in the initial inspection set-up. Hole #1 was chosen as the origin to collect data. The data is recorded in the chart below. In this early evaluation, holes #3 and #4 are shown out of position. See chart below. When the holes are plotted and evaluated with the paper gage, it can be seen that all the holes, as a group, are within the .010 position tolerance zone. See figure 1 on the next page. The holes are within their accepted limit of location.

PRODUCED PART SHOWN BELOW
THE PRODUCT DRAWING IS SHOWN ON PREVIOUS PAGES.

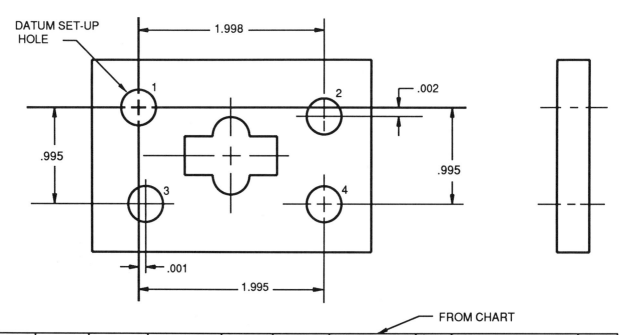

HOLE NO.	HOLE MMC	HOLE ACTUAL SIZE	POSITION TOLERANCE ALLOWED	"X" DIM.	"Y" DIM.	POSITION TOLERANCE ACTUAL	ACC or REJ	PAPER GAGE EVALUATION	ACC or REJ
1	.373	.373	.010	.000	.000	.000	A	PATTERN OF HOLES FALL WITHIN .010 ZONE AFTER PAPER GAGE EVALUATION	A
2	.373	.373	.010	-.002	-.002	.0056	A		A
3	.373	.373	.010	+.001	+.005	.0102	R		A
4	.373	.373	.010	-.005	+.005	.0141	R		A

The origin of the pattern of holes is the average of all the holes. If you notice in figure 1, the center of the round .010 tolerance zone is located X = -.002, Y = +.002 from the number 1 hole. This is approximately the origin of the pattern of holes. To make measurements to other features, reset the origin to this point.

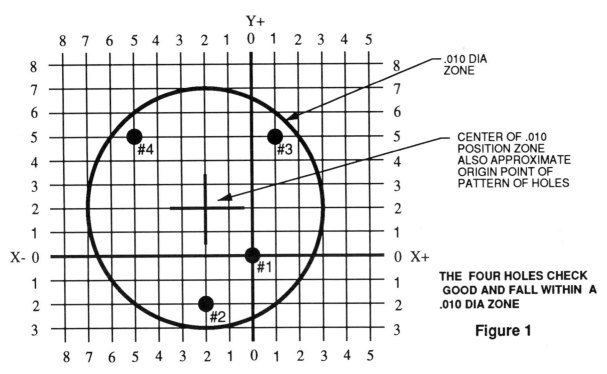

.010 DIA ZONE

CENTER OF .010 POSITION ZONE ALSO APPROXIMATE ORIGIN POINT OF PATTERN OF HOLES

THE FOUR HOLES CHECK GOOD AND FALL WITHIN A .010 DIA ZONE

Figure 1

Since the holes have size and can move around within their position limit, it is possible that the holes will depart from virtual size. If the hole pattern is referenced as a datum at MMC (virtual), then the origin may shift or be displaced by the amount the average of the holes depart from virtual. The amount of displacement depends on the size of each hole and its position. The approximate displacement of the part below can be seen in figure 2 below.

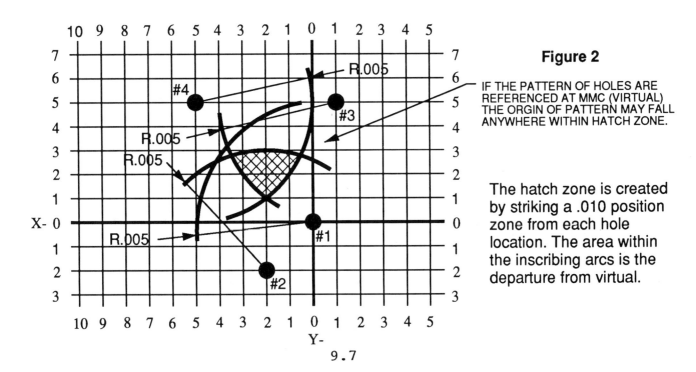

Figure 2

IF THE PATTERN OF HOLES ARE REFERENCED AT MMC (VIRTUAL) THE ORGIN OF PATTERN MAY FALL ANYWHERE WITHIN HATCH ZONE.

The hatch zone is created by striking a .010 position zone from each hole location. The area within the inscribing arcs is the departure from virtual.

COAXIAL HOLES AS DATUMS - SEAT LATCH

THIS APPLICATION

The seat latch bracket locates on a pin through the two holes and mounts inside the frame. The three slots engage in the seat rack. The adjustment lever attaches to the hole in the seat latch bracket and provides engagement and disengagement for various locked seat positions.

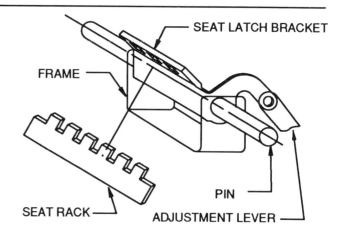

SEAT LATCH BRACKET

FRAME

PIN

SEAT RACK

ADJUSTMENT LEVER

THIS ON THE DRAWING

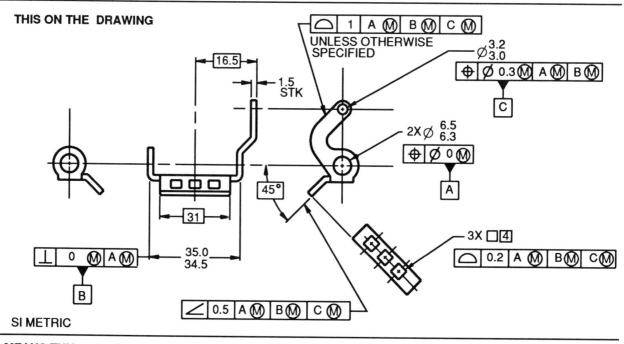

UNLESS OTHERWISE SPECIFIED

SI METRIC

MEANS THIS

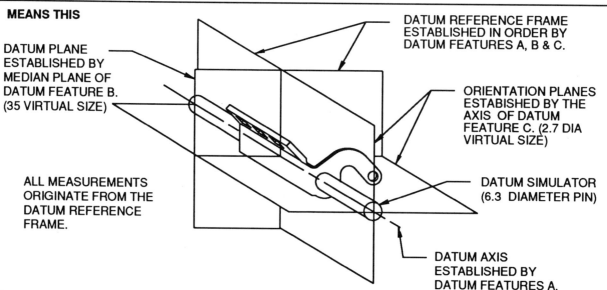

DATUM PLANE ESTABLISHED BY MEDIAN PLANE OF DATUM FEATURE B. (35 VIRTUAL SIZE)

ALL MEASUREMENTS ORIGINATE FROM THE DATUM REFERENCE FRAME.

DATUM REFERENCE FRAME ESTABLISHED IN ORDER BY DATUM FEATURES A, B & C.

ORIENTATION PLANES ESTABLISHED BY THE AXIS OF DATUM FEATURE C. (2.7 DIA VIRTUAL SIZE)

DATUM SIMULATOR (6.3 DIAMETER PIN)

DATUM AXIS ESTABLISHED BY DATUM FEATURES A.

INTERRELATIONSHIP OF MULTIPLE DATUM REFERENCE FRAMES

Datum reference frames (DRF) are selected based on function. In most cases, the establishment of one DRF on a part is sufficient to define part function. It depends on how the part works. If the part is one where many components are attached to it, as in a gearbox housing or an engine block, the part may have many different datum reference frames. If the part is one that attaches to the gear box housing or engine block as in a cover or oil pan, only one DRF may be required. In other cases, even though the part may have many mounting surfaces, only one DRF is established by engineering. This is often done because the tolerance is large, it is relatively unimportant, and it certainly keeps the part definition simple. The establishment of multiple datum reference frames on a part should only be done where it is necessary for fit or function.

The assembly below shows an adaptor that may be considered for the establishment of two datum reference frames. Notice that the adaptor mounts on the bottom surface and a pilot diameter. It is held in place with four bolts. Another component mounts to the upper surface of the adaptor. It mounts on the upper face and pilot and is fastened with bolts. The relationship between the individual features on the bottom surface are important to each other. The relationship between the individual features are also important to each other. But the relationship between the two groups of features are less important to each other. It is important, however, to relate the two groups of features so calculations or verifications can be made between the two groups of features.

APPLICATION

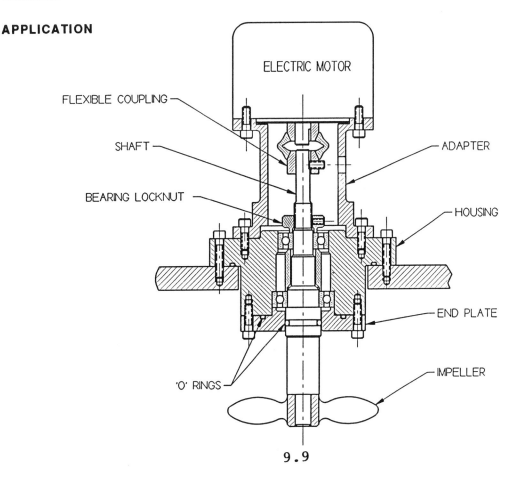

The drawing below illustrates the adaptor which has two datum reference frames established. The two datum reference frames on the drawing also have an established relationship to each other. On the drawing a DRF, we will call AB, has been established on the bottom surface and pilot. The datum features in this AB - DRF are related to each other with the flatness and perpendicularity specifications. The upper surface has been located up and down with a profile specification and the orientation has been refined with a parallelism tolerance. The upper pilot has been located with a position tolerance. The pilot and upper surface are now related to the AB - DRF.

A new datum reference frame, CD, is established from these upper features. The CD - DRF may tilt, skew or be displaced relative to the AB - DRF by the amount the upper surface and pilot is allowed to move within the profile and position tolerances. The orientation and form of the upper surface is further refined by the parallelism tolerance.

The features that establish the CD - DRF are then related to each other with the perpendicularity tolerance. The four .290 diameter holes are then related to the CD - DRF. The following illustrates a graphical representation of the interpretation of multiple datum reference frames and their relationship.

Note: In this example the holes on the upper surface of the adaptor have no relation to the holes on the bottom surface. If such a relationship is required, it must be specified.

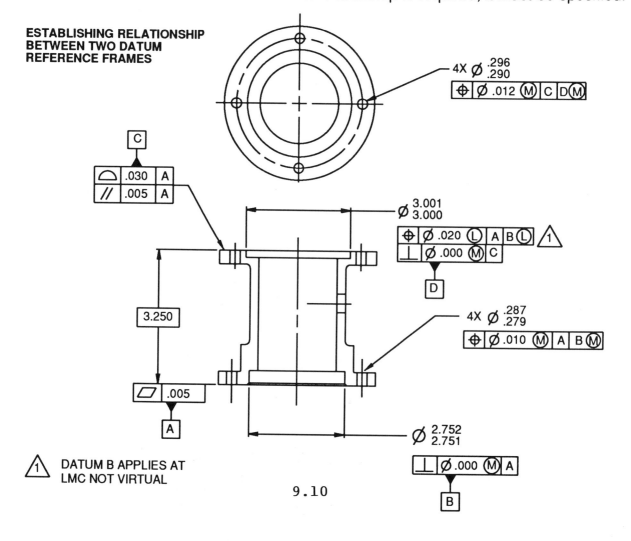

ESTABLISHING RELATIONSHIP BETWEEN TWO DATUM REFERENCE FRAMES

9.10

1. The upper face and pilot are located and oriented by the position and profile zones. The upper face orientation and flatness is further refined by the parallelism tolerance.

2. The upper face and pilot establish the CD - DRF. The CD - DRF may tilt, skew, rotate, and may be displaced relative to the AB - DRF by the amount the the upper surface and pilot is allowed to move within the confines of the position, profile and the orientation tolerance of parallelism.

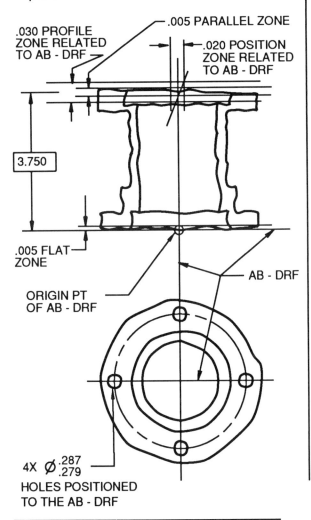

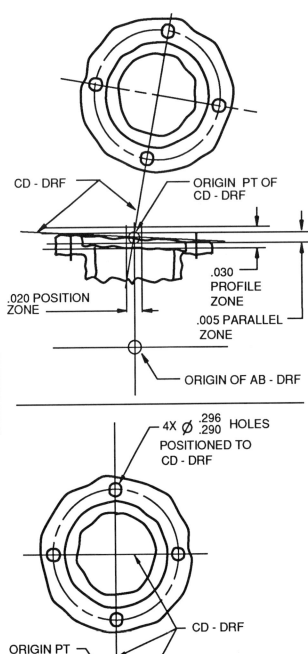

3. The features establishing the CD - DRF are related to each other by the perpendicularity tolerance. The parallelism tolerance which controls orientation of the upper surface also controls the flatness of the surface. The four holes are related to the CD - DRF.

9.11

WORKSHOP EXERCISE 9.1

1. The drawing below has individual holes selected as datums. Evaluate the geometric tolerancing and draw a simple functional gage (or mathematical 3D solid) in the space below the drawing. Label the size of the profile tolerance zone as well as the virtual sizes of the gage pins.

THIS ON THE DRAWING

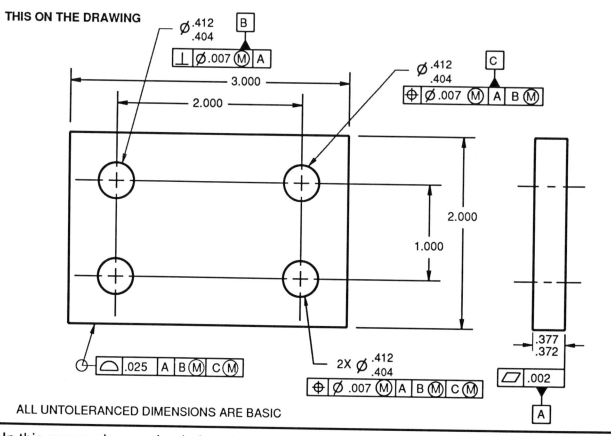

ALL UNTOLERANCED DIMENSIONS ARE BASIC

In this space, draw a simple functional gage for the part above. Label the size of the profile zone as well as the virtual size of the gage pins.

The drawing below has no geometric tolerancing applied. The part mounts on the four holes. The four holes are important to each other as the part mounts on the four holes. The outside edges are not important and are simply clearance. It is your task to apply the geometric tolerancing per the following instructions.

2. Select the back face as datum A, and make this surface flat within .002.

3. Make all the dimensions basic, except of course, the size dimensions.

4. Position the four holes to each other and datum A within .007 at MMC.

5. Since the four holes locate the part, identify the four holes as datum feature B.

6. Since the outside edge is not important, profile this surface all around within .025, and relate the profile to datum features A and B at MMC.

7. Now that the part is completely toleranced, explain the differences between a functional gage for this part and the part in problem 1 of this exercise.

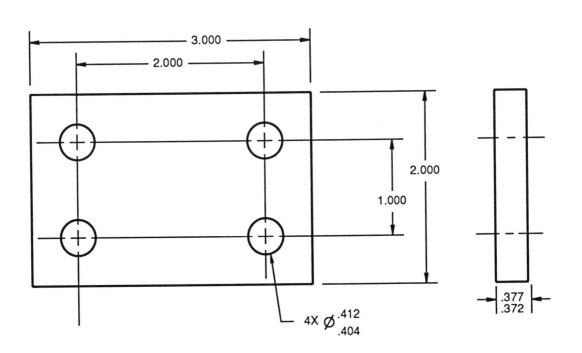

4X Ø .412 / .404

3.000
2.000
2.000
1.000
.377 / .372

8. The drawing below is a produced part from either problem 1 or problem 7 in this exercise. Take your pick. The two product drawings have identical interpretations. In the inspection procedure, the part was set-up on the back face and the hole pattern was balanced or aligned to obtain the average of the holes. Hole #1 was chosen as the initial origin in order to collect data.

Your task has two parts.

First, you must evaluate the data and find if the hole pattern is in position.

Second, you must find the approximate DRF origin point of the balanced pattern relatve to the initial set-up.

Use the paper gage concept to complete your evaluations and record your calculations in the chart below. In this example, ignore the profile specification. It is not part of this exercise. If you have trouble with this exercise, see the similar examples in this unit. The paper gage materials can be found in unit 3.

PRODUCED PART SHOWN BELOW
SEE PREVIOUS PROBLEMS FOR PRODUCT DRAWING

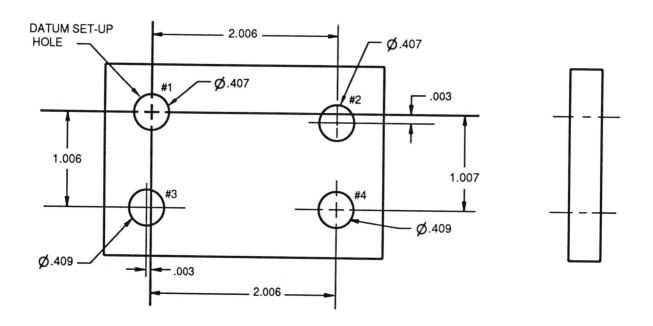

HOLE NO.	HOLE MMC	HOLE ACTUAL SIZE	POSITION TOLERANCE ALLOWED	"X" DIM.	"Y" DIM.	POSITION TOLERANCE ACTUAL	ACC or REJ	PAPER GAGE EVALUATION	ACC or REJ
1									
2									
3									
4									

UNIT 10

FORM TOLERANCES

Overview
Flatness
       Design Considerations
       Verification
Flatness Applied on a Unit Basis
Straightness - Design Considerations
Straightness - Line Elements Applied to a Flat Surface
Straightness - Line Elements, Two Directions
Straightness - Line Elements Applied to a Pin
Straightness - Axis at MMC
Straightness - Median Plane at MMC
Circularity
       Average Diameter - Free State Variation
Cylindricity
Workshop Exercise 10.1

FORM TOLERANCES

FORM TOLERANCES CONTROL THE FORM OF INDIVIDUAL FEATURES
NO DATUMS ARE ALLOWED

SYMBOL	TYPE OF TOLERANCE	SHAPE OF TOLERANCE ZONE	2D OR 3D	APP OF FEATURE MODIFIER
◻	FLATNESS	2 PARALLEL PLANES	3D	NO
—	STRAIGHTNESS CONTROLS SURFACE LINE ELEMENTS	2 PARALLEL LINES	2D	NO
—	STRAIGHTNESS CONTROLS AXIS OR MEDIAN PLANE	2 PARALLEL PLANES OR CYLINDRICAL ZONE	3D	YES
○	CIRCULARITY ROUNDNESS	2 CONCENTRIC CIRCLES	2D	NO
⌭	CYLINDRICITY	2 CONCENTRIC CYLINDERS	3D	NO

OVERVIEW:

There are four form tolerances: FLATNESS, STRAIGHTNESS, CIRCULARITY (ROUNDNESS) AND CYLINDRICITY. These form tolerances are shown in the chart above. Form tolerances control the form of an individual feature. There are no datums allowed. Note: *There are also cases where profile specified without datums can be considered a form tolerance. See the profile section for more information on this subject.*

As we know, the Taylor Principle (rule #1) states that the size of an individual feature will control the size as well as the form of that feature. In effect, this means that the size of an individual feature will control how much the form will vary.

Generally, in order to refine or tighten up the form control on size features, a form tolerance is used. As you can see from the chart above, the form controls can be either 2 dimensional (2D) or 3 dimensional (3D).

Flatness and straightness - Line elements are identical with the exception that flatness is a 3D control (tolerance zone is 2 parallel planes) and straightness - line elements is a 2D control (tolerance zone is 2 parallel lines).

Cylindricity and circularity are also identical with the exception that cylindricity is a 3D control (tolerance zone is two concentric cylinders) and circularity is a 2D control (tolerance zone is two concentric circles).

The verification of cylindricity and circularity are often difficult. The surface periphery of the part must be verified relative to a reference axis. Often an external feature is rotated in a vee bock and the surface is verified with an indicator. This is a rough check. Depending on the varying number and arrangement of lobes on the part, plus the angle of the vee bock, this verification method can give varying results. In some cases, it can show a good part bad and other times it can show a bad part good.

A more precise method of verification is through the use of a roundness machine where the part is rotated utilizing a precision spindle. A stylus reads the surface and transcribes an enlarged profile of the part periphery on a polargraph or strip chart.

The method of describing roundness and circularity in the ASME Y14.5M-1994 standard is called minimum radial separation. In the ANSI B89.3.1-1972 standard on roundness there are more sophisticated methods available to describe more specific needs. This includes methods such as the "least mean square" which determines the center of circular form based on a mathematical formula. This method is often used by coordinate measuring machines.

There is also methods such as "minimum circumscribed circle (MCC) or maximum inscribed circle (MIC)" Consult the ANSI B89.3.1-1972 standard for more specifics.

THE RENEGADE - STRAIGHTNESS AXIS OR MEDIAN PLANE

The form tolerances are fairly straight forward with one exception. The exception is straightness - axis or median plane. This form control is often referred to as "the renegade," as it reacts differently from the other form controls.

Straightness - axis or median plane is the only form tolerance that controls the axis or median plane of a feature and, therefore, can have a material condition modifier (MMC. LMC or RFS) applied. All the other form tolerances control the surface of a feature and, therefore, material condition modifiers are not applicable.

In addition, all of the form tolerances (with the exception of the circularity average diameter concept) must be within the size and form requirements of the Taylor Principle. The application of a form tolerance to a feature will tighten or restrict the form even further. In the case of straightness-axis or median plane, this control will relax the form control provisions of rule #1.

FLATNESS

Flatness is the condition of a surface having all elements in one plane. A flatness tolerance specifies a tolerance zone defined by two parallel planes within which the surface must lie.

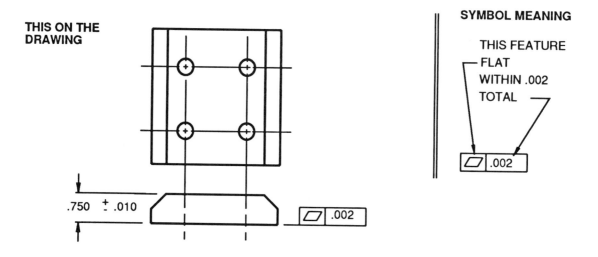

THIS ON THE DRAWING

.750 ± .010

□ .002

SYMBOL MEANING

THIS FEATURE
FLAT
WITHIN .002
TOTAL

□ .002

The surface must lie between two parallel planes .002 apart. In addition, the surface must be within the specified limits of size.

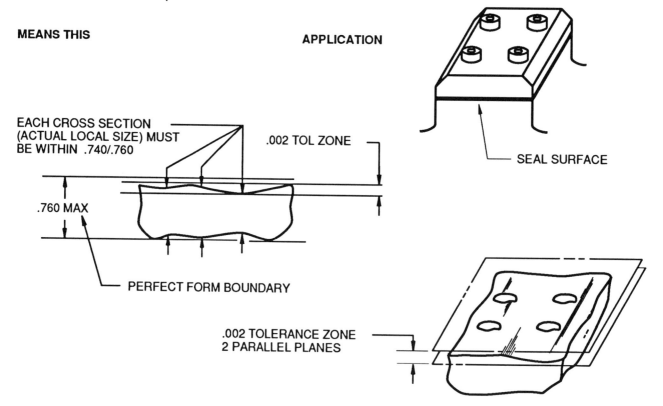

MEANS THIS

EACH CROSS SECTION
(ACTUAL LOCAL SIZE) MUST
BE WITHIN .740/.760

.002 TOL ZONE

.760 MAX

PERFECT FORM BOUNDARY

.002 TOLERANCE ZONE
2 PARALLEL PLANES

APPLICATION

SEAL SURFACE

Flatness is a form tolerance and datums are not allowed. Since flatness controls the surface, the material condition modifiers RFS, MMC and LMC are not applicable.

FLATNESS - VERIFICATION

There are many ways to check a flatness specification. Some are better than others. As with verifications for any geometric tolerance, the method or procedure used for verification will depend on many factors. How many parts are there to check? Is this the 1st part produced or the1000th? Is the tolerance well within the process capability? Are statistical process controls being done? How tight is the tolerance? What kind of equipment is available? Is it an in-process check or a final check?

All of these factors and many more will have an effect on how the part is to be verified. The procedures for verification should be recorded in the dimensional measurement plan and coordinated with anyone who is involved with the part.

The measurement procedures and techniques shown below are for illustration and background information. These procedures are intended to assist the reader in understanding the concepts. In addition to those methods shown below, there are many other ways to verify flatness, including the use of a coordinate measuring machine. (CMM)

VERIFICATION METHODS
FOR FLATNESS

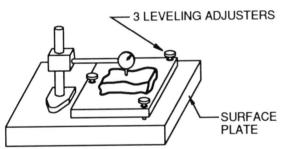

Leveling plate
Level part and move indicator over surface, readings must not exceed total flatness tolerance. Good check but may be time consuming.

Feeler gage check
Quick check. It is good for large tolerances. May miss concave variations.

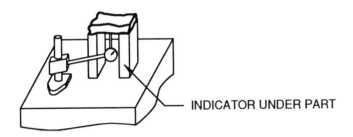

Set on gage blocks of same height, then indicate under part. Will not check surface under blocks.

Indicate thru hole in plate. Slide part over indicator. Good in process check. May misread on convex parts.

FLATNESS - DESIGN CONSIDERATIONS

The manufacturability or producibility of a flatness tolerance, or any geometric tolerance for that matter, is not just dependent on the size of the tolerance zone. The size of the controlled surface is also a factor. A large geometric tolerance in conjunction with an extremely large feature can be more difficult to produce than a small geometric tolerance on a small feature. When selecting geometric tolerances the designer should consider the size of the controlled feature as well as the size of the geometric tolerance.

Flatness is often used on datum features. Specifying a flatness control on a primary datum surface will limit the amount of rock or instability of the surface relative to the datum reference frame. The rock or instability cannot be eliminated; it can only be minimized through a flatness control. There is an old "rule of thumb" that states: Datum features should be as good as the features that are related to it. This rule still holds good today with some clarification.

It stands to reason that if features are related to a datum feature that is unstable and can rock, then the feature that is related to the unstable datum surface will rock also. But, this does not mean that if the flatness tolerance is .005 that all the related features will rock .005. When considering the flatness tolerance, the designer should also consider the size, distance and relationship geometry of the datum feature relative to the related features.

Flatness is often applied to sheet metal, flexible or flanged surfaces that will be bolted, screwed or welded together. If the parts will bend or deform when assembled, the flatness tolerance may not control rock or instability, but rather how much the parts will be deformed, bent or stressed in the assembly process.

Flatness is also used on gasket or seal surfaces. The flatness tolerance will control the size of the gap or crack when parts are assembled. A flatness on two mating surfaces of .005 each could result in a .010 total gap.

In order to meet functional requirements and still meet producibility requirements, the designer may consider alternatives to specifying tight tolerances. The notation, MUST NOT BE CONCAVE or MUST NOT BE CONVEX, may be placed under the feature control frame as applicable.

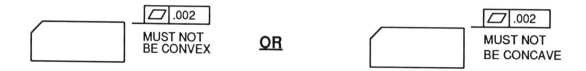

FLATNESS APPLIED ON A UNIT BASIS

Flatness may be applied on a unit basis as a means of preventing an abrupt surface variation within a relatively small area of the feature. The unit variation is used either in combination with a specified total variation or alone. The unit variation can be specified as a square zone as shown below or as a diameter zone. If the diameter zone is required the diameter symbol is entered in the feature control frame.
Note: This type of unit basis control can be used with other geometric controls to obtain similar effects.

FLATNESS SPECIFIED ON A UNIT BASIS WITH A SPECIFIED TOTAL VARIATION.

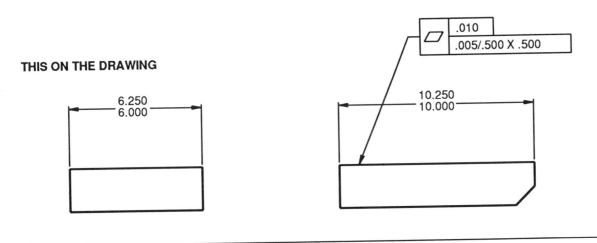

THIS ON THE DRAWING

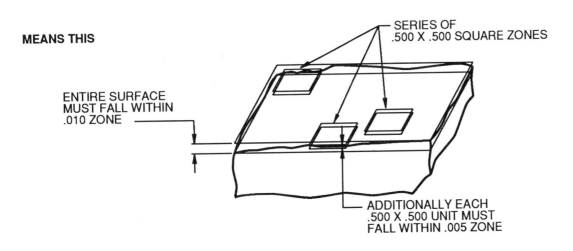

MEANS THIS

Caution should be used when applying a unit control without specifying a maximum limit. It is possible to have relatively large theoretical variations if the toleranced feature has a "bow" or is bent or warped. Each unit of the feature must be within the specified unit tolerance, but can produce a cumulative effect if allowed to continue over the entire length of the feature without a specified maximum value. This possible cumulative effect can be substantial if the overall feature is significantly larger than the unit length.

STRAIGHTNESS

There are two kinds of straightness controls. The two controls are very different and will be discussed separately. The two types of straightness use the same geometric characteristic symbol. The manner in which the feature control frame is directed to the feature determines which control applies.

1. Straightness - line elements - This is a 2D specification and controls line elements of surfaces. No material condition modifiers are allowed. This control refines and further restricts the form control provisions of the Taylor Principle. (Rule#1) The feature control frame is directed to the surface.

2. Straightness - axis or median plane - This is a 3D specification and controls the axis or median plane of features of size. This control relaxes the form control provisions of the Taylor Principle. (Rule#1) The perfect form boundary at MMC can be violated. Material condition modifiers are applicable. The feature control frame is attached to, placed under or associated with the size tolerance.

STRAIGHTNESS-LINE ELEMENTS - DESIGN CONSIDERATIONS

Straightness - line elements is a surface control; and, when applied to a feature, it will tighten or further restrict the straightness control provisions of rule #1. This type of control might be used on dowel pins or shafts where the straightness must be better than provided with the size control.

It can also be used in conjunction with, and as a refinement to, other geometric tolerances. It can be used on corrugated material where control in one direction is more important than another.

It might also be specified on features that are not controlled by the form provisions of rule #1, such as: stock sizes or commercially purchased parts.

Caution: As we know, the straightness 2D line elements apply in the view in which shown. Manufactured parts are not perfect and will have inherent variations on all the surfaces. In some instances, when the orientation of straightness tolerances or specifications are critical, it may be unclear on how to orient the part in order to make the straightness line element checks. Since straightness cannot be applied with datums, the designer might consider the profile of a line control with datums.

STRAIGHTNESS - LINE ELEMENTS APPLIED TO A FLAT SURFACE

Straightness - line elements is a condition where an element of a surface is a straight line. The tolerance specifies a tolerance zone within which the considered line element must lie. A straightness tolerance is applied to the feature in the drawing view in which it is to apply. Note the feature control frame is directed to the surface.

THIS ON THE DRAWING

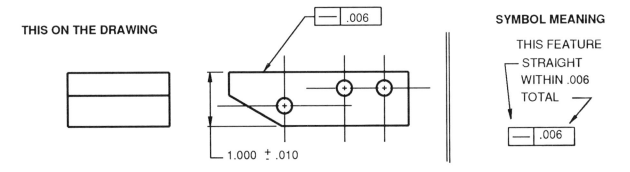

SYMBOL MEANING

THIS FEATURE
STRAIGHT
WITHIN .006
TOTAL

The surface, in the view shown, must lie between two parallel lines .006 apart. In addition, the feature must be within the limits of size.

MEANS THIS

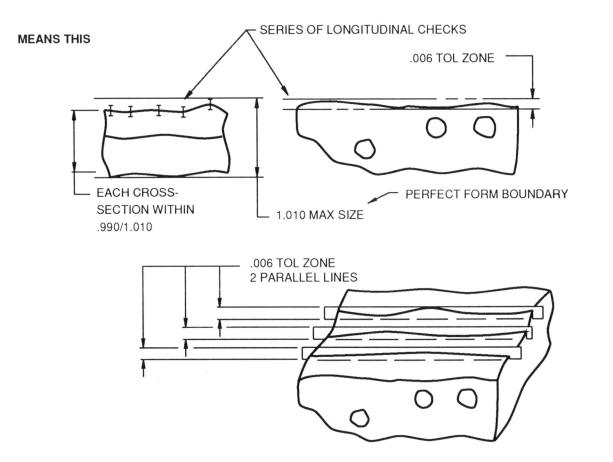

The figure above illustrates how a straightness-line elements specification controls the tolerance in 2D line elements in the view shown. If orientation of the tolerance zones are critical to other features, consider the use of profile of a line with datums.

10.9

STRAIGHTNESS - LINE ELEMENTS IN 2 DIRECTIONS

Since straightness-line elements is a 2D control, a feature can have different straightness values specified in different views as shown below. The tolerance applies in the view in which it is shown.

THIS ON THE DRAWING

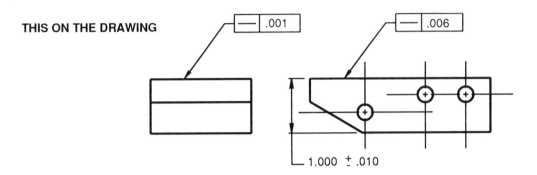

Each longitudinal element of the surface must lie between two parallel lines of .001 in the left view and .006 in the right view of the drawing.

MEANS THIS

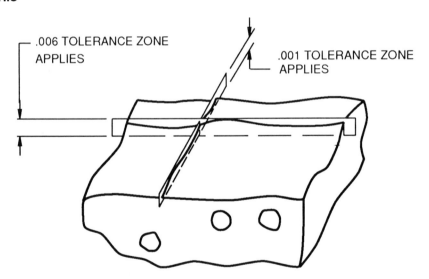

If the orientation of the straightness tolerance zones are critical to other features, consider the use of profile of a line with datums.

STRAIGHTNESS - LINE ELEMENTS

Straightness - line elements can be applied to the surface of a pin. The tolerance specifies a tolerance zone within which the considered line elements must lie. Note the feature control frame is directed to the surface.

THIS ON THE DRAWING

SYMBOL MEANING

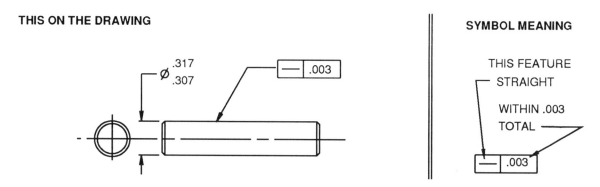

THIS FEATURE
STRAIGHT

WITHIN .003
TOTAL

Each longitudinal element of the surface must lie between two parallel lines .003 apart where the two lines and the nominal axis of the part share a common plane. In addition, the feature must be within the specified limits of size and the boundary of perfect form at MMC. Straightness - line elements will control waisted, barrelled and bent shapes. It does not control taper.

MEANS THIS

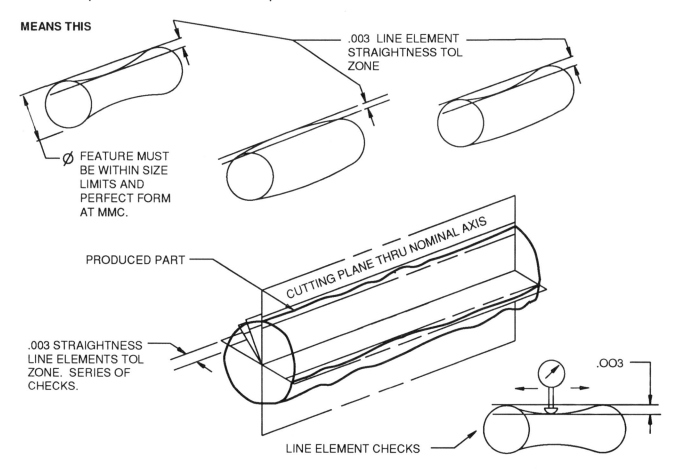

10.11

STRAIGHTNESS - AXIS AT MMC

Straightness of an axis is a condition where an axis is a straight line. The tolerance specifies a tolerance zone within which the derived median line must lie. This type of control is used where the size of the pin is important, but the pin can "bow or be bent beyond the perfect form limits of size. Note the feature control frame below is associated with the size tolerance of the feature.

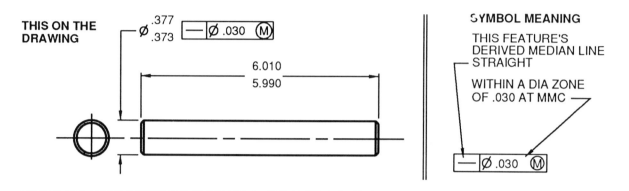

THIS ON THE DRAWING

Ø .377 / .373   — | Ø .030 (M)

6.010
5.990

SYMBOL MEANING

THIS FEATURE'S DERIVED MEDIAN LINE STRAIGHT

WITHIN A DIA ZONE OF .030 AT MMC

— | Ø .030 (M)

The derived median line of the feature's actual local sizes must lie within a cylindrical tolerance zone of .030 diameter at MMC. As each local size departs from MMC, an increase in the local diameter of the tolerance cylinder is allowed which is equal to the amount of such departure. Each circular element of the surface must be within the specified limit of size.

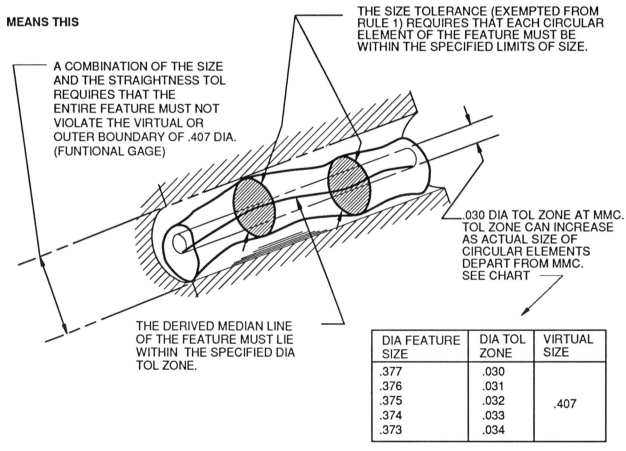

MEANS THIS

THE SIZE TOLERANCE (EXEMPTED FROM RULE 1) REQUIRES THAT EACH CIRCULAR ELEMENT OF THE FEATURE MUST BE WITHIN THE SPECIFIED LIMITS OF SIZE.

A COMBINATION OF THE SIZE AND THE STRAIGHTNESS TOL REQUIRES THAT THE ENTIRE FEATURE MUST NOT VIOLATE THE VIRTUAL OR OUTER BOUNDARY OF .407 DIA. (FUNTIONAL GAGE)

.030 DIA TOL ZONE AT MMC. TOL ZONE CAN INCREASE AS ACTUAL SIZE OF CIRCULAR ELEMENTS DEPART FROM MMC. SEE CHART

THE DERIVED MEDIAN LINE OF THE FEATURE MUST LIE WITHIN THE SPECIFIED DIA TOL ZONE.

DIA FEATURE SIZE	DIA TOL ZONE	VIRTUAL SIZE
.377	.030	
.376	.031	
.375	.032	.407
.374	.033	
.373	.034	

Straightness is a form tolerance and datums are not allowed. Since straightness of an axis controls the axis of this feature, the material condition modifiers MMC, LMC and RFS can be applied.

When a straightness of an axis control is applied to a feature of size, all circular elements of the surface must be within the specified size tolerance; however, the boundary of perfect form at MMC may be violated.

Where necessary the straightness tolerance may be greater than the size tolerance. The collective effect of size and form variation can produce a virtual condition equal to the MMC size plus the straightness tolerance. The acceptance boundary can be thought of as a 3D solid or functional gage that the feature may never violate.

STRAIGHTNESS - MEDIAN PLANE

Straightness can be applied to the median plane of a non-cylindrical feature. This is the same concept as straightness of an axis. In this case it is just applied to a median plane. This control may be applied when the local size of the feature must be maintained but the feature may be allowed to "bow or warp." This straightness median plane control will require the application of material condition modifiers. The part below is specified at MMC. Note the feature control frame is associated with the size tolerance.

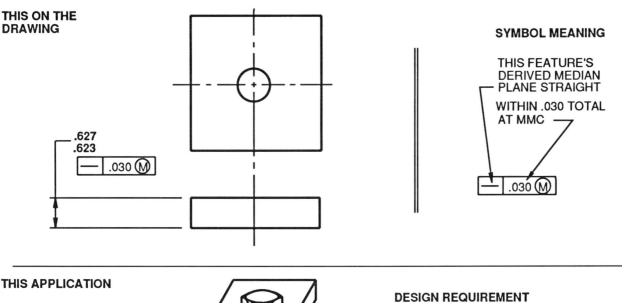

THIS ON THE DRAWING

.627
.623

| — | .030 Ⓜ |

SYMBOL MEANING

THIS FEATURE'S DERIVED MEDIAN PLANE STRAIGHT

WITHIN .030 TOTAL AT MMC

| — | .030 Ⓜ |

THIS APPLICATION

WARPED PART

DESIGN REQUIREMENT

THE SIZE OR THICKNESS IS IMPORTANT, BUT THE PART MAY BE "BOWED" OR WARPED. IT WILL STRAIGHTEN OUT WHEN BOLTED DOWN IN THE ASSEMBLY.

The derived median plane of the actual feature must lie within two planes .030 apart at maximum material condition. Additionally, each local size of the feature must fall within the specified limits of size.

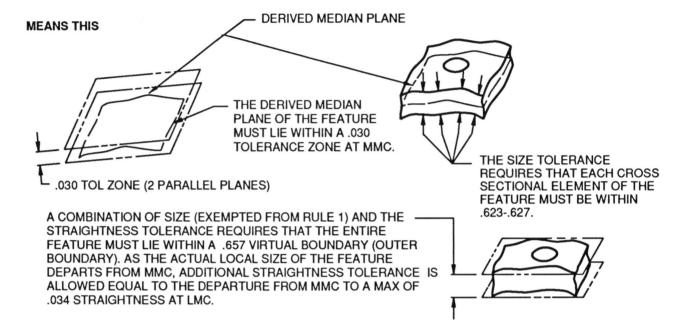

MEANS THIS

DERIVED MEDIAN PLANE

THE DERIVED MEDIAN PLANE OF THE FEATURE MUST LIE WITHIN A .030 TOLERANCE ZONE AT MMC.

.030 TOL ZONE (2 PARALLEL PLANES)

THE SIZE TOLERANCE REQUIRES THAT EACH CROSS SECTIONAL ELEMENT OF THE FEATURE MUST BE WITHIN .623-.627.

A COMBINATION OF SIZE (EXEMPTED FROM RULE 1) AND THE STRAIGHTNESS TOLERANCE REQUIRES THAT THE ENTIRE FEATURE MUST LIE WITHIN A .657 VIRTUAL BOUNDARY (OUTER BOUNDARY). AS THE ACTUAL LOCAL SIZE OF THE FEATURE DEPARTS FROM MMC, ADDITIONAL STRAIGHTNESS TOLERANCE IS ALLOWED EQUAL TO THE DEPARTURE FROM MMC TO A MAX OF .034 STRAIGHTNESS AT LMC.

STRAIGHTNESS - AXIS OR MEDIAN PLANE DESIGN CONSIDERATIONS

Straightness - axis or median plane is a form control that is often called "the renegade," as it reacts differently than the other form tolerances.

In all of the form tolerances (with the exception of circularity, average diameter concept) the feature must be within the size and form requirements of the Taylor Principle. The application of a form tolerance to a feature will tighten or restrict the form even further. Straightness-axis or median plane will relax the form control of rule #1 (Taylor Principle).

Straightness - axis or median plane is the only form tolerance that is applied to the axis or median plane of a feature. It is also the only form control that can have a modifier applied (MMC, RFS or LMC). The other form tolerances are only surface controls and, therefore, can not have material condition modifiers applied.

Straightness of an axis or median plane is used when the form requirements of rule #1 are too restrictive.

CIRCULARITY - (ROUNDNESS)

CIrcularity is a condition of a surface where:
- a. for a feature other than a sphere, all points of the surface intersected by any plane perpendicular to an axis are equidistant to that axis.
- b. for a sphere, all points of the surface intersected by any plane passing through a common center are equidistant from that center.

THIS ON THE DRAWING

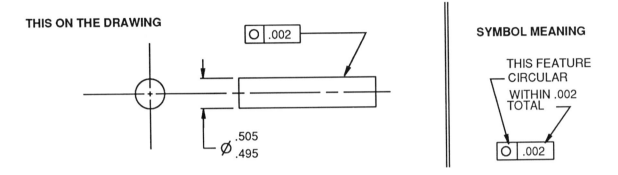

SYMBOL MEANING

THIS FEATURE
CIRCULAR
WITHIN .002
TOTAL

Each circular element of the surface in a plane perpendicular to an axis must lie between two concentric circles, one having a radius .002 larger than the other. In addition, the feature must be within the specified limits of size.

MEANS THIS

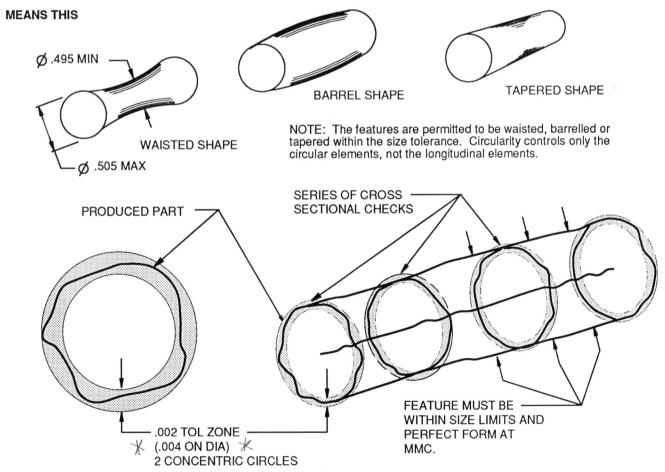

NOTE: The features are permitted to be waisted, barrelled or tapered within the size tolerance. Circularity controls only the circular elements, not the longitudinal elements.

10.15

Circularity is a 2D tolerance. It controls circular elements only, not longitudinal elements. The value of the tolerance zone is the width of the band between the two concentric circles. Circularity is a form tolerance and, therefore, datums are not allowed. Circularity is a surface control; thus the modifiers MMC, LMC and RFS are not applicable.

Circularity tolerance is used to refine the size requirements and is always less than the size tolerance, except in the case of parts subject to free state variation. Circularity is used on round or spherical features. Usually the size and form requirements of the Taylor Principle are sufficient to control the form of the feature.   There are situations where the circularity of a feature must be refined better than the requirements of the Taylor Principle. This is an application where circularity may be applied.

Examples of use: ball bearings, tubes, pipes, and circular elements of tapered, barreled or waisted parts such as nose or tail cones, seals, valves, etc. It can also be used as a 2D refinement on cylindricity.

AVERAGE DIAMETER - FREE STATE VARIATION

There are cases, especially with diameters subject to free state variation, where the circularity tolerance (and possibly other geometric tolerances) may exceed the size and form requirement of the Taylor Principle. This is often the case of flexible aircraft rings that go "out of round" after fabrication.

If a diameter is qualified with the abbreviation AVG (average dia), the size of this diameter may exceed the form provisions of the Taylor Principle. The size of the feature is the average of several diametral measurements taken across a circular or cylindrical feature. Usually enough measurements (at least four) are taken to assure the establishment of an average diameter. The average diameter designation will negate the form provisions of the Taylor Principle requirements.

Specifying a circularity requirement on an average diameter of a nonrigid part is often necessary, as it will place a limit on the maximum out of roundness. This will ensure the actual diameter of the feature can be restrained to the desired shape at assembly and not be bent or deformed beyond its elastic limit. It is in this case where the circularity tolerance (and, as an extension of the principles, possibly some other geometric tolerances) may exceed the size tolerance.

CYLINDRICITY

Cylindricity is the condition of a surface of revolution in which all points of the surface are equidistant from a common axis. The tolerance zone is two concentric cylinders within which the surface must lie.

THIS ON THE DRAWING

⌭ | .0002

Ø .5005 / .4995

SYMBOL MEANING

THIS FEATURE
CYLINDRICAL
WITHIN .0002
TOTAL

⌭ | .0002

THE SIZE OF THE BEARINGS CAN VARY WITHIN THE SPECIFIED LIMITS, BUT THE BEARINGS MUST BE CYLINDRICAL.

APPLICATION

The cylindrical surface must lie between two concentric cylinders, one having a radius of .0002 larger than the other. In addition, the feature must be within the specified limits of size.

MEANS THIS

PRODUCED PART

.0002 TOL ZONE
(.0004 ON DIA)
2 CONCENTRIC CYLINDERS

THE SIZE TOLERANCE REQUIRES
THAT THE FEATURE MUST BE
WITHIN THE LIMITS OF SIZE AND
THE PERFECT FORM
REQUIREMENT AT MMC.

Cylindricity is a 3D tolerance. It controls both the circular and longitudinal elements of the feature. It includes circularity, straightness and taper. Cylindricity is a form tolerance, therefore datums are not allowed. Cylindricity is a surface control; thus the modifiers, MMC, LMC and RFS are not applicable. Cylindricity is used to refine the size requirements and is always a smaller value than the size tolerance. Examples of use: bearings, bearing journals, cylinders and the qualification of datum features. Cylindricity can be compared to flatness wrapped around a journal.

WORKSHOP EXERCISE 10.1

1. In the chart below, list the four classic form tolerances, along with their symbols and common shapes of their tolerance zones.

SYMBOL	TYPE OF TOLERANCE	SHAPE OF TOLERANCE ZONE
⧫	Flatness	2 // planes
—	Straightness	2 // lines/planes or cyl. zone
○	Circularity Roundness	2 concentric circles
⌀	Cylindricity	2 concentric cylinders

2. Can any of the form tolerances be applied with datums? If so, which ones?

No

3. Can any of the form tolerances be applied with feature modifiers? If so, which ones?

Yes, straightness

4. List the two dimensional (2D) form tolerances.

Straightness, circularity roundness

5. List the three dimensional (3D) form tolerances.

flatness, straightness, cylindricity

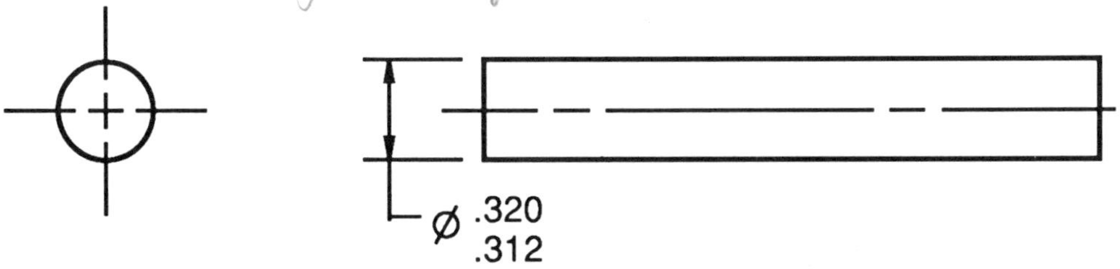

⌀ .320
.312

6. In the illustration above, what is the maximum line element straightness of the pin before any geometric controls are applied?

.320

7. In the illustration above, what is the maximum "bow" of the pin before any geometric controls are applied?

8. What rule in geometric tolerancing defines the size and form limits of an individual feature such as in the pin above?

straightness

10.19

Ø .320
.312

⌓ .003

9. Apply a .003 straightness line element control to the surface of the part above and show the resultant tolerance zone on the produced part below.

—.003

⌓ .003

10. Can the MMC, LMC or RFS modifier be applied to this control? Why or why not?

Yes

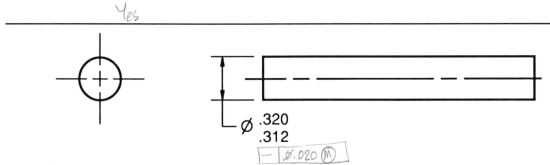

Ø .320
.312

— Ø.020 Ⓜ

11. Apply a .020 dia at MMC straightness control to the axis of the pin above and show the resultant tolerance zone and virtual size on the produced part below.

Virtual Size: .332 Ø

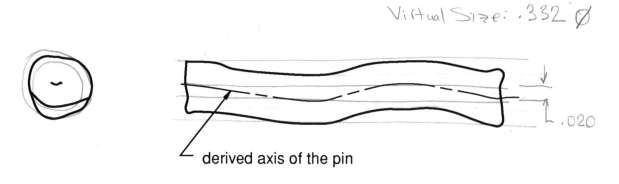

.020

derived axis of the pin

12. Complete the chart below illustrating the effect of MMC on the straightness tolerance.

PRODUCED SIZE OF PIN	ALLOWABLE DIA STRAIGHTNESS TOL ZONE	VIRTUAL SIZE
.313	.027	
.315	.025	.332
.318	.022	
.320	.020	

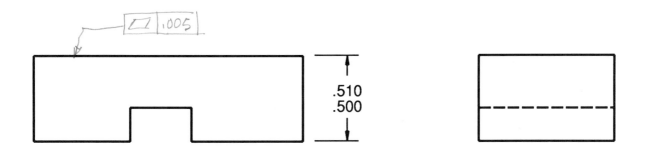

13. Apply a .005 flatness control to the top surface of the part above.

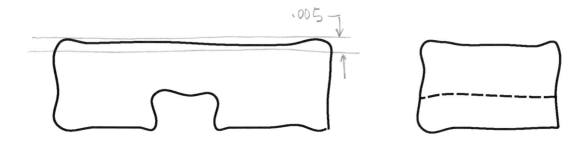

14. On the produced part above, show a simple sketch on how the part may possibly be checked and how the resultant tolerance zone is defined.

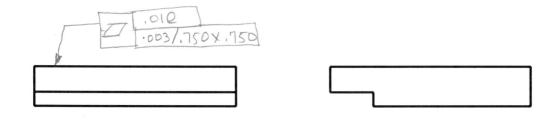

15. The entire top surface on the part above must be flat within .010 but each .750 X .750 unit must be flat within .003. Apply the necessary geometric controls to achieve that result.

16. What is the maximum circularity on the part below before any geometric symbols are applied?

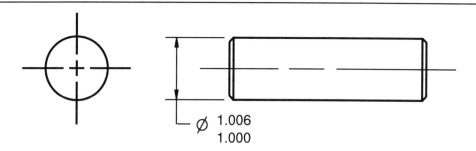

Ø 1.006
 1.000

17. Place a .0005 circularity control on the part above and show the resultant tolerance zone on the produced part below.

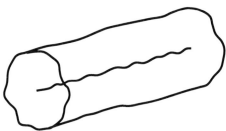

18. What is the maximum cylindricity on the part below before any geometric symbols are applied?

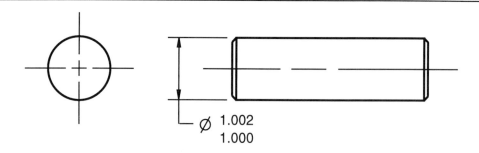

Ø 1.002
 1.000

19. Place a .0002 cylindricity control on the part above and show the resultant tolerance zone on the produced part below.

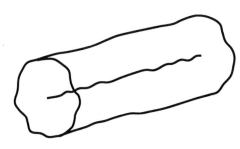

20. What is the main difference between circularity and cylindricity?

_____

_____

21. Is the circularity and cylindricity tolerance specified on a radius or diameter?

_____

22. Is a circulartiy and cylindricity tolerance additive or a refinement of the size tolerance?

_____

23. A designer has a base sealing plate and is worried that the part may come in warped or bent and will rock or be unstable in the assembly. What form tolerance would you suggest be considered? Why?

_____

_____

24. A designer has a shaft that will have a bearing pressed on a journal. The size tolerance on the journal is a tight tolerance to accept the press fit of the bearing race. Ovality on the journal could deform the bearing race. The bearing journal ovality must be refined even tighter than the size tolerance, but in this case the taper is not as important. Which form tolerance would you consider applying to the bearing journal?

_____

_____

25. Considering the same problem as above but this time the taper on the journal is also important. Which form tolerance would you consider using? Why?

_____

_____

26. The thickness on a machined plate is very important. After machining, the plate will enter a heat treating process and warpage is a problem. Manufacturing is asking for a relaxation of the tolerances. The plate is thin enough that the bolts in the assembly process will straighten the warpage. What form tolerance would you consider applying to the plate to allow this relaxation?

_____

27. A piece of corrugated material requires the form of the longitudinal elements of the surface to be controlled in one direction, but not the other. What form tolerance would you consider using?

_____

10.23

28. On this geometric overview chart, fill in the required information for only the form tolerances. You will return later to complete this chart as you progress through other units in the workbook.

GEOMETRIC CHARACTERISTIC OVERVIEW

DATUMS	TYPE OF TOLERANCE	CHARACTERISTIC	SYMBOL	2D OR 3D	CONTROLS — AXIS OR MEDIAN PLANE	CONTROLS — SURFACE	APPLICABILITY OF FEATURE MODIFIERS	APPLICABILITY OF DATUM MODIFIERS	COMMON SHAPES OF TOLERANCE ZONE
DATUMS NOT ALLOWED	FORM	STRAIGHTNESS LINE ELEMENT							
		STRAIGHTNESS AXIS OR MEDIAN PLANE							
		FLATNESS							
		CIRCULARITY (ROUNDNESS)							
		CYLINDRICITY							
DATUMS REQUIRED	ORIENTATION	ANGULARITY		SEE NOTE 3					
		PERPENDICULARITY		SEE NOTE 3					
		PARALLELISM		SEE NOTE 3					
	RUNOUT SEE NOTE 1	CIRCULAR RUNOUT							
		TOTAL RUNOUT							
	PROFILE (LOCATION OF SURFACES)	PROFILE OF A LINE							
		PROFILE OF A SURFACE							
DATUMS REQUIRED SEE NOTE 2	LOCATION OF FEATURES OF SIZE	POSITION				SEE NOTE 5			
		CONCENTRICITY			SEE NOTE 4				
		SYMMETRY			SEE NOTE 4				

NOTES:

1. CAN CONTROL FORM, ORIENTATION AND LOCATION.

2. THERE ARE SPECIAL CASES WHERE POSITION AND PROFILE MAY NOT REQUIRE DATUMS

3. THESE CHARACTERISTICS CAN BE MADE 2D BY WRITING "LINE ELEMENTS" UNDER THE FEATURE CONTROL FRAME.

4. THESE CHARACTERISTICS CONTROL OPPOSING MEDIAN POINTS.

5. CAN ALSO CONTROL SURFACE BOUNDARY.

UNIT 11

ORIENTATION TOLERANCES

ORIENTATION TOLERANCES

ORIENTATION TOLERANCES CONTROL THE ORIENTATION OF INDIVIDUAL FEATURES
DATUMS ARE REQUIRED

SYMBOL	TYPE OF TOLERANCE	SHAPE OF TOLERANCE ZONE	2D OR 3D	APP OF FEATURE MODIFIER
⊥	PERPENDICULARITY	2 PARALLEL LINES	2D OR 3D*	YES IF FEATURES HAVE SIZE
//	PARALLELISM	2 PARALLEL PLANES		
∠	ANGULARITY	CYLINDRICAL		

* Default for orientation tolerances is 3D. They can be made 2D by writing LINE ELEMENTS under the feature control frame.

OVERVIEW:

There are three orientation tolerances: PARALLELISM, ANGULARITY, AND PERPENDICULARITY. These orientation tolerances are shown in the chart above. Orientation tolerances control the orientation of individual features. The orientation control of a surface will also control the form of that surface if no form tolerance is specified. Note: *There are also some cases where profile specified with datums can be considered an orientation tolerance. See the profile section for more information on this subject.*

The orientation tolerances are all 3D controls. If it is necessary to control a feature in 2D, a note, such as EACH ELEMENT" or "EACH RADIAL ELEMENT," is placed under the feature control frame. In some cases, it may be necessary to control only the tangent plane of a feature. In this situation, the T in a circle designation is placed in the feature control frame.

Datum(s) are required on all orientation tolerances. It is quite common to relate an orientation tolerance to one or more datums. This requirement is specified if it is necessary to stabilize the tolerance zone relative to the datum reference frame.

The orientation tolerances will control features without size (plane surfaces) as well as features with size. (holes, slots, tabs, pins etc.) If the orientation tolerances are applied to plane surfaces or a tangent plane, then feature modifiers (MMC, RFS or LMC) are not allowed. If the orientation tolerances are applied to the axis or median plane of a feature, then feature modifiers are required.

As we know, the size requirements of rule #1 (Taylor Principle) will control the form and size of an individual feature but do not control the relationship between individual features. In fact, the ASME Y14.5M, 1994 standard states, "Features shown perpendicular, coaxial, or symmetrical to each other must be controlled for location or orientation to avoid incomplete drawing requirements." The orientation tolerances will control the orientation of features to a datum reference frame.

The reason we still have the three symbols is simply because of past history. Mathematically the three orientation tolerances are identical. They control orientation. The only difference between the three controls is the angle of orientation to the primary datum. If the feature is 90 degrees from the primary datum, then the perpendicularity symbol is used. If the feature is 180 degrees from the primary datum, then the parallelism symbol is used. If the feature is any thing other than 90 or 180 degrees from the primary datum, then the angularity symbol is used. In reality, we could use the angularity symbol in place of the parallelism and perpendicularity symbol and it would all have the same meaning. In many places in the text, the word orientation is used in place of parallelism, angularity and perpendicularity to help us think generically about orientation tolerances. In fact, this is the way it is explained in the ASME Y14.5.1M, 1994 Mathematical Definition of Dimensioning and Tolerancing Principles Standard.

As you review the orientation tolerances in this chapter, you will find that in some complex examples it is difficult to determine whether the specification should be perpendicularity, parallelism or angularity. Remember, they are all orientation tolerances. An orientation tolerance will just control the orientation of an individual feature at the specified angle and to the specified datums. Orientation tolerances will not locate features. In order to locate features, a profile or position tolererance is used. Generally, profile is used to locate surfaces and position is used to locate features of size.

PARALLELISM

Parallelism is the condition of a surface or center plane, equidistant at all points from a datum plane or an axis, equidistant along its length from one or more datum planes or a datum axis.

THIS ON THE DRAWING

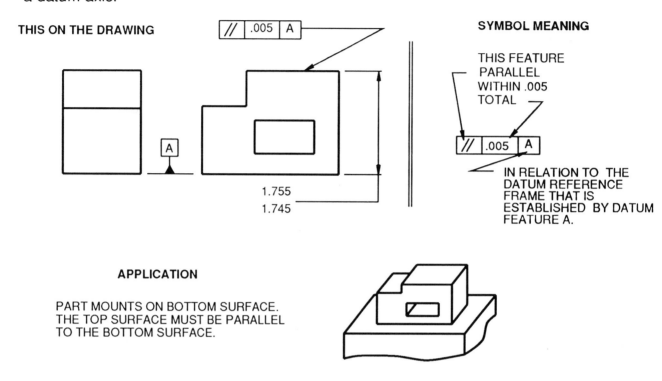

SYMBOL MEANING

THIS FEATURE PARALLEL WITHIN .005 TOTAL

IN RELATION TO THE DATUM REFERENCE FRAME THAT IS ESTABLISHED BY DATUM FEATURE A.

APPLICATION

PART MOUNTS ON BOTTOM SURFACE. THE TOP SURFACE MUST BE PARALLEL TO THE BOTTOM SURFACE.

The surface must lie between two parallel planes .005 apart which are parallel to the datum plane. In addition, the surface must be within the limits of size or profile.

MEANS THIS

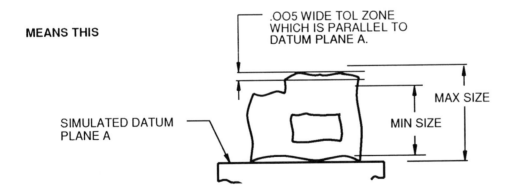

The above parallelism specification is a surface control. The MMC, LMC and RFS modifiers are not applicable. Parallelism tolerance applied to a plane surface also controls the flatness of that surface. Thus, if no flatness tolerance is specified, the flatness tolerance will be at least as close as the parallelism requirement.

11.4

MEANS THIS

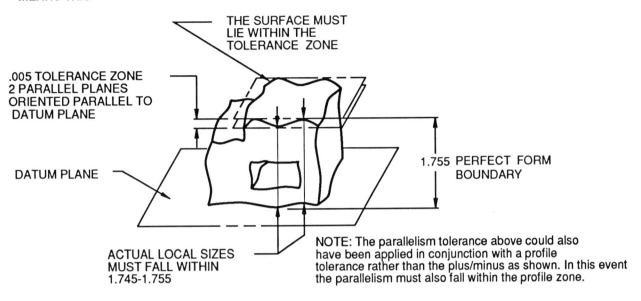

THE SURFACE MUST
LIE WITHIN THE
TOLERANCE ZONE

.005 TOLERANCE ZONE
2 PARALLEL PLANES
ORIENTED PARALLEL TO
DATUM PLANE

DATUM PLANE

1.755 PERFECT FORM
BOUNDARY

ACTUAL LOCAL SIZES
MUST FALL WITHIN
1.745-1.755

NOTE: The parallelism tolerance above could also
have been applied in conjunction with a profile
tolerance rather than the plus/minus as shown. In this event
the parallelism must also fall within the profile zone.

The figure above illustrates the interpretation of the prior parallelism specification. Notice that the part must be within the limits of size and perfect form at MMC. The plus/minus tolerance on this part is for illustrative purposes only and can be replaced with a profile tolerance.

If a parallelism tolerance is applied in conjunction with a profile tolerance, the parallelism tolerance zone must also fall within the profile zone just as it did with the plus/minus tolerance.

SAMPLE INSPECTION

PART IS MOUNTED ON DATUM A. THE
ENTIRE SURFACE IS INDICATED AND
MUST FALL WITHIN A .005 TOL ZONE.

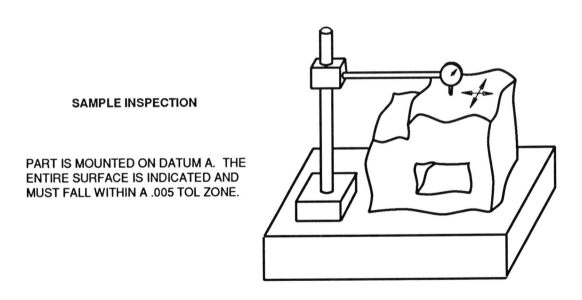

The example above shows a possible inspection procedure for verifying a parallelism specification. A CMM or other inspection devices can be used as well.

ANGULARITY - SURFACE

Angularity is the condition of a surface, center plane, or axis at a specified angle (other than 90° or 180°) from a DRF. The example below illustrates a surface located with a profile tolerance and then refined for orientation with an angularity tolerance.

THIS ON THE DRAWING

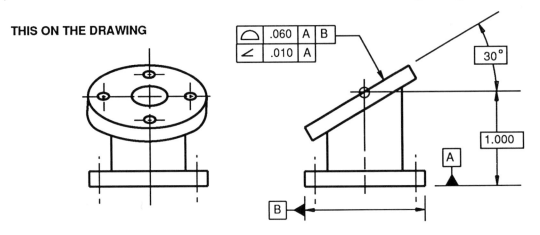

The surface must lie between two parallel planes .010 apart which are inclined at 30° to datum plane A. In addition, the surface must also fall within the profile zone.

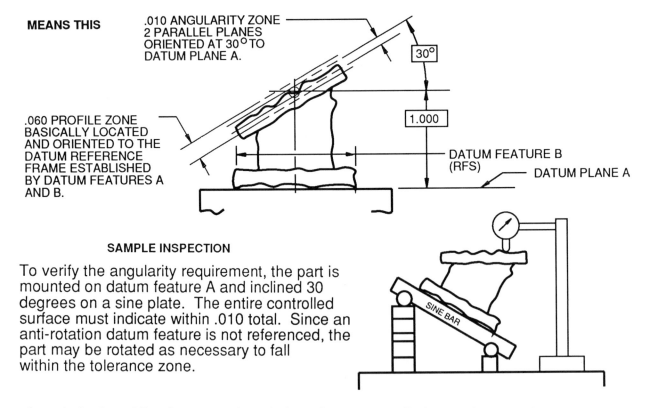

MEANS THIS

.010 ANGULARITY ZONE
2 PARALLEL PLANES
ORIENTED AT 30° TO
DATUM PLANE A.

.060 PROFILE ZONE
BASICALLY LOCATED
AND ORIENTED TO THE
DATUM REFERENCE
FRAME ESTABLISHED
BY DATUM FEATURES A
AND B.

DATUM FEATURE B (RFS)

DATUM PLANE A

SAMPLE INSPECTION

To verify the angularity requirement, the part is mounted on datum feature A and inclined 30 degrees on a sine plate. The entire controlled surface must indicate within .010 total. Since an anti-rotation datum feature is not referenced, the part may be rotated as necessary to fall within the tolerance zone.

Angularity is a 3D tolerance. Angularity tolerance applied to a plane surface also controls the flatness of that surface. The surface on the part above is not controlled for rotation relative to the DRF. If rotation control is necessary, additional datums (possibly the pattern of holes on the bottom surface) can be specified.

PERPENDICULARITY - SURFACE

Perpendicularity is the condition of a surface, median plane or an axis at a right angle to the datum plane(s) or axes.

THIS ON THE DRAWING

SYMBOL MEANING

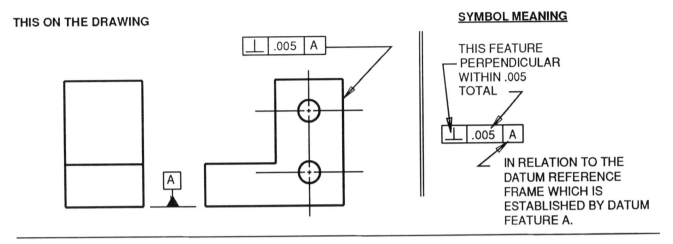

THIS FEATURE
PERPENDICULAR
WITHIN .005
TOTAL

IN RELATION TO THE
DATUM REFERENCE
FRAME WHICH IS
ESTABLISHED BY DATUM
FEATURE A.

APPLICATION

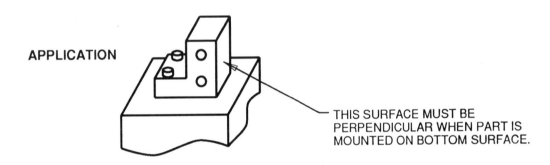

THIS SURFACE MUST BE
PERPENDICULAR WHEN PART IS
MOUNTED ON BOTTOM SURFACE.

The surface must lie between two parallel planes .005 apart. In addition, the feature must be within the limits of size or location.

MEANS THIS

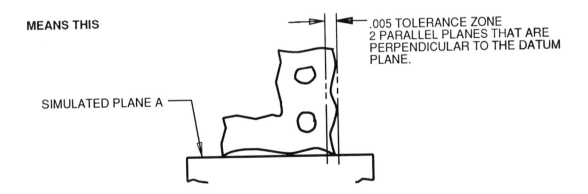

.005 TOLERANCE ZONE
2 PARALLEL PLANES THAT ARE
PERPENDICULAR TO THE DATUM
PLANE.

SIMULATED PLANE A

Perpendicularity is a 3D tolerance. Perpendicularity tolerance applied to a plane surface also controls the flatness of that surface. Thus, if no flatness is specified, the flatness tolerance will be as close as the perpendicularity requirement.

MEANS THIS

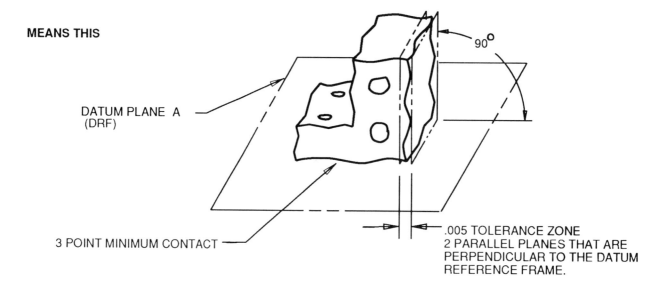

DATUM PLANE A
(DRF)

90°

3 POINT MINIMUM CONTACT

.005 TOLERANCE ZONE
2 PARALLEL PLANES THAT ARE
PERPENDICULAR TO THE DATUM
REFERENCE FRAME.

The figure above illustrates in 3D the perpendicularity requirement to one datum feature. The surface must fall between two parallel planes. The two parallel planes are oriented perpendicular to the datum reference frame. In this case, the datum reference frame consists of only one plane.

In the perpendicularity example, the variations of the datum feature are not controlled. The datum feature will just contact the datum plane and the assembly on the high points of the surface. In a practical example, a flatness specification might be used to control the variations of the datum feature.

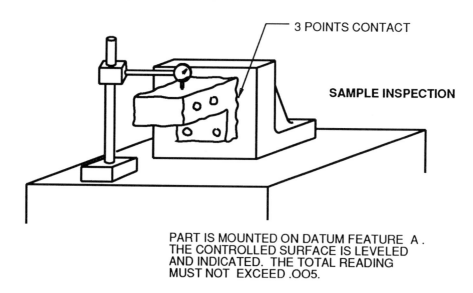

3 POINTS CONTACT

SAMPLE INSPECTION

PART IS MOUNTED ON DATUM FEATURE A.
THE CONTROLLED SURFACE IS LEVELED
AND INDICATED. THE TOTAL READING
MUST NOT EXCEED .005.

The figure above illustrates a possible inspection procedure for the perpendicularity requirement. Note that the datum reference frame established in the inspection set-up coincides with the datum reference frame established on the part.

PERPENDICULARITY - TWO DATUM FEATURES

Perpendicularity is an orientation tolerance and can be applied to two or more datum features. The example below illustrates perpendicularity to two datum features.

THIS ON THE DRAWING

SYMBOL MEANING

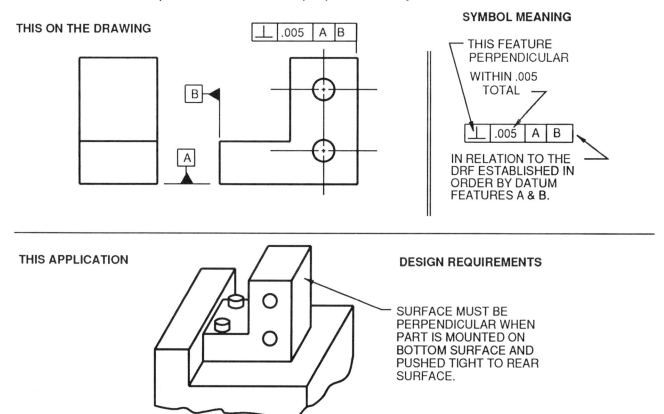

THIS APPLICATION

DESIGN REQUIREMENTS

SURFACE MUST BE PERPENDICULAR WHEN PART IS MOUNTED ON BOTTOM SURFACE AND PUSHED TIGHT TO REAR SURFACE.

The example below illustrates in 3D how the perpendicularity requirement is related to two datum features. Notice how the datum selection and function of the part shown above correlates with the datum set-up and interpretation shown below.

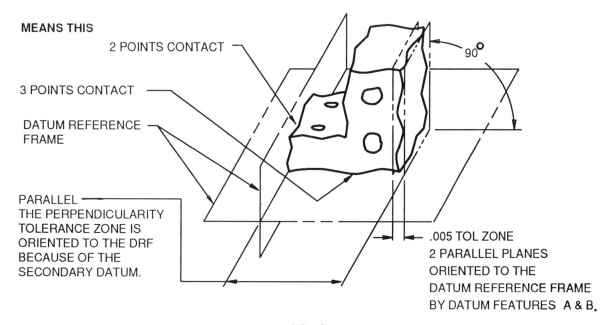

MEANS THIS

2 POINTS CONTACT

3 POINTS CONTACT

DATUM REFERENCE FRAME

PARALLEL
THE PERPENDICULARITY TOLERANCE ZONE IS ORIENTED TO THE DRF BECAUSE OF THE SECONDARY DATUM.

90°

.005 TOL ZONE
2 PARALLEL PLANES ORIENTED TO THE DATUM REFERENCE FRAME BY DATUM FEATURES A & B.

11.9

Perpendicularity, or any orientation tolerance to two or more datums, can be very confusing if you do not clearly understand the datum reference frame concept. It is important to remember that the datums only set up the part. The part, in this case, is set up with 3 points on A and 2 points on B. The orientation or perpendicularity tolerance zones are oriented to the datum reference frame and not the datum features. The surface must then fall within the tolerance zone.

The perpendicularity specification reads: This feature is perpendicular to the datum reference frame which is established, in order, by datum features A and B.

A separate perpendicularity and a separate parallelism specification, instead of the applied two datum perpendicularity requirement, will yield entirely different results. The parallelism and perpendicularity requirement would have two different primary datums and, as a result, would require two datum set ups. This will not match the functional requirements of the part.

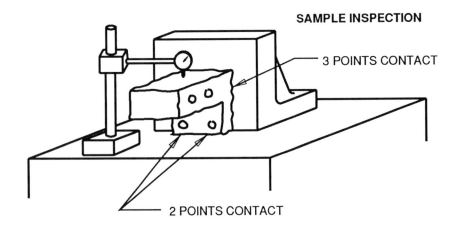

SAMPLE INSPECTION

3 POINTS CONTACT

2 POINTS CONTACT

PART IS MOUNTED ON DATUM FEATURE A AND ORIENTED TO DATUM FEATURE B. THE ENTIRE CONTROLLED SURFACE MUST INDICATE WITHIN .OO5 MAX.

The above inspection set-up is shown to help illustrate a possible procedure to verify a perpendicular requirement to two datum features. Note the inspection set up matches the functional requirements as the stated interpretation.

PARALLELISM - TWO DATUM FEATURES

The part in this example is fully toleranced with form, orientation, profile and position tolerances. The dimensions were left off for clarity. In fact, the part could be a solid model and have all the dimensions defined in a CAD data base. In any event, all the dimensions are basic and were left off for clarity.

This example was fully toleranced to give the reader a feel as to how simple and straight forward it can be to apply geometric tolerances. The important thing on a part like this is to establish a datum reference frame based on functional requirements and then relate everything to that reference frame. The primary purpose of this example is to show the application of orientation tolerances to two datum features. For additional information on the interpretations of the other geometric tolerances, see the appropriate sections in this text.

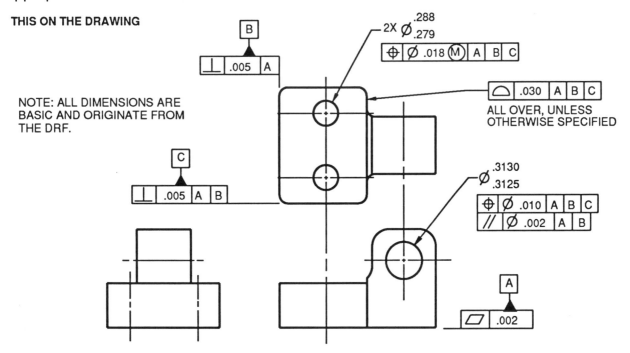

The following figure shows how the part mounts in the assembly. Datums and geometric tolerances are selected based on functional requirements. The part mounts and is aligned on the three surfaces shown. It is then fastened with two screws. The clearance holes are positioned at MMC. The outside contour is not important and is defined with a large profile tolerance. The datum reference frame is established and the datum features are qualified with flatness and perpendicularity.

The location of the large hole is not important and is given a .010 dia position RFS. If no other controls are specified, the position requirement also controls the parallelism of the hole within .010 dia to the DRF. Since the orientation of the hole is more critical, a parallelism is specified and it refines the orientation of the hole relative to the DRF. All of the applicable geometric tolerances, except the clearance hole position, are applied at RFS, as this is functionally required.

THIS APPLICATION

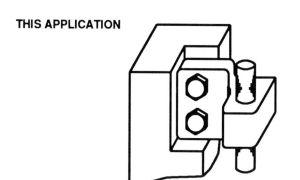

DESIGN REQUIREMENT

The part mounts on the 3 surfaces as shown. The location of the large bore must be within a .010 dia. The bore, however, must be oriented (parallel) relative to the mounting within a .002 dia.

The 3D figure shown below illustrates mathematically how the parallelism tolerance zone is related to the datum reference frame. It is important to remember that all the tolerance zones are oriented or located to the datum reference frame and not the datum features. The datums specified in the feature control frame only tell how to establish the DRF.

On a part like this, where it mounts in only one position, it is important that only one datum reference frame is established, as this is the way the part functions. One datum reference frame also makes the part much easier to understand, design, inspect and manufacture.

MEANS THIS

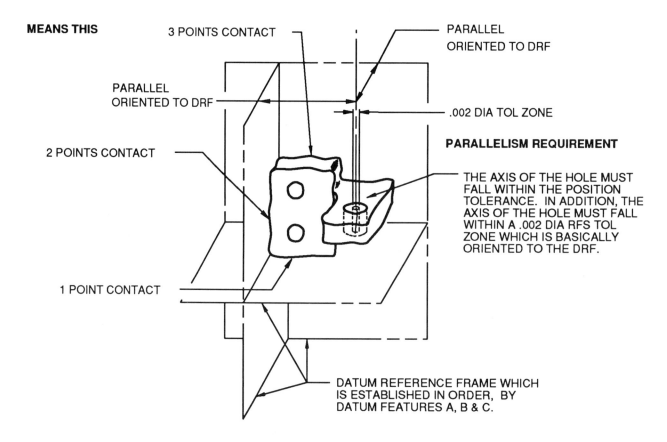

3 POINTS CONTACT

PARALLEL ORIENTED TO DRF

PARALLEL ORIENTED TO DRF

.002 DIA TOL ZONE

2 POINTS CONTACT

PARALLELISM REQUIREMENT

THE AXIS OF THE HOLE MUST FALL WITHIN THE POSITION TOLERANCE. IN ADDITION, THE AXIS OF THE HOLE MUST FALL WITHIN A .002 DIA RFS TOL ZONE WHICH IS BASICALLY ORIENTED TO THE DRF.

1 POINT CONTACT

DATUM REFERENCE FRAME WHICH IS ESTABLISHED IN ORDER, BY DATUM FEATURES A, B & C.

PARALLELISM - PATTERN OF HOLES

The part below illustrates an orientation tolerance (in this case parallelism) to a pattern of holes. This example is an extension of the principles shown in the last problem with parallelism to two datum features. The dimensions are left off for clarity. Part definition could also be defined in a CAD data base. If dimensions are applied, they will originate from one of the holes in the pattern. All dimensions are basic.

The primary reason for this example is to reinforce the idea that tolerances are related to the datum reference frame (DRF) and not the datum features. This example will concentrate on the parallelism specification. For additional information on the interpretations of other geometric tolerances, see the appropriate sections in this text.

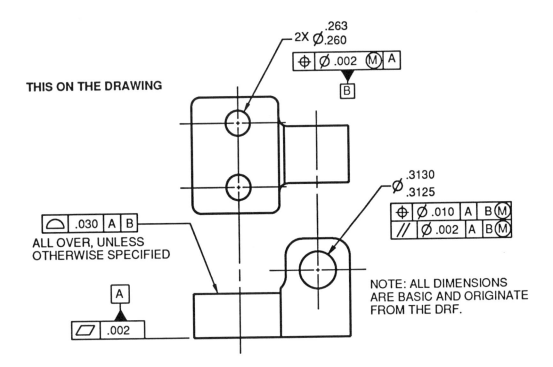

The following figure shows how the part mounts in the assembly. Datums and geometric tolerances are selected based on functional requirements. The part mounts and is aligned by two screws in the pattern of holes. The size and the position tolerance between the clearance holes is rather tight as these features locate and orient the part in the assembly. The outside contour is not important and is defined with a large profile tolerance.

The location of the large hole is not important and is given a .010 dia position RFS. If no other controls are specified, this position tolerance will also control the parallelism of the hole within .010 dia to the DRF. In this case, the orientation of the hole is important and a parallelism tolerance is specified which refines the orientation of the hole within .002 dia relative to the DRF.

THIS APPLICATION

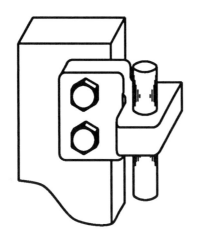

APPLICATION

The part mounts on the bottom surface and is located and oriented in the assembly by the two screws. The location of the bore must be within .010 dia. The orientation, however, must be within .002 dia.

The example below shows how the datum reference frame is established by the position tolerance cylinders set at basic dimensions. The actual axis of the holes must fall within these cylinders. The designer should recognize that the size and clearance of the mounting holes can have an end effect on the parallelism tolerance. Even though the parallelism tolerance on the large hole is held to .002 dia RFS, the total parallelism tolerance stack-up in the final assembly could be larger because the part mounts on the holes and these clearance hole sizes and location can vary. This example is shown with only two holes as a datum; but, in effect, this same concept can be applied to a pattern of many holes.

CREATING THE DATUM REFERENCE FRAME (DRF)

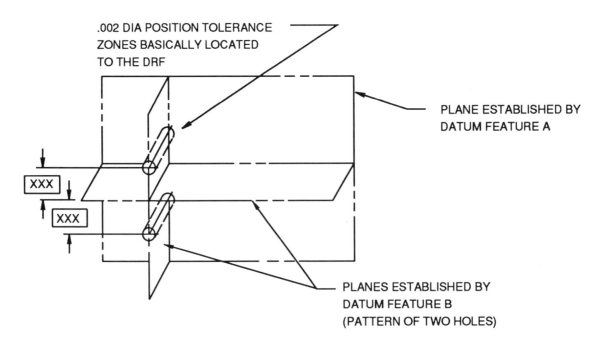

.002 DIA POSITION TOLERANCE ZONES BASICALLY LOCATED TO THE DRF

PLANE ESTABLISHED BY DATUM FEATURE A

XXX

XXX

PLANES ESTABLISHED BY DATUM FEATURE B (PATTERN OF TWO HOLES)

The following 3D figure shows mathematically how the parallelism tolerance zone is related to the DRF. It is important to remember that the datums in the feature control frame only set the part up in the DRF. The tolerance zones are all relative to the DRF. The part features must then fall in the tolerance zones.

This part is a good example to show how simple it is to tolerance a part if you understand the DRF concept. Establish the datum reference frame based on function and then relate everything to the DRF.

This parallelism specification also reinforces the discussion in the overview in the beginning of this orientation tolerances chapter. As the orientation requirements on the part become more complex and must be related to more than one datum, it is difficult to decide if the specification should be a parallelism, perpendicularity or angularity control. In the final analysis, it probably does not matter which symbol is used. The orientation tolerances only orient the features. Therefore, any of the symbols will do the job.

PARALLELISM TOLERANCE ZONE IN RELATION TO THE DATUM REFERENCE FRAME

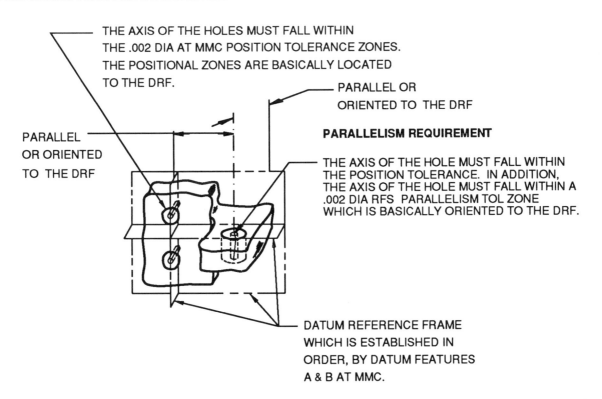

THE AXIS OF THE HOLES MUST FALL WITHIN
THE .002 DIA AT MMC POSITION TOLERANCE ZONES.
THE POSITIONAL ZONES ARE BASICALLY LOCATED
TO THE DRF.

PARALLEL OR
ORIENTED TO THE DRF

PARALLEL
OR ORIENTED
TO THE DRF

PARALLELISM REQUIREMENT

THE AXIS OF THE HOLE MUST FALL WITHIN
THE POSITION TOLERANCE. IN ADDITION,
THE AXIS OF THE HOLE MUST FALL WITHIN A
.002 DIA RFS PARALLELISM TOL ZONE
WHICH IS BASICALLY ORIENTED TO THE DRF.

DATUM REFERENCE FRAME
WHICH IS ESTABLISHED IN
ORDER, BY DATUM FEATURES
A & B AT MMC.

The verification process on this part is relatively simple. The part is mounted on datum feature A, and the two holes are verified to each other. If the holes check good, then a DRF has been established on the part. The remaining geometric tolerances are checked relative to this set-up. This is a simple task for a CMM, open set-up or a functional gage.

PERPENDICULARITY - ZERO TOLERANCING AT MMC

The two examples below represent two mating assemblies. Example #1 specifies a .002 at MMC perpendicularity requirement. Example #2 specifies a zero perpendicularity at MMC requirement. Notice that example #2 provides more tolerance (greater variation on size) while still maintaining the same virtual size as example #1. The zero tolerancing concept can be applied to other geometric tolerances as well.

EXAMPLE #1 - PERPENDICULARITY .002 DIA AT MMC.

Use this method if it is necessary to control the size of the feature within set limits and also control the orientation within set limits.

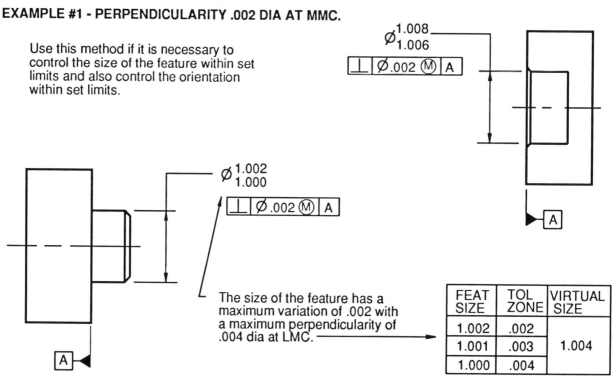

The size of the feature has a maximum variation of .002 with a maximum perpendicularity of .004 dia at LMC.

FEAT SIZE	TOL ZONE	VIRTUAL SIZE
1.002	.002	
1.001	.003	1.004
1.000	.004	

EXAMPLE #2 - PERPENDICULARITY .000 DIA AT MMC.

Use this method if it is not necessary to control the size of the feature with one value and the perpendicularity at another value. This method combines the size and orientation and simply states the virtual size of the feature. It is a common specification on pilots. It certainly provides more tolerance than example #1.

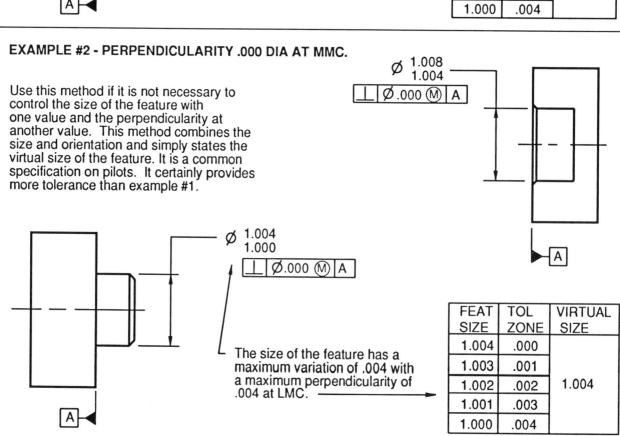

The size of the feature has a maximum variation of .004 with a maximum perpendicularity of .004 at LMC.

FEAT SIZE	TOL ZONE	VIRTUAL SIZE
1.004	.000	
1.003	.001	
1.002	.002	1.004
1.001	.003	
1.000	.004	

PERPENDICULARITY - LINE ELEMENTS

All of the orientation tolerances are by default 3D. If there is a need to specify an orientation tolerance in 2D, the words "EACH ELEMENT" are placed under the feature control frame. The direction of the tolerance zone applies in the view where it is specified.

THIS ON THE DRAWING

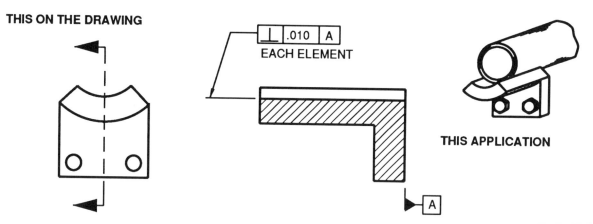

THIS APPLICATION

Each line element of the surface must fall between two parallel lines .010 apart which are oriented perpendicular to the specified datum reference frame. In addition, the surface must be within the applicable size or profile zone.

MEANS THIS

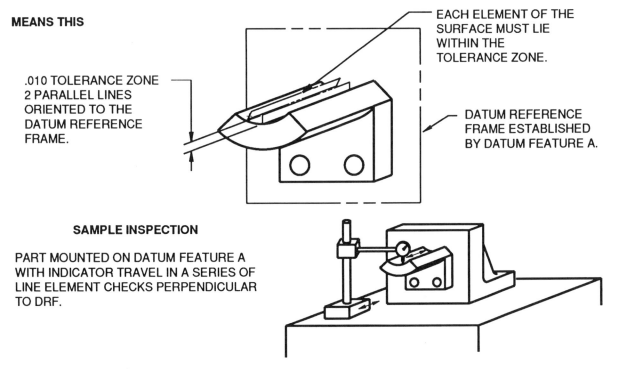

EACH ELEMENT OF THE SURFACE MUST LIE WITHIN THE TOLERANCE ZONE.

.010 TOLERANCE ZONE 2 PARALLEL LINES ORIENTED TO THE DATUM REFERENCE FRAME.

DATUM REFERENCE FRAME ESTABLISHED BY DATUM FEATURE A.

SAMPLE INSPECTION

PART MOUNTED ON DATUM FEATURE A WITH INDICATOR TRAVEL IN A SERIES OF LINE ELEMENT CHECKS PERPENDICULAR TO DRF.

Line element tolerance applied to a surface also controls the straightness of that surface. Thus, if no straightness is specified, the straightness will be at least as close as the line element requirement. The line element concept above can be expanded to control other 2D requirements. Notations such as, "EACH RADIAL ELEMENT" or "EACH CIRCULAR ELEMENT" may also be used.

PARALLELISM - TANGENT PLANE

The tangent plane symbol can be applied to many of the geometric characteristics. The tangent plane symbol is shown below with a parallelism specification. A plane contacting the high points of the surface must lie within the parallelism tolerance. The tangent plane application does not control flatness of the surface. If flatness of the surface is of concern, a separate control must be specified.

THIS ON THE DRAWING

SYMBOL MEANING

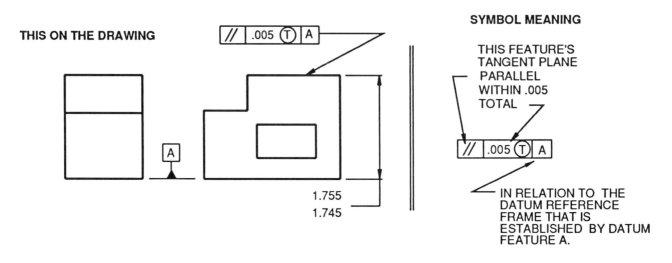

THIS FEATURE'S
TANGENT PLANE
PARALLEL
WITHIN .005
TOTAL

IN RELATION TO THE
DATUM REFERENCE
FRAME THAT IS
ESTABLISHED BY DATUM
FEATURE A.

The tangent plane must lie between two parallel planes .005 apart which are parallel to datum plane A. In addition, the surface must be within the specified limits of size or profile.

MEANS THIS

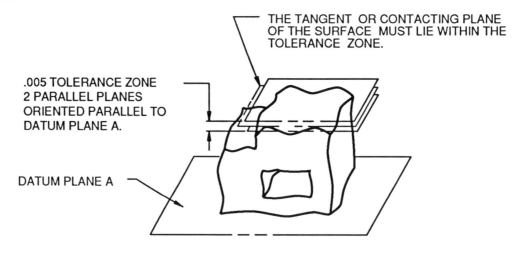

THE TANGENT OR CONTACTING PLANE
OF THE SURFACE MUST LIE WITHIN THE
TOLERANCE ZONE.

.005 TOLERANCE ZONE
2 PARALLEL PLANES
ORIENTED PARALLEL TO
DATUM PLANE A.

DATUM PLANE A

SAMPLE INSPECTION

The part is mounted on datum A, and a parallel bar is placed covering the entire top surface. The simulated contacting plane of the parallel bar must lie within the .005 tolerance zone.

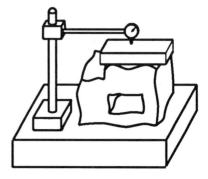

WORKSHOP EXERCISE 11.1

1. Name the three orientation tolerances.

_____

2. Are datums always required with orientation tolerances?

_____

3. On the part above, what is the maximum variation in parallelism before any geometric tolerancing is applied?

_____

4. On the part above, specify a parallelism tolerance to make the top surface parallel to the bottom surface within a total of .006. Be sure to properly identify the datum feature.

5. Can the MMC modifier be applied to the feature tolerance or the datum feature? Why or why not?

_____

6. What can we expect the flatness tolerance on the top surface to be?

_____

7. On the produced part below show a simple sketch on how the part would be checked. Make sure to dimension and show the tolerance zones.

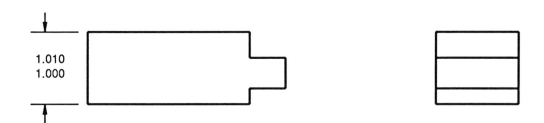

8. On the part above make the tangent plane of the top surface parallel to the bottom surface within .002.

9. Explain the difference between the parallelism specification in problem 4 and the tangent plane specification above.

_____

_____

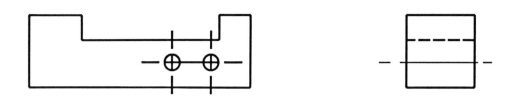

ALL ANGLES ± 3°

10. On the part above, specify a geometric control to make the left hand surface in the front view perpendicular to the bottom surface within .005 total.

11. Can a datum or feature modifier be applied to the parallelism specification? Why or why not?

_____

_____

12. What is the flatness on the left hand surface of the part above?

_____

13. On the part above, there is no perpendicularity requirement between the right hand surface and the bottom surface. What perpendicularity is implied?

_____

_____

14. On the part below, sketch the perpendicularity tolerance zone and show an example of how it may be verified.

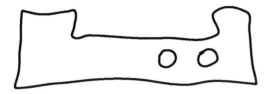

15. It is possible that orientation tolerances can be applied to two or more datums. On the part below, show an example of perpendicularity to two datums. Select your own datums and tolerance.

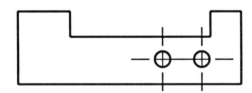 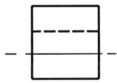

16. On the produced part below, show by hand sketch the specified perpendicularity tolerance with two datums and an example of how it may be checked.

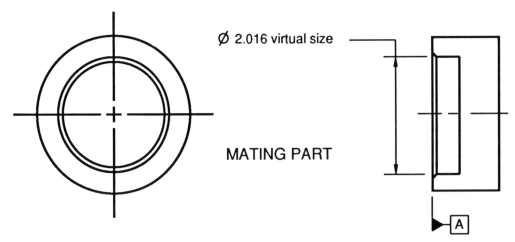

MATING PART

Ø 2.016 virtual size

A

17. The pilot part below mates with the part above. There is a design clearance of .002 between the virtual size fit of the two parts. Complete the chart below showing the amount of size tolerance and perpendicularity tolerance allowed with the specification below.

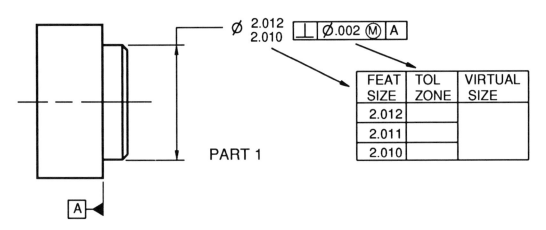

Ø 2.012 / 2.010   ⊥ | Ø.002 Ⓜ | A

FEAT SIZE	TOL ZONE	VIRTUAL SIZE
2.012		
2.011		
2.010		

PART 1

A

18. The pilot part belows mates with the mating part above. There is a design clearance of .002 between the virtual size fit of the two parts. Complete the chart below showing the amount of size tolerance and perpendicularity tolerance allowed with the specification below.

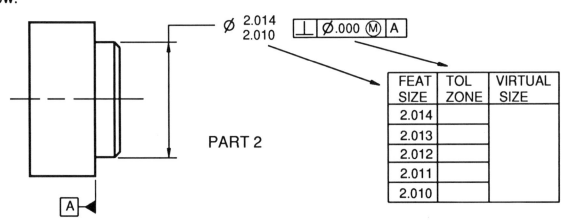

Ø 2.014 / 2.010   ⊥ | Ø.000 Ⓜ | A

FEAT SIZE	TOL ZONE	VIRTUAL SIZE
2.014		
2.013		
2.012		
2.011		
2.010		

PART 2

A

19. Which part provides the most manufacturing tolerance and is the easiest to work with, part 1 or part 2? Why?

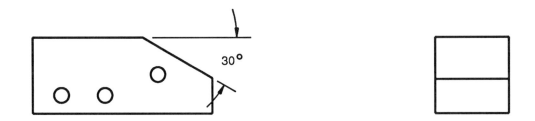

20. On the part above, show an angulartiy specification to make the angle surface in relation to the top surface within .010 total.

21. Should the 30 degree angle above have a plus/minus tolerance applied or should it be applied as basic? Complete this requirement on the drawing above.

22. What is the flatness tolerance on the angled surface above?

---

23. Draw a simple sketch below on how the produced part may be checked and the shape of the resultant angularity tolerance zone.

24. Turn back the pages to the geometric overview chart on the last page in unit 10. Complete the chart, filling in the required information on the orientation tolerances.

UNIT 12

PROFILE TOLERANCES

PROFILE TOLERANCES

PROFILE TOLERANCES CONTROL THE LOCATION AND/OR ORIENTATION, AND/OR FORM, AND/OR SIZE OF A FEATURE MAY OR MAY NOT HAVE DATUMS

SYMBOL	TYPE OF TOLERANCE	SHAPE OF TOLERANCE ZONE	2D OR 3D	APP OF FEATURE MOD
⌒	PROFILE OF A LINE	2 DIMENSIONAL UNIFORM BOUNDARY 	2D	NO
⌓	PROFILE OF A SURFACE	3 DIMENSIONAL UNIFORM BOUNDARY 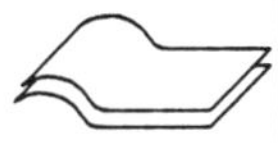	3D	NO

OVERVIEW:

There are two types of profile tolerances: PROFILE OF A SURFACE and PROFILE OF A LINE. These profile tolerances are shown in the chart above. Datums may or may not be applied. Profile of a surface is a 3D control, and profile of a line is a 2D control.

Profile of a surface is probably the most powerful control in the entire geometric tolerancing system. It can be used to control the size, form, orientation or location of a feature. In the past, the position symbol seemed to be the most important. But, as we begin to understand the geometric system and work with CAD data bases, solid models, CMM's and CNC, we find that profile of a surface is probably the most powerful tool we have.

The plus/minus tolerancing system we have used in the past to describe parts is subject to many interpretations. All of the 90 degree angles shown on the drawing have angle tolerance. The plus/minus tolerances have only implied datums. The plus/minus tolerances are only 2D, and all of our parts are 3D. It is for these reasons that profile tolerancing is replacing the plus/minus tolerancing on surfaces.

Profile tolerance is often compared to position tolerance in that position is used to locate features of size (holes, slots, tabs, pins etc.), and profile tolerance is used to locate features that have no size (surfaces). It is a lot easier to use a profile tolerance relative to a DRF than try to figure out all the plus/minus variations with implied datums.

In industry today, most companies have and use coordinate measuring machines (CMM's) and computer aided design (CAD). These systems require that the parts are defined mathematically in 3D relative to a coordinate system. Profile tolerancing defines parts in 3D relative to a coordinate system.

In some industries, especially automotive sheet metal, profile tolerance is used exclusively; and very little plus/minus tolerance is used. The object of this chapter is to allow you feel more comfortable with profile tolerancing. If you are now using a small amount of profile on the drawings, you may wish to use more in the future.

Profile tolerance is also very versatile. The classic profile specification is usually applied with datums and basic dimensions. But profile can also be applied with plus/minus tolerances. If it is applied with plus/minus tolerances, depending on the situation, it will react like a form or orientation tolerance and must lie within the confines of the plus/minus tolerance.

Profile is easy to verify. It is just like a plus/minus tolerance. The difference is that profile clearly defines design intent in 3D. It is impossible to check all the points in a profle tolerance. It is also impossible to check all the points with plus/minus tolerancing. It is also impossible to check all the points when verifying a surface finish requirement. Defining a part using profile tolerance does not mean it is more important or a tigher tolerance. The size of profile tolerance value is what will determine if it is a tight tolerance. Profile is simply another clearer method for defining design intent.

The use of profile tolerancing does not require any exotic verification procedures. Depending on the situation, profile may be verified with a CMM, micrometer, calipers, gages, optical comparator, indicators or visually. How many points, how often to verify, and the method of verification will be determined by the quality engineer and defined in the dimensional measurement plan. These decisions are based on the number of parts to be checked, acceptable risk, the variability of the process, availability of equipment and many other factors.

PROFILE OF A SURFACE - BILATERAL

Profile tolerancing specifies a uniform boundary along the true profile within which the elements of the surface must lie. Profile tolerancing can be applied on a bilateral, unilateral or unequal distribution basis. Below is shown the bilateral distribution. A single arrow from the feature control frame points directly at the surface.

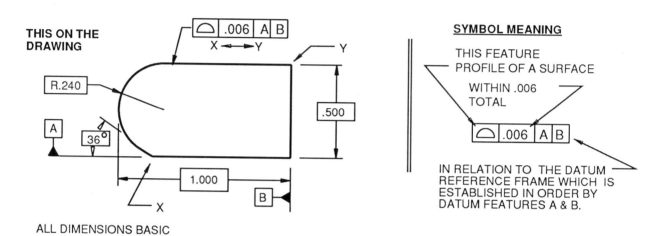

The tolerance zone established by the profile of a surface control is three dimensional, extending along the full length and width of the considered feature.

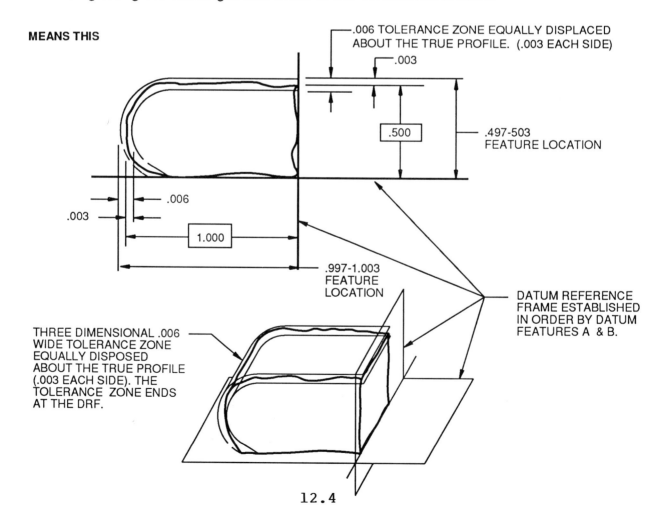

PROFILE OF A SURFACE - UNILATERAL AND BILATERAL UNEQUAL

Profile can also be applied unequally about the true profile. If an unequal distribution is required, two arrows are used to define the width or direction of the tolerance zone.

THIS ON THE DRAWING

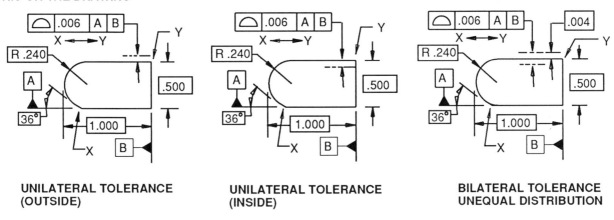

UNILATERAL TOLERANCE (OUTSIDE)

UNILATERAL TOLERANCE (INSIDE)

BILATERAL TOLERANCE UNEQUAL DISTRIBUTION

NOTE: TRUE PROFILE IS DEFINED BY BASIC DIMENSIONS

MEANS THIS

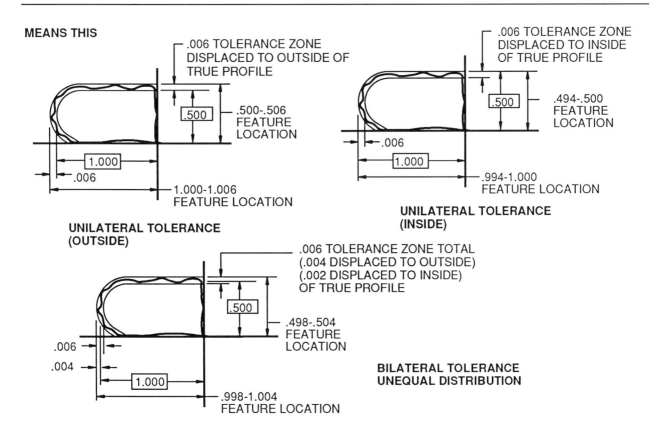

UNILATERAL TOLERANCE (OUTSIDE)

UNILATERAL TOLERANCE (INSIDE)

BILATERAL TOLERANCE UNEQUAL DISTRIBUTION

Where it is not clear as to the type of profile or direction of the tolerance zone, a note such as UNILATERAL OUT or UNILATERAL IN, may be placed under the feature control frame. This method may be necessary when working with solid models, untoleranced drawings, sheet metal parts or where drawing designation is unclear.

PROFILE OF A LINE - BILATERAL

The concepts for profile of a surface and profile of a line are identical, with the exception that profile of a surface is three dimensional and profile of a line is two dimensional. Profile of a line is often used in conjunction with profile of a surface as shown below. Profile of a surface defines the shape or location of a feature, and profile of a line is used to refine the feature in one direction, as in extruded parts. The line elements apply in the view in which the feature control frame is directed.

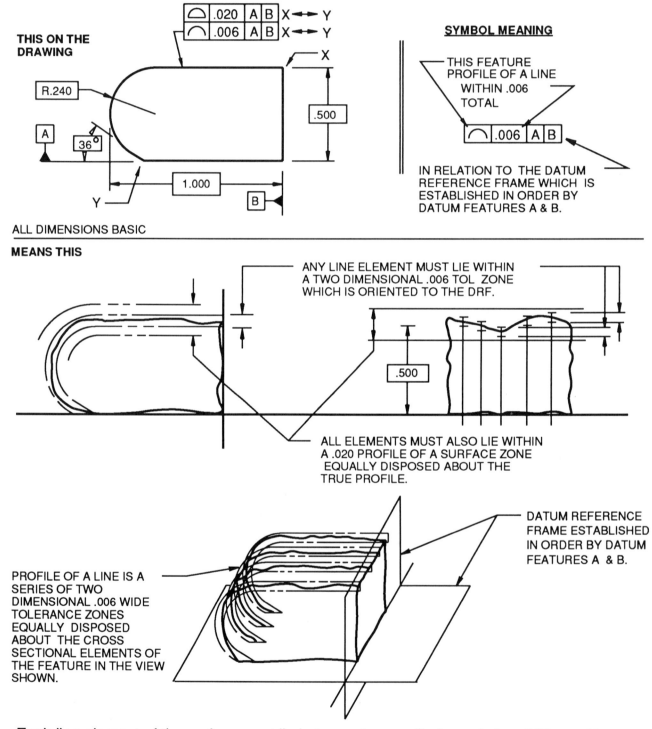

Each line element of the surface must lie between two profile boundaries .006 apart in relation to the datum reference frame. When used in conjunction with a size tolerance or a profile of a surface control, the surface must also fall within these specified limits.

PROFILE OF A SURFACE - CAM WHEEL

Profile tolerancing is a very effective method for controlling contoured or flat surfaces. The cam wheel below has two profile controls applied. The upper control is a bilateral profile which specifies a zone that is equally disposed about the true profile. The lower control is a unilateral profile which specifies a zone that is displaced to the inside of the true profile. The specified surfaces of the part must fall within these zones.

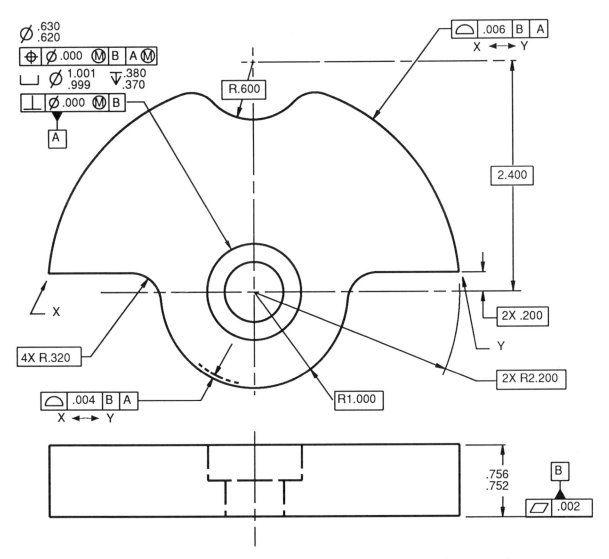

The MMC, LMC and RFS modifiers can not be applied to the surface, as they are non-size features. The MMC, LMC and RFS modifiers can, however, be applied to referenced datums that have size. The MMC or LMC modifier can be applied to datum A in the part above. If either of these modifiers are applied, and datum A departed from the specified material condition, the profile zone width remains the same size. However, it will allow a shift of the profile zone relative to the datum axis by the amount of the datum feature departure. Datum A on the cam wheel above is implied RFS, therefore no profile shift is allowed.

PROFILE APPLICATIONS - FUEL LINE BRACKET ASSEMBLY

The assembly below is a bracket and clamp that secures a fuel line in position relative to the engine. As you can see from the assembly drawing, the bracket mounts to the engine by screws in the two holes. The periphery of the bracket must clear the engine housing in one area; the rest of the contour is unimportant. The elongated holes on the bracket are for adjustment and must line up with the clamp holes. The clamp profile is relatively unimportant except where it mounts to the bracket and the fuel line.

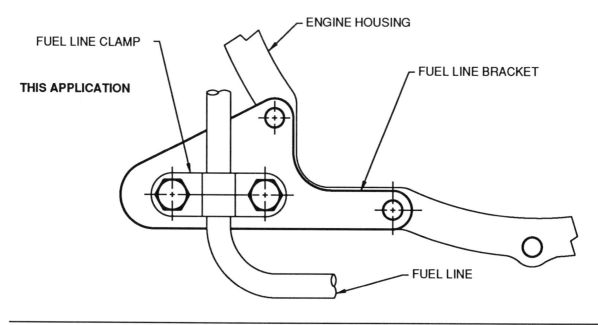

THIS APPLICATION

FUEL LINE CLAMP

ENGINE HOUSING

FUEL LINE BRACKET

FUEL LINE

THIS ON THE DRAWING

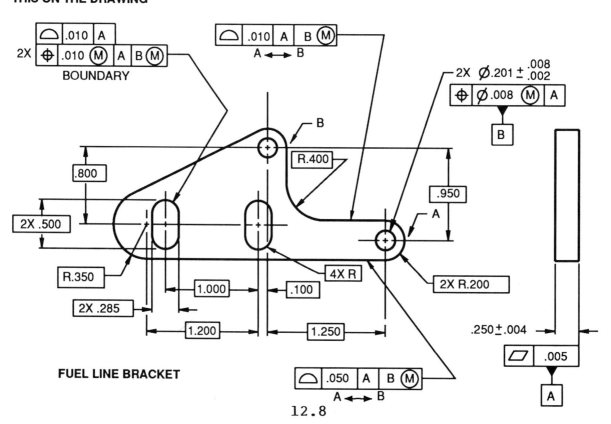

FUEL LINE BRACKET

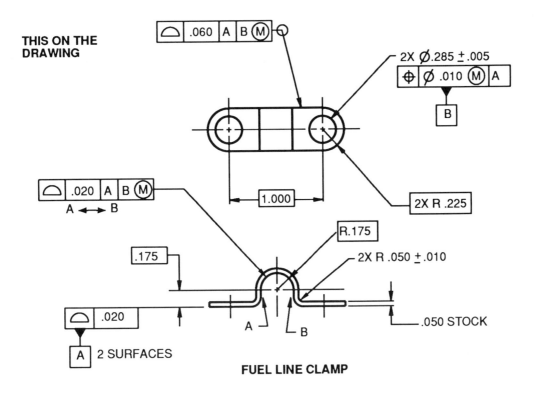

THIS ON THE DRAWING

FUEL LINE CLAMP

The drawings of the bracket and clamp illustrate a simple application of geometric tolerancing reflecting functional requirements. A datum reference frame is established on the bracket using the face and the two holes. The datum features are related to the DRF and all the remaining features are related to this DRF. The profile tolerance on outside contour is held closer in the area that clears the engine housing. The remaining outside contour has a larger profile tolerance. The slotted holes are profiled for size and then positioned with the boundary concept. As an alternative, both the slotted hole size and location could have been controlled with a profile tolerance.

The datum reference frame on the clamp is established by the two mounting surfaces and the two mounting holes. The two datum surfaces are specified with profile to control co-planarity. (A flatness specification does not control co-planarity between surfaces. Flatness is only a form tolerance.) The two holes are positioned to each other and datum A. The two surfaces and the pattern of holes establish the DRF. All features are related to this DRF. The profile in the area of the tube mounting is held a little closer than the rest of the contour as it interfaces to the tube.

1. On the cam wheel below, calculate the min and max of the designated dimension.

MIN 2.4-.003+.6+1 = 2.997÷.002
     ÷2.995

MAX 2.4+.003-.6+1+.002 = 2.805

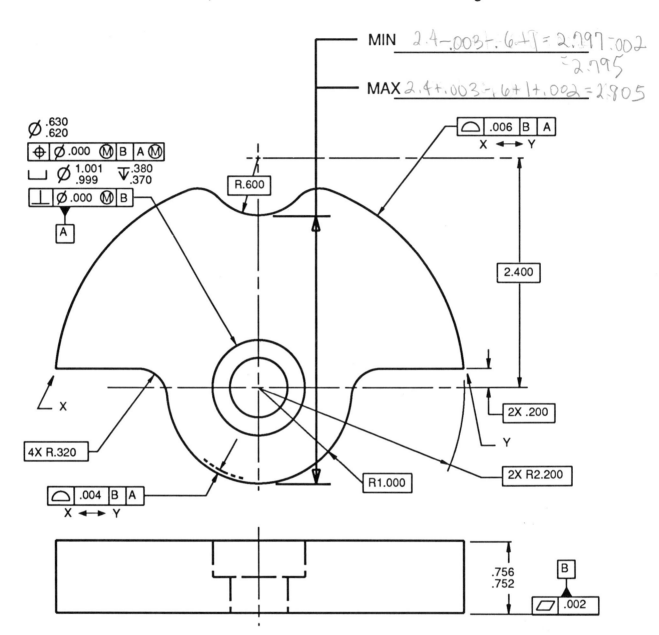

2. What is the size of the datum feature.

3. What is the virtual size of the datum feature.

4. The upper profile tolerance is called a bilateral profile. What is the bottom profile called?

   Unilateral - inside

5. What is the depth of the counterbore?

   .37-.38

6. Often, in the verification of parts, we want to simply monitor a few dimensions on the part. On the part below calculate the min and max of the designated dimensions.

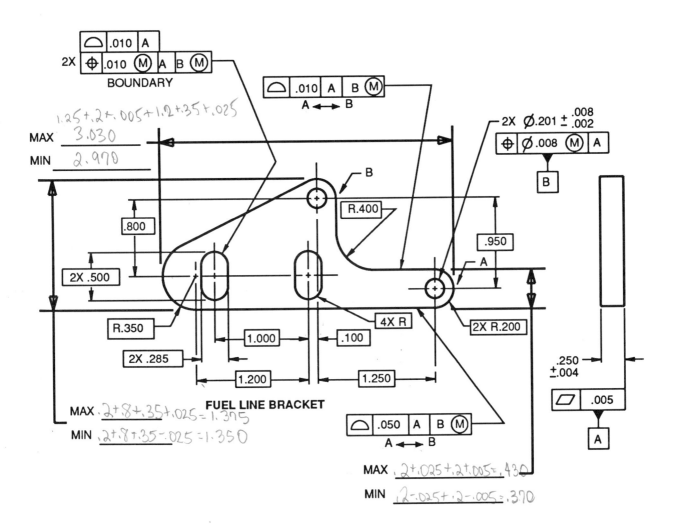

FUEL LINE BRACKET

MAX 1.25+.2+.005+1.2+.35+.025
MAX 3.030
MIN 2.970

MAX .2+.8+.35+.025=1.375
MIN .2+.8+.35-.025=1.350

MAX .2+.025+.2+.005=.430
MIN .2-.025+.2-.005=.370

7. What is the virtual size of datum features B?

_____

8. The MMC modifier is applied to datum feature B in all the feature control frames. Does this allow each profile and position specification to shift independently, or are all the profile and position considered one composite pattern? Why or why not?

_____

_____

PROFILE TOLERANCING - HAT BRACKET

The sheet metal hat bracket drawing below is a sample application of geometric tolerancing.
The object is to keep the tolerancing simple while still preserving functional requirements

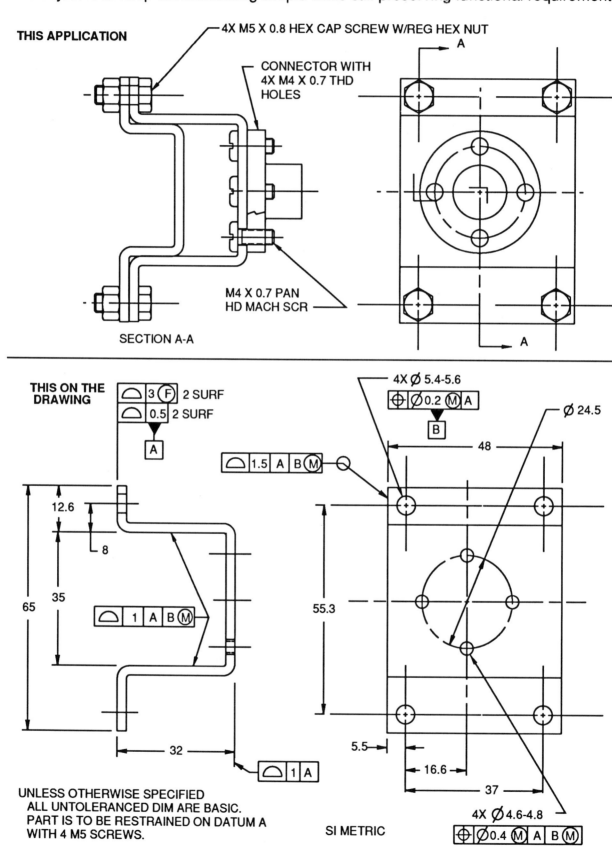

THIS APPLICATION

4X M5 X 0.8 HEX CAP SCREW W/REG HEX NUT

CONNECTOR WITH
4X M4 X 0.7 THD
HOLES

M4 X 0.7 PAN
HD MACH SCR

SECTION A-A

THIS ON THE DRAWING

4X Ø 5.4-5.6

Ø 24.5

UNLESS OTHERWISE SPECIFIED
ALL UNTOLERANCED DIM ARE BASIC.
PART IS TO BE RESTRAINED ON DATUM A
WITH 4 M5 SCREWS.

SI METRIC

4X Ø 4.6-4.8

12.12

The hat bracket mounts on the two bottom surfaces and 4 hole pattern of holes and then clears the mating part. A profile is applied to the two bottom surfaces to ensure co-planarity and is then defined as datum feature A. The pattern of holes is selected as the secondary datum. The designer had concerns about the flexibility of the part and applied a note that requires all tolerances, except those noted with the free state symbol, to be verified in the restrained condition.

The figure below illustrates the inner and outer boundary of the profile and position tolerances in the restrained condition. Note how simple it is to do a stack-up between the various features. The verification procedure is equally as simple. The pattern of holes are verified for conformance to the position requirement. If the holes conform, then retain this same DRF set-up, and everything else is verified from this established DRF. An optical comparator, CMM or functional gage also verifies this part very easily.

HAT BRACKET - INNER AND OUTER BOUNDARIES IN RESTRAINED CONDITION

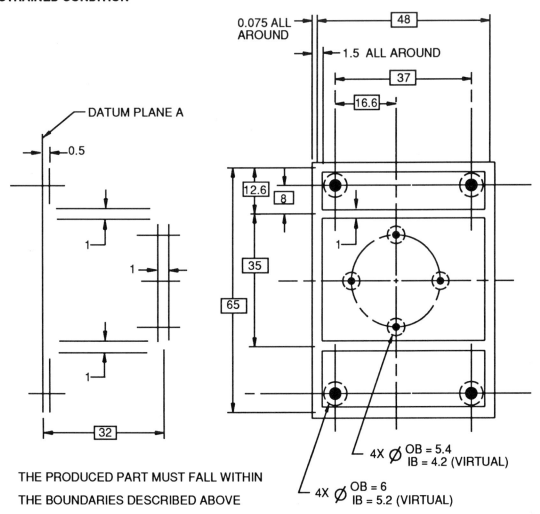

THE PRODUCED PART MUST FALL WITHIN

THE BOUNDARIES DESCRIBED ABOVE

12.13

COMPOSITE PROFILE TOLERANCING OF AN IRREGULAR FEATURE

Composite profile may be used to locate irregular shaped features as shown on the accompanying figure. The feature is defined with basic dimensions. A composite profile is applied to the feature.

Composite profile is special. (It works in the same manner as composite position.) The profile symbol is entered once and is applicable for both the upper and lower segment. The upper segment of the feature control frame, in effect, locates the feature to the specified datums, and the lower segment of the feature control frame specifies the size, form or shape and the orientation (perpendicularity in following example) of the feature to the specified datums.

To help solidify the composite tolerancing concept in your thoughts, consider the following scenarios. In the lower entry, if, in addition to datum A, a datum B is also specified, the rotation or orientation of the shape relative to the datums also is controlled. A third datum specified in the lower segment will not add any control, as two datums are sufficient to control the orientation. If there are no datums specified in the lower segment, then only the size, form or shape of the feature is controlled as there are no datums to control orientation.

Composite profile used for locating irregular features is similar to the position boundary concept for locating irregular features. The difference is that the position boundary concept establishes a virtual condition boundary that the feature must clear when applied with MMC or LMC modifiers.

The composite profile concept can not be applied with feature modifiers and provides both an inner and outer boundary in which the feature must lie. The composite profile control on an irregular feature is a more restrictive control than the position boundary concept. See also the position boundary concept later in the text.

COMPOSITE PROFILE TOLERANCING OF AN IRREGULAR FEATURE

THIS ON THE DRAWING

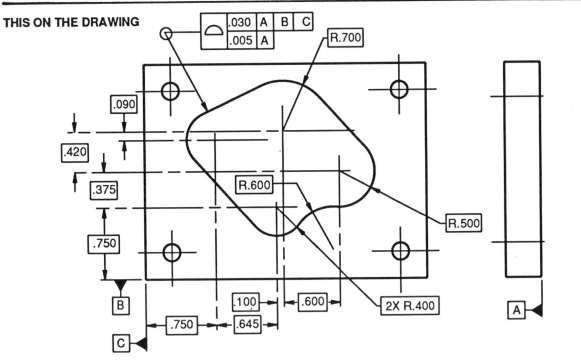

MEANS THIS

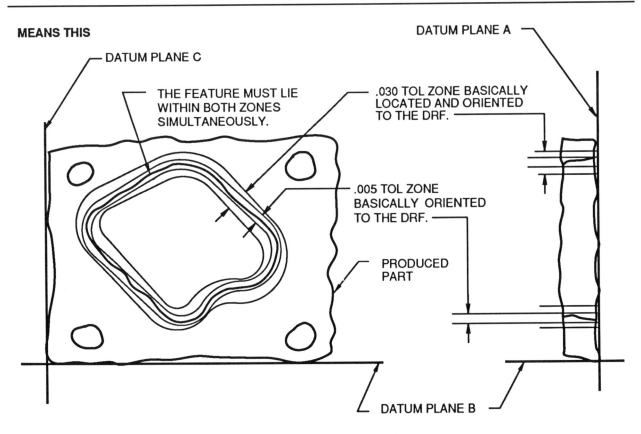

Composite profile of a surface can be used to locate irregular shaped features. The feature must lie within a .030 tol zone that is located and oriented to the specified DRF established by datum features A, B and C. In addition, the feature must also lie within a .005 tol zone that is basically oriented (perpendicular) to datum plane A. The .005 zone "floats" within the .030 zone.

COMPOSITE TOLERANCING IS SPECIAL

Composite tolerancing is special. It can be used with profile and position tolerances. The symbol is entered once and is applicable to both horizontal entries. It is different than two single segmented feature control frames.

The upper segment on a composite profile feature control frame controls location, orientation, form and in some cases, size to the specified datums. The lower segment does not control location to the specified datums, but does control orientation, form and in some cases, the size of the feature to the specified datums.

In contrast, all single segment profile frames control the location, orientation, form and in some cases, size between the features as well as to the specified datums. This type of control is interpreted simply as two profile requirements.

COMPOSITE PROFILE FRAME
    (SPECIAL TYPE OF PROFILE)

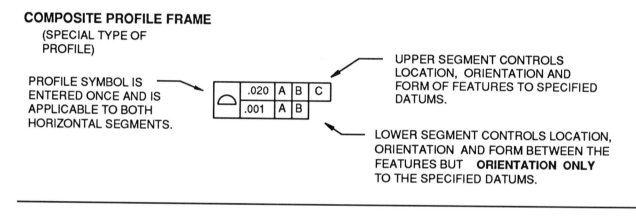

PROFILE SYMBOL IS ENTERED ONCE AND IS APPLICABLE TO BOTH HORIZONTAL SEGMENTS.

UPPER SEGMENT CONTROLS LOCATION, ORIENTATION AND FORM OF FEATURES TO SPECIFIED DATUMS.

LOWER SEGMENT CONTROLS LOCATION, ORIENTATION AND FORM BETWEEN THE FEATURES BUT **ORIENTATION ONLY** TO THE SPECIFIED DATUMS.

TWO SINGLE SEGMENTED FRAMES
    (SIMPLY TWO PROFILE REQUIREMENTS)

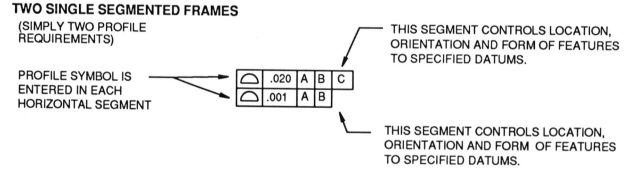

PROFILE SYMBOL IS ENTERED IN EACH HORIZONTAL SEGMENT

THIS SEGMENT CONTROLS LOCATION, ORIENTATION AND FORM OF FEATURES TO SPECIFIED DATUMS.

THIS SEGMENT CONTROLS LOCATION, ORIENTATION AND FORM OF FEATURES TO SPECIFIED DATUMS.

The difference between the terms location and orientation should be clear. Location locates features and is associated with basic linear dimensions. It can also include orientation. Orientation, on the other hand, is not associated with location or with basic linear dimensions, only basic angles. Orientation is usually thought of as parallelism, perpendicularity or angularity. The composite tolerancing concepts explained above apply to both position and profile tolerances. See the position section of this text for a more complete explanation of composite tolerancing.

COMPOSITE PROFILE - ONE DATUM FEATURE

The upper entry on composite tolerancing controls location to the DRF. The lower entry on composite tolerancing controls size/shape and orientation (perpendicularity) to the DRF established by the specified datum.

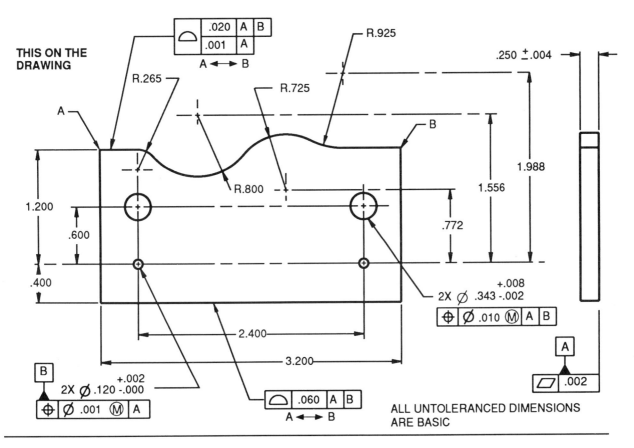

THIS ON THE DRAWING

R.925

R.265

R.725

R.800

R.260 (A B) .020 / .001 / A

A ↔ B

.250 ±.004

A

1.200

.600

.400

1.988

1.556

.772

B

2X Ø .343 +.008/-.002

⊕ Ø .010 Ⓜ A B

A

▱ .002

2.400

3.200

B

2X Ø .120 +.002/-.000

⊕ Ø .001 Ⓜ A

▱ .060 A B

A ↔ B

ALL UNTOLERANCED DIMENSIONS ARE BASIC

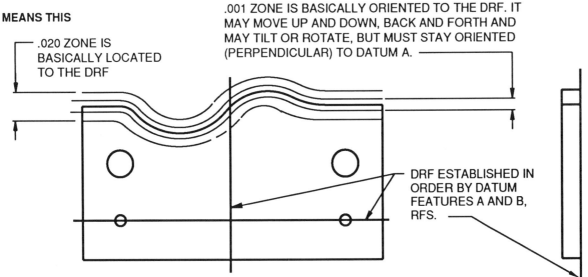

MEANS THIS

.001 ZONE IS BASICALLY ORIENTED TO THE DRF. IT MAY MOVE UP AND DOWN, BACK AND FORTH AND MAY TILT OR ROTATE, BUT MUST STAY ORIENTED (PERPENDICULAR) TO DATUM A.

.020 ZONE IS BASICALLY LOCATED TO THE DRF

DRF ESTABLISHED IN ORDER BY DATUM FEATURES A AND B, RFS.

The composite tolerance specification above allows the .001 zone to "float" up and down and back and forth and tilt or rotate within the confines allowed by the .020 zone. The .001 tolerance zone, however, must stay perpendicular to datum A. The datum in the lower entry controls the orientation (perpendicularity) of the .001 zone relative to the DRF established by datum feature A. The surface of the part must lie in both zones simultaneously and meet both requirements.

COMPOSITE PROFILE - TWO DATUM FEATURES - CAM BLOCK

The following application is a cam block. The part mounts on dowels in two .120 diameter holes and is fastened down with screws in the two .343 diameter clearance holes. The contoured surface on top is a cam surface on which a cam follower rides and provides motion for another component.

The contour and the orientation of the contour is important as the follower rides on the contour. The follower, however, can be adjusted in the up and down direction. The functional design requirements allow the top contoured surface to move up and down in a large tolerance but the form and orientation (perpendicularity and parallelism) of the contour must be held at a closer requirement relative to the mounting.

The drawing has geometric tolerancing applied relative to functional requirements. The face and the two dowel holes are specified as datum features and are related to the datum reference frame with a flatness and a position tolerance. The area that the cam follower travels is identified as the surface between points A and B.

A composite profile is applied to this surface. The upper entry, in effect, locates the surface within .020 to the DRF; and the lower entry controls the form and the orientation of the surface to the DRF. Remember, the specified datums in the lower entry of a composite feature control frame controls only orientation (perpendicularity, parallelism or angularity) to the specified datums, not location.

The composite tolerancing specification has two requirements. The specified part feature must meet both requirements. The upper entry specifies a .020 zone in which the feature must lie. This upper zone is basically located and oriented to the DRF. The lower entry specifies a smaller .001 zone within which the feature must lie. This zone is basically oriented to the DRF. The specified feature must lie in both zones simultaneously.

The remaining features are also located to the specified DRF. The outside edges are not important and are profiled with a large tolerance.

Notice that the datums are selected relative to function. The square edges, although convenient, are not functional. It is possible, however, that manufacturing might select these edges to produce the part. It is their choice. Manufacturing is aware that if the dowels and contour are machined in the same set-up, the functional requirements will be met.

Note:
On the preceeding page is another example of the same part with only one datum specified in the lower entry of the composite feature control frame. This part has a different function than described above. Since only one datum is specifed in the lower entry of the composite feature control frame, orientation (perpendicularity) of the contour on this part is only controlled to datum feature A.

COMPOSITE PROFILE - TWO DATUM FEATURES

The upper entry on composite tolerancing controls location to the DRF. The lower entry on composite tolerancing controls size/shape and orientation (perpendicularity, parallelism or angularity) to the DRF established by the specified datums.

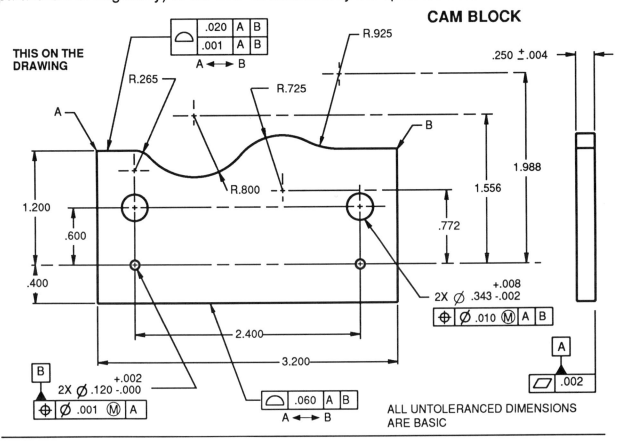

CAM BLOCK

ALL UNTOLERANCED DIMENSIONS ARE BASIC

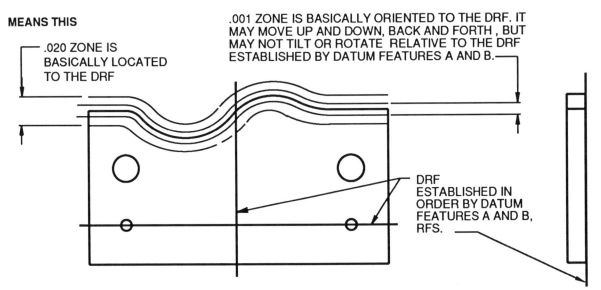

MEANS THIS

.020 ZONE IS BASICALLY LOCATED TO THE DRF

.001 ZONE IS BASICALLY ORIENTED TO THE DRF. IT MAY MOVE UP AND DOWN, BACK AND FORTH , BUT MAY NOT TILT OR ROTATE RELATIVE TO THE DRF ESTABLISHED BY DATUM FEATURES A AND B.

DRF ESTABLISHED IN ORDER BY DATUM FEATURES A AND B, RFS.

The composite tolerance specification above allows the .001 zone to "float" up and down and back and forth within the confines allowed by the .020 zone. The .001 tolerance zone, however, may not tilt or rotate. The datums in the lower entry control the orientation (parallelism/ perpendicularity) of the .001 zone relative to the DRF established by datum features A and B. The surface of the part must lie in both zones simultaneously and meet both requirements.

PROFILE TOLERANCING - MULTIPLE CONTROLS

The part below has multiple profile controls applied to illustrate how adding and removing datums in the feature control frame can effect a profile (or position) control. Notice, with the addition or removal of datums, a control can be changed from location, and/or orientation and/or form and/or size.

THIS ON THE DRAWING

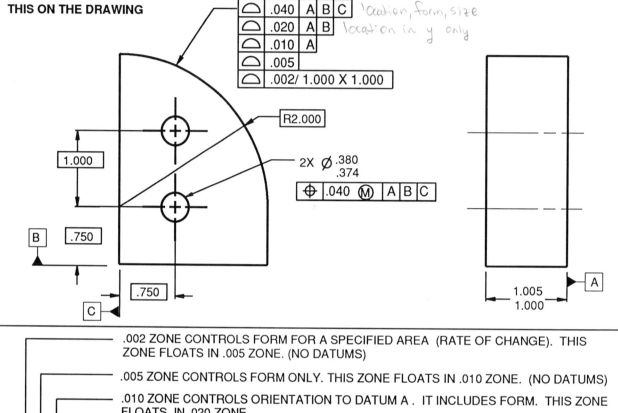

Handwritten notes:
- location, form, size
- location in y only

Feature control frames:
- ⌒ .040 A B C
- ⌒ .020 A B
- ⌒ .010 A
- ⌒ .005
- ⌒ .002/ 1.000 X 1.000

R2.000

2X ⌀ .380 / .374

⊕ | .040 Ⓜ | A | B | C

1.000

B .750

.750

C

1.005 / 1.000 ▶ A

.002 ZONE CONTROLS FORM FOR A SPECIFIED AREA (RATE OF CHANGE). THIS ZONE FLOATS IN .005 ZONE. (NO DATUMS)

.005 ZONE CONTROLS FORM ONLY. THIS ZONE FLOATS IN .010 ZONE. (NO DATUMS)

.010 ZONE CONTROLS ORIENTATION TO DATUM A . IT INCLUDES FORM. THIS ZONE FLOATS IN .020 ZONE.

.020 ZONE CONTROLS LOCATION TO DATUMS A AND B. IT INCLUDES ORIENTATION AND FORM. THIS ZONE FLOATS FORE AND AFT IN .040 ZONE.

.040 ZONE CONTROLS LOCATION TO DATUMS A, B AND C. IT INCLUDES ORIENTATION, FORM AND SIZE.

MEANS THIS

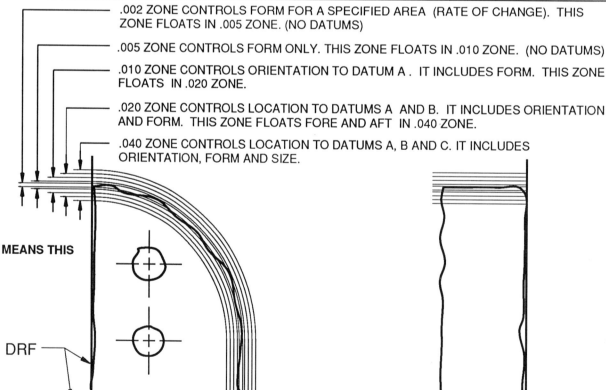

DRF

THE PART MUST MEET ALL REQUIREMENTS. EACH CONTROL STARTING FROM THE BOTTOM UP RELAXES THE TOLERANCE.

NONRIGID PARTS - FREE STATE CONDITION

Unless otherwise specified, all dimensioning and tolerancing applies in a free state condition with no restraint. Some parts, such as sheet metal, thin metal, plastics and rubber, are nonrigid in nature. These parts are naturally somewhat flexible or wobbly. Commonly, it may be necessary to specify design requirements on the part in a natural or free state as well as in a restrained condition. The restrained condition usually requires some restraint or force to insure functional requirements. The restraint or force on the nonrigid parts is usually applied in such a manner to resemble or approximate the functional or mating requirements. Sometimes, it may also be necessary to specify the direction of the force of gravity.

Nonrigid parts can fall into two categories:

1. Inherently flexible parts - These parts are very flexible and wobbly. The force of gravity alone can cause distortion in inspection and lead to the misreading of design requirements. Parts fitting in this category are thin metal rings, rubber, gasket material, etc. These inherently flexible parts are very floppy or wobbly. To ensure design intent, these parts will require the application of forces or restraints to hold or bring them into shape. It might also be necessary to specify the direction or force of gravity.

2. Inherently rigid parts - These parts are rigid, but due to the imposition or release of internal stresses resulting from the manufacturing process, these parts can distort beyond the specified design requirements. Parts fitting in this category are sheet metal stampings, flexible welded or riveted assemblies, plastic parts etc. These inherently rigid parts may require the application of certain restraints to force them into the functional design requirements.

The dimensioning and tolerancing of the inherently flexible parts and the inherently rigid parts are usually very similar. To ensure functional design requirements, it may be necessary to assess these parts both in the natural or free state and in a restrained condition.

Example: An inherently rigid part such as a sheet metal stamping may have some surfaces that have spring-back or draft due to the stamping operation. These surfaces may be used as datum features. These datum features are bolted, screwed or otherwise restrained in the assembly process. The parts in their free state may be in or out of tolerance. If the part is restrained in the assembly, this assembly restraint may have an effect on the final outcome of the tolerances. The restraint conditions can change the physical characteristics of the part. To ensure functional requirements, the designer may specify the datum surfaces restrained in such a way to resemble the bolted or screwed assembly condition. All of the characteristics related to the datums are verified in the specified restrained condition.

As you might imagine, the method of restraining the datum features can vary widely depending on the particular part. There is no standard note or symbol available that will cover all the possible restraint conditions. The usual method for specification of restrained conditions is done with a note on the drawing. If all the parts are similar or there is a standard restraint procedure, this information may be specified in a company standard specification. This company standard specification may be referenced on the drawing.

This note or specification should explain how the part is restrained and the force required to facilitate the restraint. A sample note may be found on the drawing below.

TOLERANCES APPLIED IN RESTRAINED CONDITION

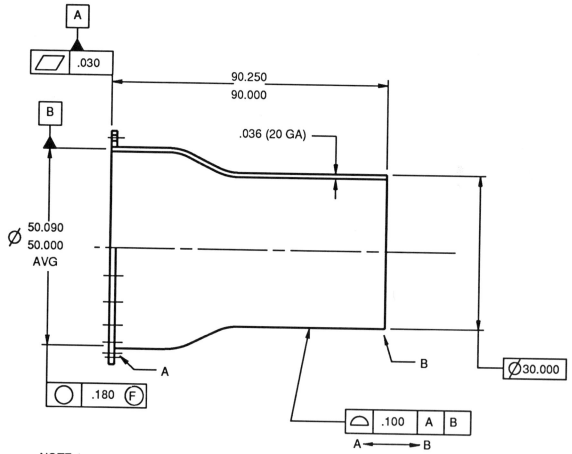

NOTE 1

UNLESS OTHERWISE SPECIFIED, ALL DIMENSIONS AND TOLERANCES APPLY IN THE RESTRAINED CONDITION. MOUNT THE PART ON THE PRIMARY DATUM FEATURE A USING 24, .500-13 UNC BOLTS TORQUED TO 100 TO 120 INCH POUNDS, WHILE RESTRAINING THE SECONDARY DATUM FEATURE B AT ITS SPECIFIED LIMIT.

In some cases it may be necessary for a designer to specify some dimensions and tolerances to the part in the free state condition. This may be required because, even though the part may meet the design requirements in the restrained condition, there are certain design requirements that must also be met in the free state condition. The free state requirements may require that before the part is restrained it should be checked to make sure it is not bent beyond its elastic limit or that it is not bent so much that the assembly process will require too much force as to restrict assembly or insert screws. In another case, the part may be bent so much that the forces required to restrain it may be great enough that it will deform the part to which it is attached. These are but a few of the many reasons that a designer may require the part to be checked both in the free and restrained state.

If any of the part specifications are to be verified in the free state, the designer may specify this requirement with the words FREE STATE or the free state symbol. The free state symbol is an F in a circle. This free state symbol means that dimensions and tolerances that have the free state symbol applied are checked in the free state and not the restrained condition. The free state condition is the condition of a part when it is only subjected to the forces of gravity. The free state symbol is applied by placing it next to or associating it with the required dimensions and tolerances. If it is applied in a feature control frame, it always follows the feature tolerance and any modifiers.

In theory, there should not be any assumptions. It is important to specify all the design requirements on the drawing to ensure functional requirements. But, is there a point where we can apply too many specifications? Nonrigid parts have been built for a long time. There are many established procedures that manufacturing and quality already use to ensure product integrity. It is generally understood that certain nonrigid or flexible parts are subject to variation and must be clamped or restrained. If an organization is in the business of working with these parts on a regular basis, there are often some written or unwritten rules than will determine how a part shall be verified.

There may be some instances where, rather than specifying restraint requirements, the designer might let standard shop procedures or "common sense" apply. It should be understood that there is some risk when it is assumed that everyone understands all the parameters in which tolerance is applied and verified. It might be wise for a designer to consult with manufacturing and quality to determine what, if any, standard procedures may apply or are already in place.

In many cases, the part may be of significant importance to require the specification of restraint forces to insure functional requirements. In this case, it may be absolutely necessary that the restraint forces and methods of application are specified in great detail.

Where the line is drawn to specify or not specify restraints will depend on the part, the number built, common sense, tolerance limits, part size, material, stock size and past history, to name a few. The designer must make the final decision on the specifications.

There is additional information on nonrigid parts in the ASME Y14.5M, 1994 standard and the ISO 10579-NR, 1993 (E) standard. If the designation ISO 10579-NR is found on ISO drawings in or near the title block, the conditions under which the part shall be restrained may be found in the notes. If there are any geometric variations allowed in the free state, it will be designated with the modifying symbol F in a circle. This means that all geometric tolerances apply in the restrained condition outlined in the note, and all geometric tolerances modified with the F in a circle will apply in the free state. The free state condition is the condition when parts are only subjected to the force of gravity. If the direction of gravity is important, it shall also be designated on the drawing.

The drawing below is an example of a sheet metal bracket in which the tolerances were applied in the restrained condition. A note on the drawing states the method of restraint. The free state symbol designates the features which are to be verified in the free state.

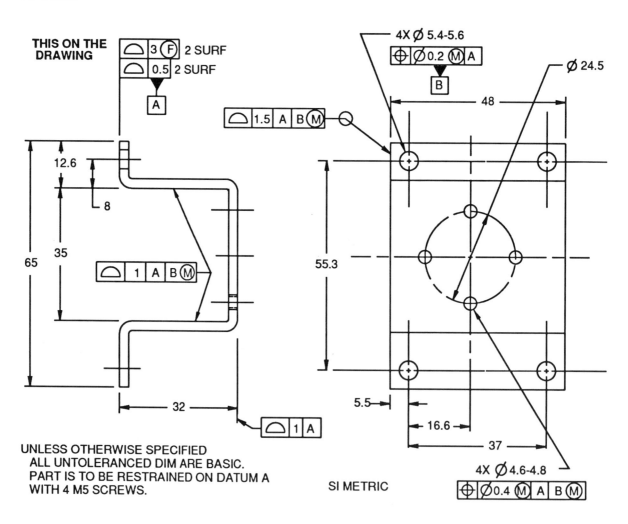

UNLESS OTHERWISE SPECIFIED
ALL UNTOLERANCED DIM ARE BASIC.
PART IS TO BE RESTRAINED ON DATUM A
WITH 4 M5 SCREWS.

SI METRIC

12.24

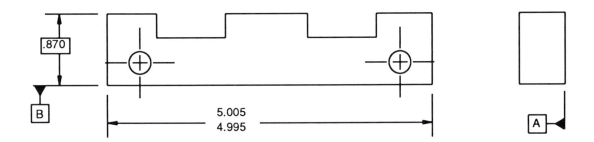

1. On the part above, apply a profile tolerance to locate the three top surfaces within .020 total in relation to datums A and B.

2. Next, apply a profile tolerance to orient the three top surfaces and make them coplanar within .010 in relation to datums A and B.

3. Make the flatness of each top surface within .005.

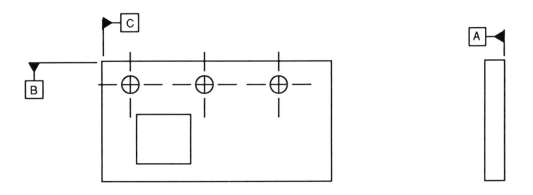

4. On the part above, the size, shape and orientation of the rectangular hole in relation to datum A must be held within .005 total all around. In addition, the location of the hole must be held within .020 in relation to datum features A, B and C. Specify a profile specification to achieve that result. For this example assume all dimensions are basic.

5. Turn back the pages to the geometric matrix chart on the last page in unit 10. Complete the chart, filling in the required information for the profile tolerances.

6. Calculate the minimum and maximum values for the indicated dimensions below. All minimum and maximum dimensions designated with an asterisk (*) should be calculated as originating from the DRF established by the datum targets. All other minimum and maximum dimensions should be calculated as the relationship between features.

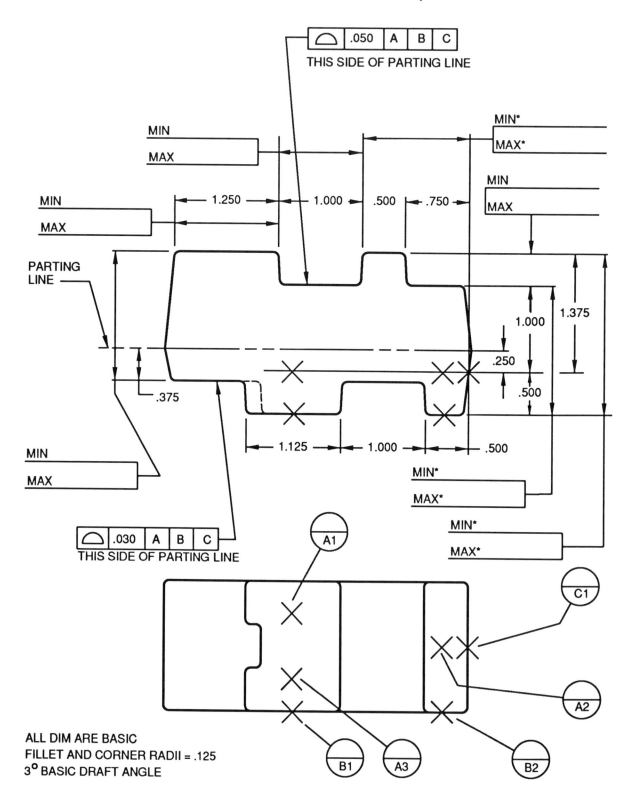

ALL DIM ARE BASIC
FILLET AND CORNER RADII = .125
3° BASIC DRAFT ANGLE

WORKSHOP EXERCISE 12.2

7. Calculate the minimum and maximum values for the indicated dimensions below. All minimum and maximum dimensions designated with an asterisk (*) should be calculated as originating from the DRF established by the datum targets. All other minimum and maximum dimensions should be calculated as the relationship between features.

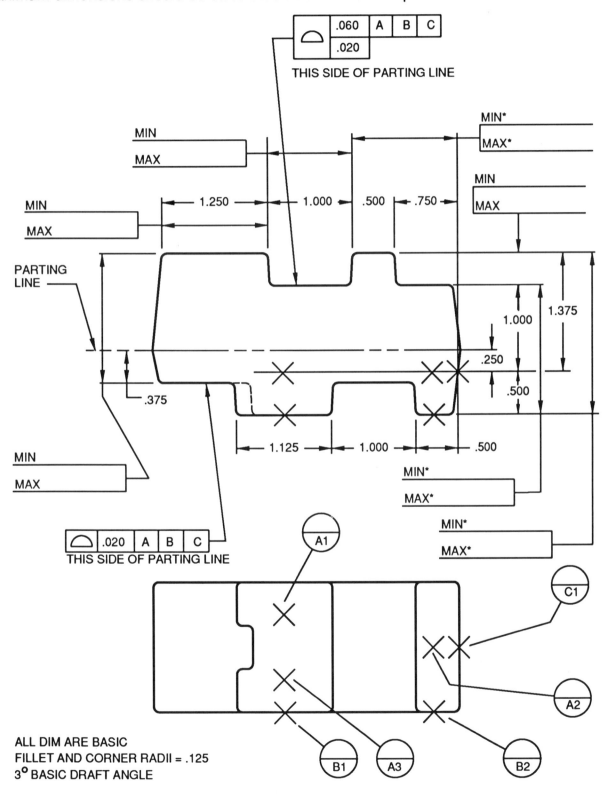

ALL DIM ARE BASIC
FILLET AND CORNER RADII = .125
3° BASIC DRAFT ANGLE

12.27

UNIT 13

POSITION TOLERANCES

Cylindrical Tolerance Zone
Rectangular Tolerance Zone
Spherical Tolerance Zone
Conical Tolerance Zone
Boundary
Composite Tolerancing is Special
Composite - One Datum Feature
Composite - Two Datum Features
Two Single Segments
Composite - Paper Gage Evaluation
Three Single Segments - Rotational Control
Coaxial Holes, Multiple Applications
Workshop Exercise 13.2

POSITION TOLERANCING

POSITION TOLERANCE CONTROLS THE LOCATION OF AN AXIS, MEDIAN PLANE OR SURFACE OF A FEATURE. DATUMS USUALLY REQUIRED.

SYMBOL	TYPE OF TOLERANCE	SHAPE OF TOLERANCE ZONE	2D OR 3D	APP OF FEATURE MODIFIER
⊕	POSITION	2 PARALLEL PLANES  CYLINDRICAL 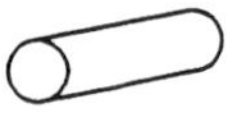 SPHERICAL 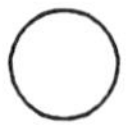 CONICAL  BOUNDARY 	3D	YES

OVERVIEW:

Position tolerancing is used for locating features of size. It defines a zone within which the axis, median plane or surface of a feature is permitted to vary from a true (theoretical exact) position.

The true position is established with basic dimensions from a datum reference frame or from other features that are related to the datum reference frame. All dimensions must be basic and are considered theoretically exact. The tolerance for the deviation of the feature from basic is found in the position feature control frame.

Position tolerancing is a three dimensional control. The common shapes of the tolerance zones are shown in the above chart. If the diameter symbol is used, the position tolerance is a cylindrical shape. If no symbol is used, the tolerance zone defaults to a total wide zone as in parallel planes for a slot.

Position tolerancing can also be used for locating irregular shaped features that have no discernable axis or median plane. In this case, the term BOUNDARY is placed under the feature control frame. The surface of the feature must not enter a theoretical boundary of identical shape located at true position.

Per the ASME Y14.5M-1994 standard, position tolerancing is applied on an MMC, LMC or RFS basis. Where MMC or LMC are required, the appropriate feature modifier symbol is applied in the feature control frame following the feature tolerance. If no modifier is specified, the default condition is implied at RFS. See modifier rules and effect of modifiers earlier in text.

Position tolerance is usually associated with datums and the datum reference frame. There are cases, though, where a group of features have an interrelationship to each other and are then used as datums. In this case, datums may not be necessary for the interrelationship of the features.

Axis versus surface interpretation

An important concept in position tolerancing is the axis interpretation versus the surface interpretation. Position tolerancing is often described as locating the axis or centerplane of a feature. The axis or centerplane of a feature is derived by using the actual mating size which contacts the high points of a feature. The location of the axis or centerplane of the actual mating size explanation is often used because of past history and because it is easy to explain. In reality, in a functional assembly it is really the surface of the feature that comes in contact with the mating part. This is especially evident when a functional gage is used to verify a position callout. The features axis or centerplane is not located, but rather the surface of the feature simply must not enter a virtual condition boundary.

This concept of axis versus surface was recognized in the ANSI Y14.5M-1982 standard. The ASME Y14.5M-1994 and ASME Y14.5.1M-1994 standards recognize that the axis versus surface interpretation can yield different results and state that, if there is a conflict between the two procedures, the surface interpretation shall prevail.

In the verification procedure the actual value of position tolerancing can be defined in terms of the axis or median plane or in terms of the surface of a feature. The axis and median plane inspection procedure is explained in unit 3. The surface interpretation is simply reported in terms of the actual virtual size of a feature.

POSITION TOLERANCING

Position tolerancing is used for locating features of size. It defines a zone within which the axis of a feature is permitted to vary from a true (theoretical exact) position. Basic dimensions establish the true position from the specified datum features as well as the interrelationship between the features.

THIS ON THE DRAWING

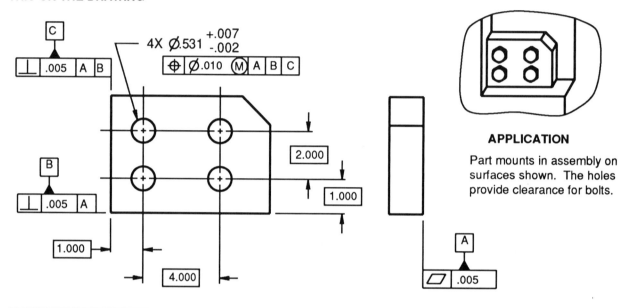

APPLICATION

Part mounts in assembly on surfaces shown. The holes provide clearance for bolts.

MEANS THIS

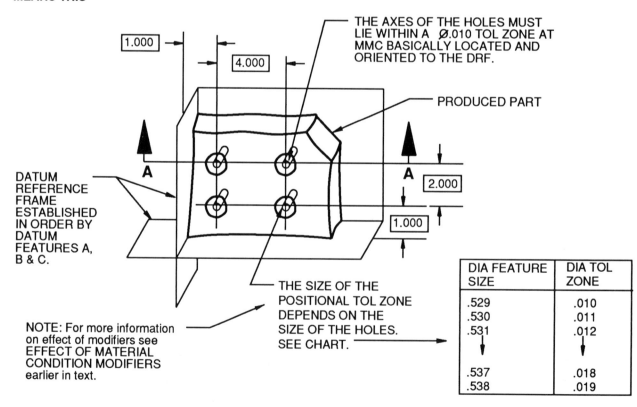

THE AXES OF THE HOLES MUST LIE WITHIN A Ø.010 TOL ZONE AT MMC BASICALLY LOCATED AND ORIENTED TO THE DRF.

PRODUCED PART

DATUM REFERENCE FRAME ESTABLISHED IN ORDER BY DATUM FEATURES A, B & C.

THE SIZE OF THE POSITIONAL TOL ZONE DEPENDS ON THE SIZE OF THE HOLES. SEE CHART.

NOTE: For more information on effect of modifiers see EFFECT OF MATERIAL CONDITION MODIFIERS earlier in text.

DIA FEATURE SIZE	DIA TOL ZONE
.529	.010
.530	.011
.531	.012
↓	↓
.537	.018
.538	.019

13.4

POSITION TOLERANCING

Position tolerancing related to datums controls the location as well as the orientation to the datums. The axis of the holes (simulated by a pin) may shift and/or tilt within the stated position zone.

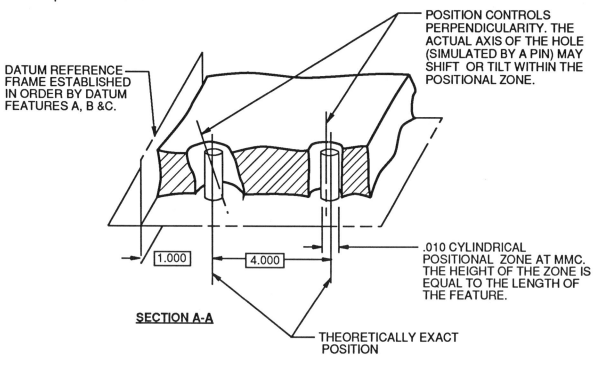

DATUM REFERENCE FRAME ESTABLISHED IN ORDER BY DATUM FEATURES A, B &C.

POSITION CONTROLS PERPENDICULARITY. THE ACTUAL AXIS OF THE HOLE (SIMULATED BY A PIN) MAY SHIFT OR TILT WITHIN THE POSITIONAL ZONE.

1.000   4.000

.010 CYLINDRICAL POSITIONAL ZONE AT MMC. THE HEIGHT OF THE ZONE IS EQUAL TO THE LENGTH OF THE FEATURE.

SECTION A-A

THEORETICALLY EXACT POSITION

FUNCTIONAL GAGE FOR POSITIONAL CALL OUT

$4X \, \emptyset.531 \, ^{+.007}_{-.002}$

| ⊕ | ∅ .010 Ⓜ | A | B | C |

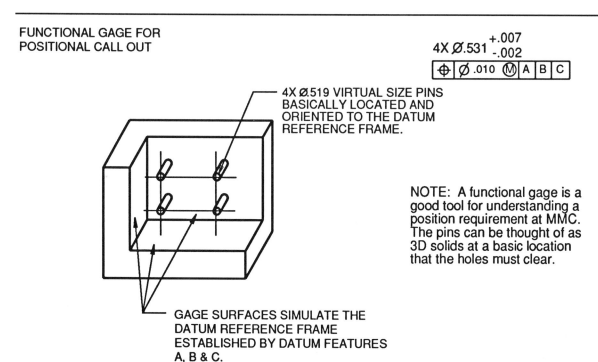

4X ∅.519 VIRTUAL SIZE PINS BASICALLY LOCATED AND ORIENTED TO THE DATUM REFERENCE FRAME.

NOTE: A functional gage is a good tool for understanding a position requirement at MMC. The pins can be thought of as 3D solids at a basic location that the holes must clear.

GAGE SURFACES SIMULATE THE DATUM REFERENCE FRAME ESTABLISHED BY DATUM FEATURES A, B & C.

Shown above is a sample functional gage or 3D solid that will verify the position specification. The holes must also be verified for size. The functional gage is shown for information only. Position can be verified with other methods as well.

POSITION TOLERANCING - RECTANGULAR TOLERANCE ZONES

In some cases it may be necessary to locate or orient features with a greater tolerance in one direction than another. This may be accomplished by the method shown below. Separate feature control frames are used to indicate the direction and magnitude of each position tolerance relative to the datum reference frame. Note that the diameter symbol is not present in the feature control frames indicating a distance between two parallel planes.

THIS ON THE DRAWING

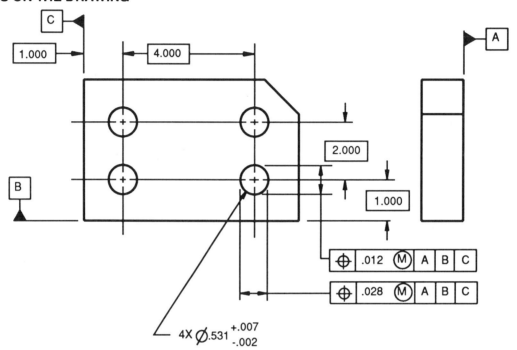

MEANS THIS

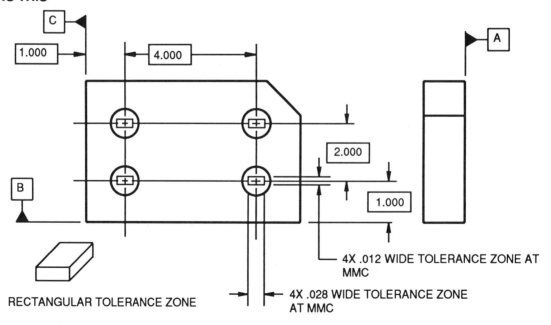

The axes of the holes must lie within the .012 X .028 rectangular tolerance zones basically located to the specified datum reference frame.

POSITION TOLERANCING - SPHERICAL ZONE

Position tolerancing may be used to control the location of a spherical feature. The symbol for spherical diameter precedes the size dimension and the feature tolerance in the feature control frame. This indicates the position tolerance is a spherical tolerance zone.

THIS ON THE DRAWING

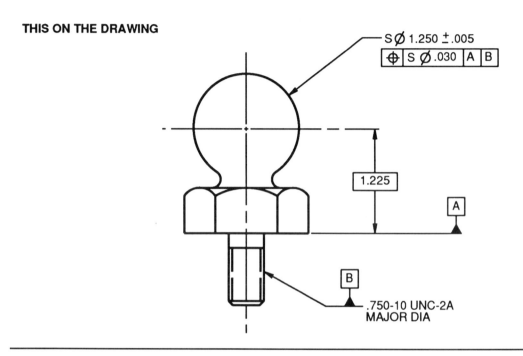

MEANS THIS

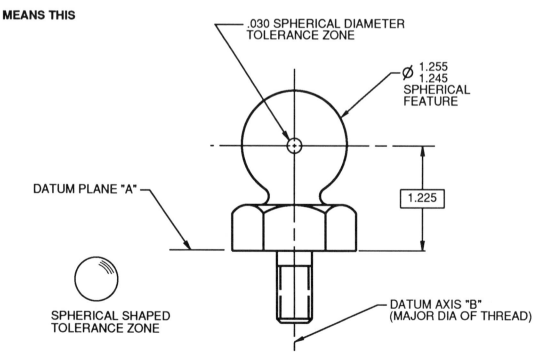

The center point of the spherical diameter must lie within a spherical diameter zone of .030 RFS, which is basically located to the datum reference frame. (DRF)

POSITION TOLERANCING - CONICAL TOLERANCE ZONE

Position tolerancing may be used to control a feature such as a deep drilled hole, closer at one surface than another. The result is a conical shaped tolerance zone.

APPLICATION

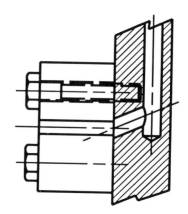

THIS ON THE DRAWING

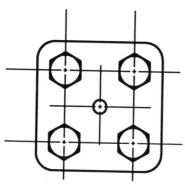

4X Ø .285 / .280 [B]

⊕ | Ø .006 Ⓜ A

⌓ .060 | A | B Ⓜ

[A]

▱ .001

SURFACE Y

□ 1.800

1.000

SURFACE X

Ø .197 / .185

⊕ | Ø .060 Ⓜ A B Ⓜ
AT SURFACE X

⊕ | Ø .010 Ⓜ A B Ⓜ
AT SURFACE Y

1.000

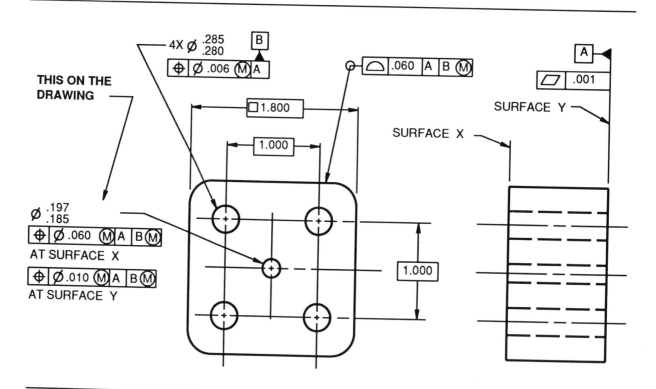

MEANS THIS

CONICAL SHAPED
TOLERANCE ZONE

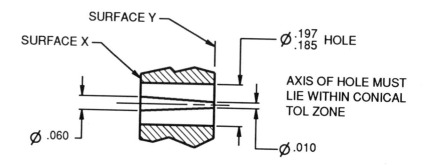

SURFACE Y

SURFACE X

Ø .197 / .185 HOLE

AXIS OF HOLE MUST
LIE WITHIN CONICAL
TOL ZONE

Ø .060

Ø .010

13.8

POSITION - BOUNDARY

Position may be used to locate irregular features. The term BOUNDARY is placed under the feature control frame. Rather than locating the axis or median plane of the feature, a virtual condition boundary is established. In the illustration below a datum reference frame is established by the flat surface and the height and width. The profile tolerance on the irregular opening defines the size, shape and orientation of the feature. The position boundary tolerance defines a boundary in which no element of the feature may lie. The feature modifiers MMC or LMC may be applied.

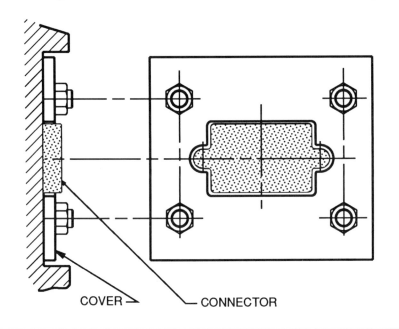

THIS APPLICATION

Cover must fit in recess, clear connector and accept 4 bolts.

COVER ⟶   ⎯ CONNECTOR

THIS ON THE DRAWING

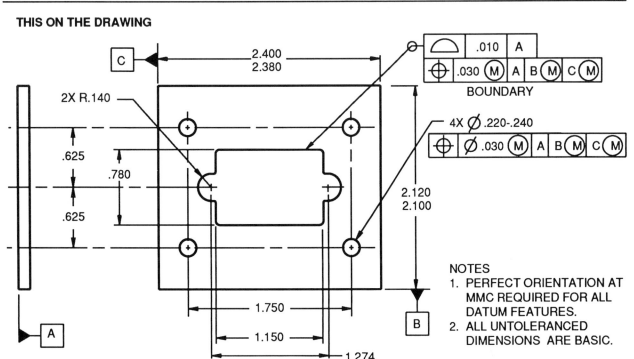

NOTES
1. PERFECT ORIENTATION AT MMC REQUIRED FOR ALL DATUM FEATURES.
2. ALL UNTOLERANCED DIMENSIONS ARE BASIC.

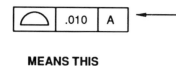

THE PROFILE REQUIREMENT SPECIFIES THE SIZE/ORIENTATION FOR THE FEATURE.

MEANS THIS

The surface, all around, must lie between two profile boundaries .010 apart equally disposed about the true profile.

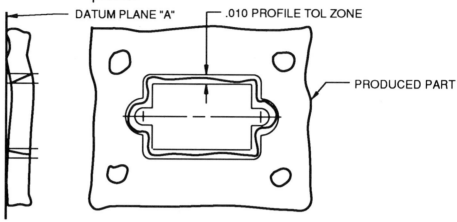

DATUM PLANE "A"   .010 PROFILE TOL ZONE

PRODUCED PART

BOUNDARY

THE POSITION BOUNDARY SPECIFICATION LOCATES THE FEATURE TO THE DATUM REFERENCE FRAME.

No portion of the surface may be permitted to lie within the boundary of MMC contour minus the position tolerance when positioned with respect to the DRF established by datum fetures A, B at MMC, and C at MMC.

MEANS THIS

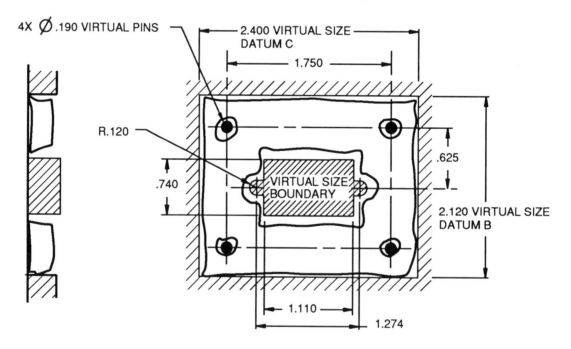

4X Ø.190 VIRTUAL PINS

2.400 VIRTUAL SIZE
DATUM C

1.750

R.120

VIRTUAL SIZE
BOUNDARY

.740

.625

2.120 VIRTUAL SIZE
DATUM B

1.110

1.274

The position boundary concept is a similar, but a somewhat different, concept than the composite profile approach to locating irregular features. Position boundary, when applied with MMC or LMC modifiers, defines only an inner or outer virtual boundary in which no element of the feature must lie. Composite profile establishes both an inner and outer boundary in which the feature must lie. See also composite profile in this text.

COMPOSITE TOLERANCING IS SPECIAL

Composite positional tolerancing provides for the location of feature patterns as well as the features within these patterns. These requirements are stated by the use of a composite feature control frame. The position tolerance symbol is entered once and is applicable to all horizontal segments. Each complete horizontal segment in the feature control frame may be verified separately, but the lower segment is always a subset of the upper segment. The composite tolerancing concepts can also be applied to profile.

COMPOSITE POSITION TOLERANCING FEATURE CONTROL FRAME

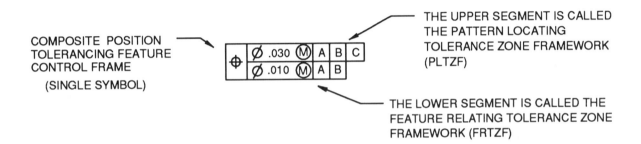

The upper segment of a composite position feature control frame is called the pattern locating tolerance zone framework (PLTZF). The PLTZF is located to applicable datums with basic dimensions. It specifies the larger positional tolerance for the location of the pattern of features as a group. Applicable datums are entered in the upper segment in their order of precedence. The upper segment locates and orients the features to each other as well as the specified datums.

The lower segment of a composite feature control frame is called the feature relating tolerance zone framework (FRTZF). The FRTZF specifies a smaller position tolerance for the feature to feature relationship within the pattern. Basic dimensions apply between the features but do not apply to the datums. The lower entry is orientation only to the specified datums and not location.

If datums are not specified in the lower segment, the FRTZF is allowed to tilt, rotate and/or shift within the confines of the PLTZF. If datums are specified in the lower segment, they govern the orientation (not location) of the FRTZF relative to the specified datums.

Note: If different datums, different datum modifiers, or the same datums in a different order of precedence are specified, this constitutes a different datum reference frame and design requirements. This requirement is not to be specified using the composite positional tolerancing method, since such a requirement no longer represents a liberation within the given limits of the FRTZF. A separately specified feature relating tolerance, using a second segment feature control frame, should be used, including applicable datums as an independent requirement.

Composite position vs two single segmented position feature control frames

Composite position tolerancing is special.   (The position symbol is entered once and is applicable to both horizontal entries.)   It is different than two single segmented feature control frames.

The upper segment on a composite feature control frame controls location and orientation to the datums. The lower segment on a composite controls location between the features but only orients (not location) the features to the specified datums.

In contrast, all single segment position frames control both the location and orientation between the features as well as the specified datums. This type of control is interpreted simply as two position requirements.

DIFFERENCE BETWEEN A COMPOSITE TOLERANCE AND 2 SINGLE SEGMENTED FEATURE CONTROL FRAMES

COMPOSITE POSITION FRAME
  (SPECIAL TYPE OF
  POSITION)

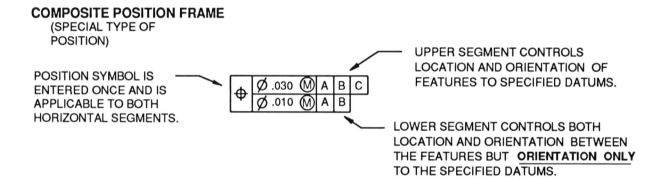

POSITION SYMBOL IS
ENTERED ONCE AND IS
APPLICABLE TO BOTH
HORIZONTAL SEGMENTS.

UPPER SEGMENT CONTROLS
LOCATION AND ORIENTATION OF
FEATURES TO SPECIFIED DATUMS.

LOWER SEGMENT CONTROLS BOTH
LOCATION AND ORIENTATION BETWEEN
THE FEATURES BUT **ORIENTATION ONLY**
TO THE SPECIFIED DATUMS.

TWO SINGLE SEGMENTED FRAMES
  (SIMPLY TWO POSITION
  REQUIREMENTS)

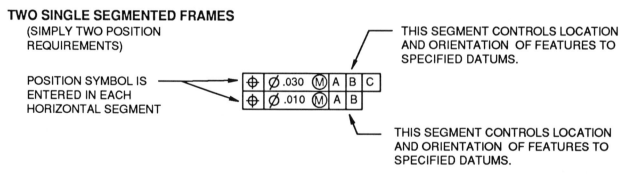

POSITION SYMBOL IS
ENTERED IN EACH
HORIZONTAL SEGMENT

THIS SEGMENT CONTROLS LOCATION
AND ORIENTATION OF FEATURES TO
SPECIFIED DATUMS.

THIS SEGMENT CONTROLS LOCATION
AND ORIENTATION OF FEATURES TO
SPECIFIED DATUMS.

The difference between the terms position and orientation should be clear. Location locates features and is associated with basic linear dimensions. It can also include orientation. Orientation, on the other hand, is not associated with location or with basic linear dimensions, only basic angles. Orientation is usually thought of as parallelism, perpendicularity or angularity. The composite tolerancing concepts explained above apply to both position and profile tolerances.

In some particular cases, either the application of a composite tolerance or two single segmented frames to a group of features can yield the same result. This is especially true when only a primary datum plane is established perpendicular to the features and is entered in the lower segment. The lower entry on a composite is only orientation to the datums. Although all the entries on single segmented frames are location, only orientation can be established to the perpendicular plane. See the four hole plate composite position examples in this text for an illustration.

The concepts concerning composite tolerancing (both position and profile) have been enhanced and clarified in the ASME Y14.5M, 1994 standard. The composite enhancement happens when more than one datum is entered in the lower segment of a composite tolerance or the primary datum is established from a feature other than perpendicular to the positioned features. The ANSI Y14.5M, 1982 standard did not cover this topic in detail. This expanded interpretation of composite tolerancing is applicable to both position and profile tolerancing.

Composite tolerancing is very useful for providing a loose location but a more restrictive orientation of the features. A simple example might be a specification for a pattern of holes that locate an emblem or name plate. The holes are important to each other but not important to the datums. The orientation of the pattern, however, is important as the emblem or name plate is not allowed to tilt or mount crooked.

Composite position tolerancing is used and appropriate datums are specified in the lower segment to control orientation. Orientation tolerances such as parallelism, perpendicularity and angularity will not work as they can not control orientation between groups of features, only the orientation of individual features.

The name plate is only a simple example. There are many more examples where a designer may allow a relaxation for the location of a group of features, if the orientation of the features can be controlled. Composite tolerancing increases manufacturing tolerance by providing the designer with the tools to specify only the requirements needed. If location is required to the specified datums, use single segmented feature control frames. If a large location to the datums, but close orientation is required, use composite tolerancing.

The following examples in the text illustrate composite position tolerancing. These examples can be compared to the examples of single segmented feature control frames for additional clarification of principles. See also composite profile tolerancing.

COMPOSITE POSITION TOLERANCING - 1 DATUM FEATURE

Composite positional tolerancing is a special method of locating features of size. The position symbol is entered once and is applicable to both horizontal segments. In the example below, the upper segment locates and orients the pattern of holes to the specified datums. The lower segment locates the holes to each other and orients the hole to hole requirement to the specified datums.

THIS ON THE DRAWING

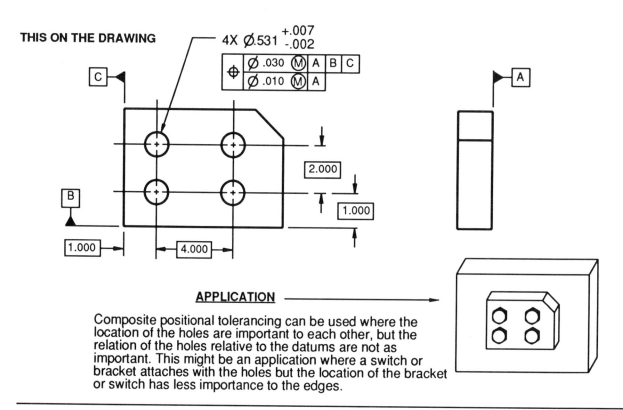

APPLICATION

Composite positional tolerancing can be used where the location of the holes are important to each other, but the relation of the holes relative to the datums are not as important. This might be an application where a switch or bracket attaches with the holes but the location of the bracket or switch has less importance to the edges.

MEANS THIS

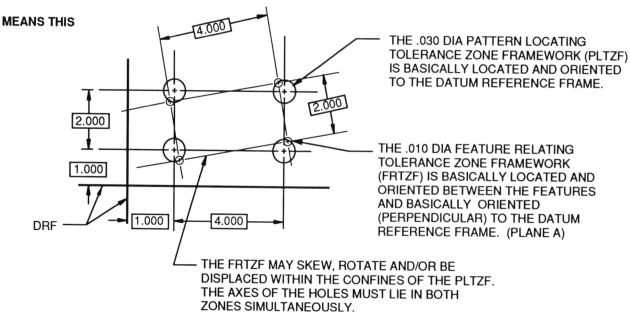

THE .030 DIA PATTERN LOCATING TOLERANCE ZONE FRAMEWORK (PLTZF) IS BASICALLY LOCATED AND ORIENTED TO THE DATUM REFERENCE FRAME.

THE .010 DIA FEATURE RELATING TOLERANCE ZONE FRAMEWORK (FRTZF) IS BASICALLY LOCATED AND ORIENTED BETWEEN THE FEATURES AND BASICALLY ORIENTED (PERPENDICULAR) TO THE DATUM REFERENCE FRAME. (PLANE A)

THE FRTZF MAY SKEW, ROTATE AND/OR BE DISPLACED WITHIN THE CONFINES OF THE PLTZF. THE AXES OF THE HOLES MUST LIE IN BOTH ZONES SIMULTANEOUSLY.

COMPOSITE POSITION TOLERANCING - 1 DATUM FEATURE

⊕	⌀.030 Ⓜ	A	B	C

◀── THE UPPER SEGMENT SPECIFIES THE PATTERN LOCATION TO THE SPECIFIED DATUMS.

THE HOLES MUST MEET BOTH REQUIREMENTS

1. PATTERN REQUIREMENT

FIRST:
THE AXES OF THE HOLES MUST LIE WITHIN A .030 DIA AT MMC PATTERN LOCATION ZONE THAT IS BASICALLY LOCATED TO THE DATUM REFERENCE FRAME

DRF ESTABLISHED BY DATUM FEATURES A, B & C.

2.000

1.000

4.000

1.000

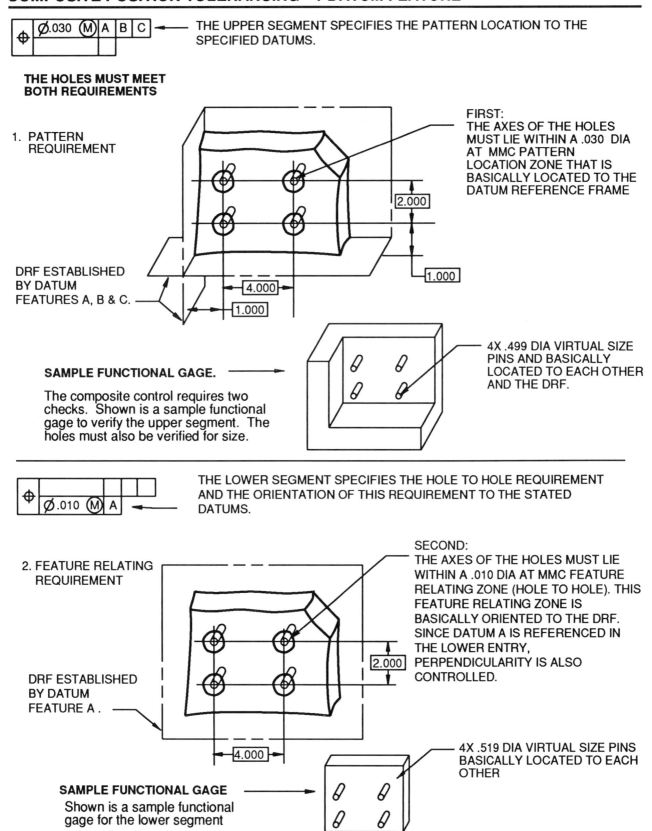

4X .499 DIA VIRTUAL SIZE PINS AND BASICALLY LOCATED TO EACH OTHER AND THE DRF.

SAMPLE FUNCTIONAL GAGE. ───▶

The composite control requires two checks. Shown is a sample functional gage to verify the upper segment. The holes must also be verified for size.

⊕	⌀.010 Ⓜ	A		

◀── THE LOWER SEGMENT SPECIFIES THE HOLE TO HOLE REQUIREMENT AND THE ORIENTATION OF THIS REQUIREMENT TO THE STATED DATUMS.

2. FEATURE RELATING REQUIREMENT

SECOND:
THE AXES OF THE HOLES MUST LIE WITHIN A .010 DIA AT MMC FEATURE RELATING ZONE (HOLE TO HOLE). THIS FEATURE RELATING ZONE IS BASICALLY ORIENTED TO THE DRF. SINCE DATUM A IS REFERENCED IN THE LOWER ENTRY, PERPENDICULARITY IS ALSO CONTROLLED.

DRF ESTABLISHED BY DATUM FEATURE A .

2.000

4.000

4X .519 DIA VIRTUAL SIZE PINS BASICALLY LOCATED TO EACH OTHER

SAMPLE FUNCTIONAL GAGE ───▶
Shown is a sample functional gage for the lower segment

Since the MMC feature modifier is specified, additional positional tolerance is available for both the pattern and hole to hole specification as the holes depart from MMC.

COMPOSITE POSITION TOLERANCING - 2 DATUM FEATURES

Composite position tolerancing can be applied with two datum features in the lower segment. This type of control is useful for locating a pattern of holes to a datum reference frame with a large tolerance while restricting the hole to hole and orientation of the hole to hole requirement to a tighter tolerance. The ability to add more than one datum to the lower segment of a composite tolerance is an enhancement of principles made available in the ASME Y14.5M, 1994 standard.

THIS ON THE DRAWING

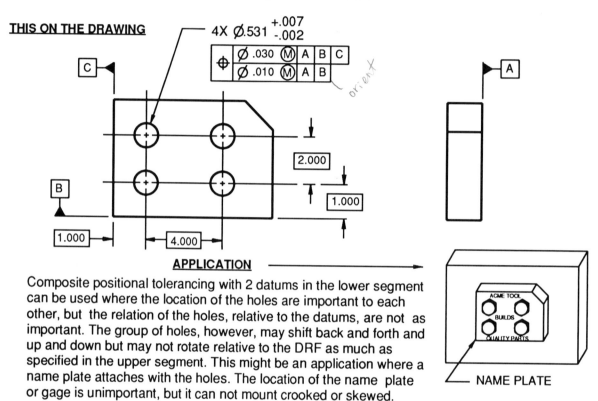

APPLICATION

Composite positional tolerancing with 2 datums in the lower segment can be used where the location of the holes are important to each other, but the relation of the holes, relative to the datums, are not as important. The group of holes, however, may shift back and forth and up and down but may not rotate relative to the DRF as much as specified in the upper segment. This might be an application where a name plate attaches with the holes. The location of the name plate or gage is unimportant, but it can not mount crooked or skewed.

NAME PLATE

MEANS THIS

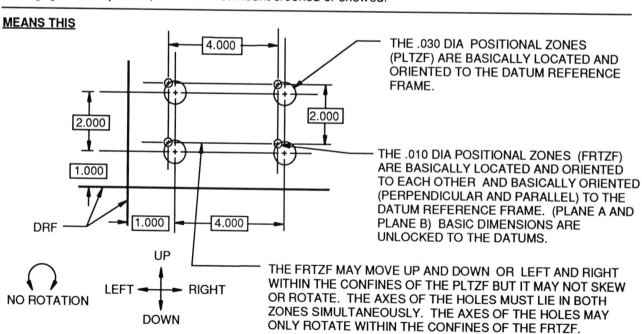

THE .030 DIA POSITIONAL ZONES (PLTZF) ARE BASICALLY LOCATED AND ORIENTED TO THE DATUM REFERENCE FRAME.

THE .010 DIA POSITIONAL ZONES (FRTZF) ARE BASICALLY LOCATED AND ORIENTED TO EACH OTHER AND BASICALLY ORIENTED (PERPENDICULAR AND PARALLEL) TO THE DATUM REFERENCE FRAME. (PLANE A AND PLANE B) BASIC DIMENSIONS ARE UNLOCKED TO THE DATUMS.

THE FRTZF MAY MOVE UP AND DOWN OR LEFT AND RIGHT WITHIN THE CONFINES OF THE PLTZF BUT IT MAY NOT SKEW OR ROTATE. THE AXES OF THE HOLES MUST LIE IN BOTH ZONES SIMULTANEOUSLY. THE AXES OF THE HOLES MAY ONLY ROTATE WITHIN THE CONFINES OF THE FRTZF.

13.16

COMPOSITE POSITION TOLERANCING - 2 DATUM FEATURES

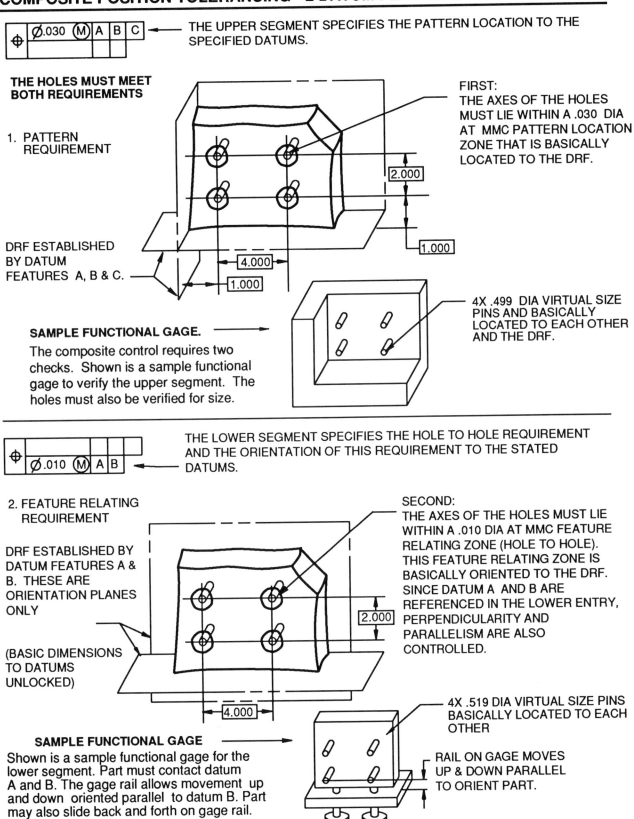

⊕ | ⌀.030 Ⓜ | A | B | C ◄——— THE UPPER SEGMENT SPECIFIES THE PATTERN LOCATION TO THE SPECIFIED DATUMS.

THE HOLES MUST MEET BOTH REQUIREMENTS

1. PATTERN REQUIREMENT

DRF ESTABLISHED BY DATUM FEATURES A, B & C.

FIRST:
THE AXES OF THE HOLES MUST LIE WITHIN A .030 DIA AT MMC PATTERN LOCATION ZONE THAT IS BASICALLY LOCATED TO THE DRF.

2.000

1.000

4.000

1.000

4X .499 DIA VIRTUAL SIZE PINS AND BASICALLY LOCATED TO EACH OTHER AND THE DRF.

SAMPLE FUNCTIONAL GAGE.

The composite control requires two checks. Shown is a sample functional gage to verify the upper segment. The holes must also be verified for size.

⊕ | ⌀.010 Ⓜ | A | B ◄——— THE LOWER SEGMENT SPECIFIES THE HOLE TO HOLE REQUIREMENT AND THE ORIENTATION OF THIS REQUIREMENT TO THE STATED DATUMS.

2. FEATURE RELATING REQUIREMENT

DRF ESTABLISHED BY DATUM FEATURES A & B. THESE ARE ORIENTATION PLANES ONLY

(BASIC DIMENSIONS TO DATUMS UNLOCKED)

SECOND:
THE AXES OF THE HOLES MUST LIE WITHIN A .010 DIA AT MMC FEATURE RELATING ZONE (HOLE TO HOLE). THIS FEATURE RELATING ZONE IS BASICALLY ORIENTED TO THE DRF. SINCE DATUM A AND B ARE REFERENCED IN THE LOWER ENTRY, PERPENDICULARITY AND PARALLELISM ARE ALSO CONTROLLED.

2.000

4.000

4X .519 DIA VIRTUAL SIZE PINS BASICALLY LOCATED TO EACH OTHER

RAIL ON GAGE MOVES UP & DOWN PARALLEL TO ORIENT PART.

SAMPLE FUNCTIONAL GAGE

Shown is a sample functional gage for the lower segment. Part must contact datum A and B. The gage rail allows movement up and down oriented parallel to datum B. Part may also slide back and forth on gage rail.

Note: If datum C were also entered in the lower segment, the interpretation would be the same. Since the lower entry is orientation only to the DRF, datum A and B are enough to establish complete orientation. The addition of datum C would be redundant. Since the MMC modifier was specified, additional positional tolerance is allowed for both the pattern and hole to hole specification as the features depart from MMC.

POSITION TOLERANCING - TWO SINGLE SEGMENTED FRAMES

In some cases, there may be a need to apply two single segmented position feature control frames to a feature. The part below illustrates this application. Single segmented position frames simply control position and orientation to the specified datums. Multiple frames may be used and datums are added or deleted as required.

THIS ON THE DRAWING

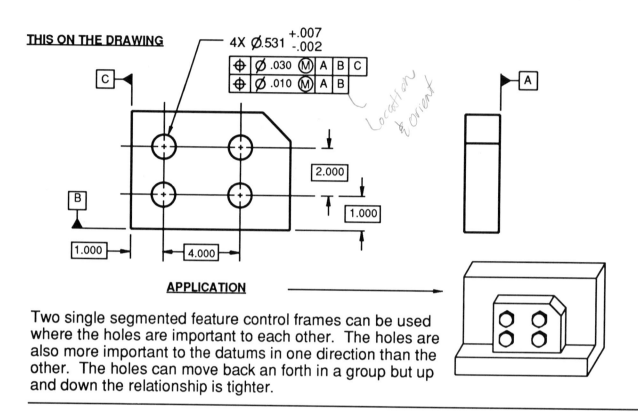

APPLICATION

Two single segmented feature control frames can be used where the holes are important to each other. The holes are also more important to the datums in one direction than the other. The holes can move back an forth in a group but up and down the relationship is tighter.

MEANS THIS

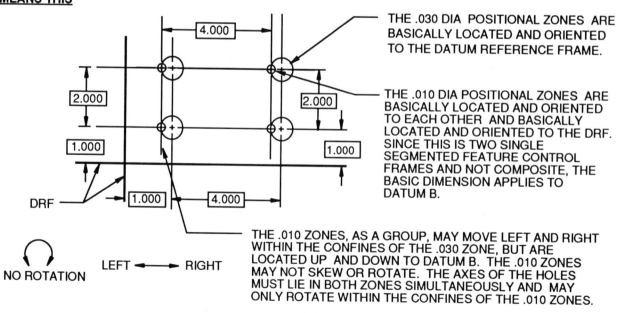

THE .030 DIA POSITIONAL ZONES ARE BASICALLY LOCATED AND ORIENTED TO THE DATUM REFERENCE FRAME.

THE .010 DIA POSITIONAL ZONES ARE BASICALLY LOCATED AND ORIENTED TO EACH OTHER AND BASICALLY LOCATED AND ORIENTED TO THE DRF. SINCE THIS IS TWO SINGLE SEGMENTED FEATURE CONTROL FRAMES AND NOT COMPOSITE, THE BASIC DIMENSION APPLIES TO DATUM B.

THE .010 ZONES, AS A GROUP, MAY MOVE LEFT AND RIGHT WITHIN THE CONFINES OF THE .030 ZONE, BUT ARE LOCATED UP AND DOWN TO DATUM B. THE .010 ZONES MAY NOT SKEW OR ROTATE. THE AXES OF THE HOLES MUST LIE IN BOTH ZONES SIMULTANEOUSLY AND MAY ONLY ROTATE WITHIN THE CONFINES OF THE .010 ZONES.

POSITION TOLERANCING - TWO SINGLE SEGMENTED FRAMES

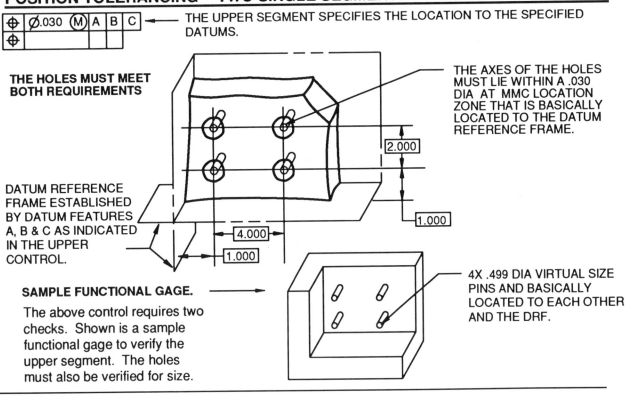

⊕	Ø.030 Ⓜ	A	B	C
⊕				

THE UPPER SEGMENT SPECIFIES THE LOCATION TO THE SPECIFIED DATUMS.

THE HOLES MUST MEET BOTH REQUIREMENTS

THE AXES OF THE HOLES MUST LIE WITHIN A .030 DIA AT MMC LOCATION ZONE THAT IS BASICALLY LOCATED TO THE DATUM REFERENCE FRAME.

DATUM REFERENCE FRAME ESTABLISHED BY DATUM FEATURES A, B & C AS INDICATED IN THE UPPER CONTROL.

SAMPLE FUNCTIONAL GAGE.

The above control requires two checks. Shown is a sample functional gage to verify the upper segment. The holes must also be verified for size.

4X .499 DIA VIRTUAL SIZE PINS AND BASICALLY LOCATED TO EACH OTHER AND THE DRF.

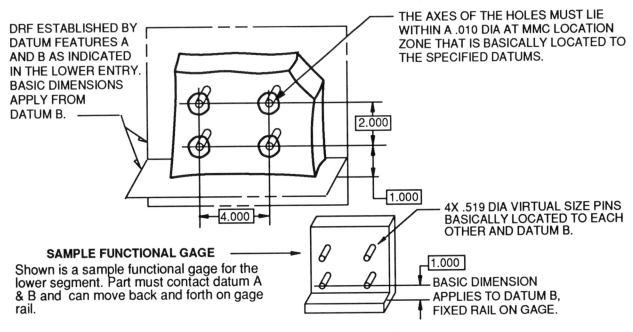

⊕				
⊕	Ø .010 Ⓜ	A	B	

THE LOWER SEGMENT SPECIFIES A LOCATION REQUIREMENT TO THE STATED DATUMS.

THE AXES OF THE HOLES MUST LIE WITHIN A .010 DIA AT MMC LOCATION ZONE THAT IS BASICALLY LOCATED TO THE SPECIFIED DATUMS.

DRF ESTABLISHED BY DATUM FEATURES A AND B AS INDICATED IN THE LOWER ENTRY. BASIC DIMENSIONS APPLY FROM DATUM B.

4X .519 DIA VIRTUAL SIZE PINS BASICALLY LOCATED TO EACH OTHER AND DATUM B.

SAMPLE FUNCTIONAL GAGE

Shown is a sample functional gage for the lower segment. Part must contact datum A & B and can move back and forth on gage rail.

BASIC DIMENSION APPLIES TO DATUM B, FIXED RAIL ON GAGE.

Note: If datum C were also entered in the lower segment, there would be a conflict. Since both the upper and lower segment are location to the datums, there would be no need for the larger tolerance in the upper segment of the feature control frame. Since the MMC modifier is specified, additional positional tolerance is allowed for both the upper and lower segment as the features depart from MMC.

COMPOSITE POSITION - PAPER GAGE EVALUATION

The top drawing is a plate with composite position tolerancing applied. The lower drawing is the produced part. The example below and following page explain how to use the paper gage to evaluate the 3 holes. The upper segment can be calculated by using the conversion chart. The lower segment must be completed using the paper gage or with appropriate CMM software. The numbers may vary slightly, as the answers will be estimates. The final outcome of accept or reject, however, will be identical. See the chart below for the record of calculations.

AS DRAWN

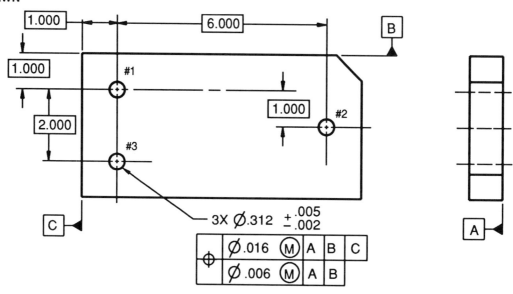

PRODUCED PART

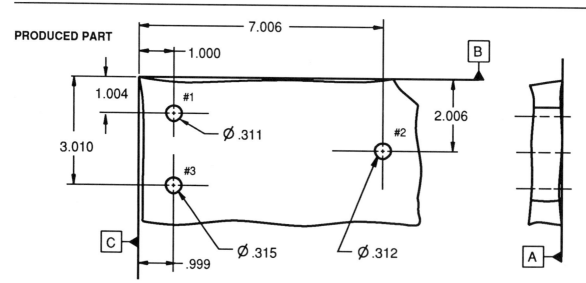

UPPER SEGMENT CALCULATIONS ⟶   LOWER SEGMENT CALCULATIONS ⟶

HOLE NO.	HOLE MMC	HOLE ACTUAL SIZE	PATTERN TOLERANCE ALLOWED	PATTERN TOLERANCE ACTUAL	ACC	REJ	HOLE TO HOLE TOLERANCE ALLOWED	HOLE TO HOLE TOLERANCE ACTUAL	ACC	REJ
1	.310	.311	.017	.008	X		.007	.007	X	
2	.310	.312	.018	.017	X		.008	.008	X	
3	.310	.315	.021	.021	X		.011	.011	X	

The upper entry on the composite position can be calculated by using the conversion chart or with the paper gage as shown below. The axis of the holes must lie within the allowable pattern tolerance. If the paper gage is used, the actual pattern tolerance is estimated with the position rings.

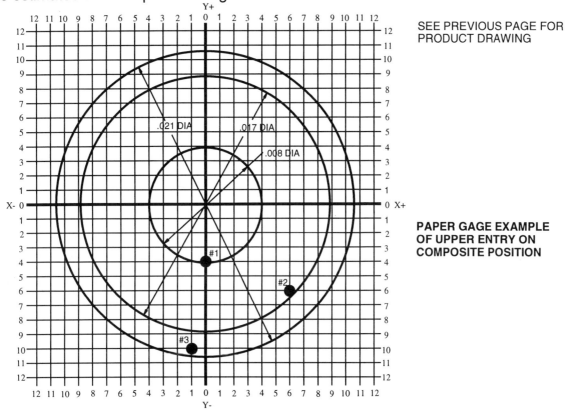

SEE PREVIOUS PAGE FOR PRODUCT DRAWING

PAPER GAGE EXAMPLE OF UPPER ENTRY ON COMPOSITE POSITION

The lower segment on the composite position tolerance can be calculated by using the paper gage. The group of four holes must lie within their allotted position tolerance. The hole to hole actual is only estimated as the position rings may be moved around to get the best fit. Individual numbers may vary slightly. The end result of pass or fail will not vary.

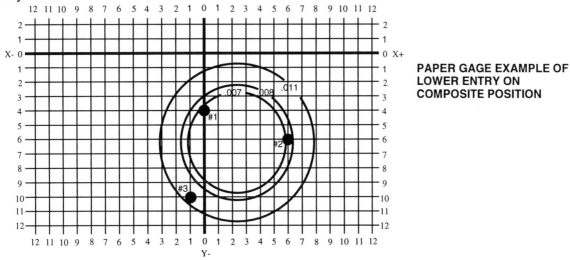

PAPER GAGE EXAMPLE OF LOWER ENTRY ON COMPOSITE POSITION

NOTE: In the above example two datums were entered in the lower entry of the composite feature control frame. This required the collection of data and verification while the part was set-up or oriented to the DRF established by datum features A and B. If datum feature B were eliminated from the lower entry, the set-up to verify the lower entry would change. The collection of data and verification would take place while the 3 holes are balanced to each other without alignment or orientation to datum feature B.

COMPOSITE POSITION APPLICATIONS - COAXIAL HOLES

Composite position tolerancing may be used to control coaxial holes as shown in the accompanying figure. In scenario #1 below, the position symbol is entered once and applies for both the upper and lower segment. The upper segment or .010 dia. tolerance zone controls the location and orientation of the four holes to the specified datum reference frame established by datum features A and B. The lower segment or .004 dia tolerance zone controls the location between the holes (coaxiality). Since datums are not entered in the lower segment, the orientation of this zone is not controlled and is free to tilt or skew within the confines of the .010 dia upper segment zone. The axes of the holes must lie in both zones simultaneously. See the following page for scenarios #2 thru 4.

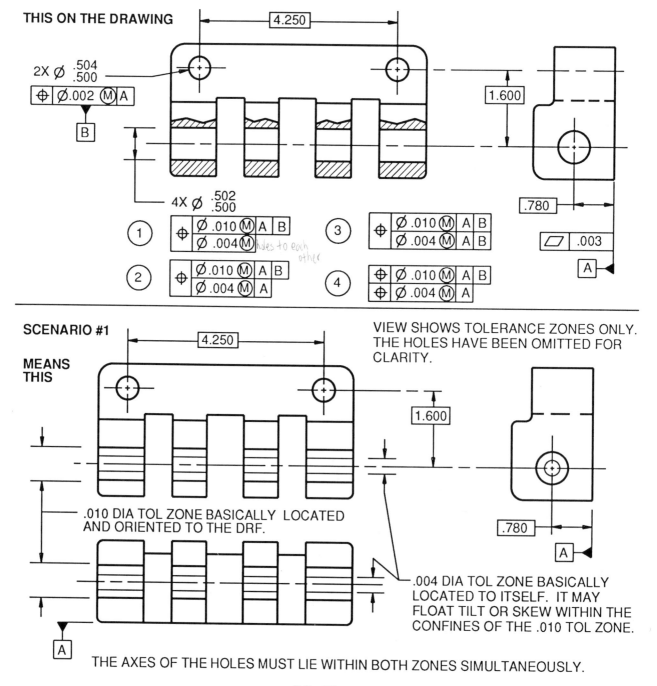

THE AXES OF THE HOLES MUST LIE WITHIN BOTH ZONES SIMULTANEOUSLY.

COMPOSITE POSITION APPLICATIONS - COAXIAL HOLES

It is very important to understand the differences between orientation, position and composite position tolerances. All of these factors, plus the application or elimination of datums, can have a substantial effect on the interpretation of specifications. It is important that the designer select a control that properly reflects design requirements. The following scenarios apply to the coaxial holes in accompanying hinge illustration.

Scenario #2 - In the accompanying figure, the composite position tolerance is revised and datum feature A is added to the lower segment. The upper segment did not change and locates and orients the holes to the specified datum reference frame. The lower segment or .004 tolerance zone controls the coaxality of the holes. In addition, since datum A is also entered in the lower segment, the orientation (parallelism) of this tolerance zone is controlled in only one direction, relative to the face or datum A. The axes of the holes must fall in both zones simultaneously. The basic dimensions to the datums are unlocked for the lower segment.

Scenario #3 - In the accompanying figure, the composite position tolerance is revised and both datum feature A and B are added to the lower segment. The upper segment did not change and locates and orients the holes to the specified datum reference frame. The lower segment or .004 tolerance zone controls the coaxality of the holes. In addition, since datum A and datum B are also entered in the lower segment, the orientation (parallelism) of this tolerance zone is controlled in all directions, (up and down as well as back and forth) relative to the datum reference frame established by datum features A and B. The axes of the holes must fall in both zones simultaneously. The basic dimensions to the datums are unlocked for the lower segment.

Note: A parallelism tolerance would not yield the same results as the composite position tolerance. Parallelism is an orientation tolerance for individual features. It will not control the position between the holes as it does in the lower segment of the composite position tolerance. A parallelism specification will require each hole to parallel but will not control the location between the holes. If there were only one hole, as shown in the example in the parallelism section of this text, a parallelism control would be used.

Scenario #4 - In the accompanying figure, the composite position tolerancing is replaced with two single segmented position controls. The upper segment has the same interpretation as the composite control and locates and orients the holes to the specified datum reference frame. The lower segment or .004 tolerance zone also controls location and orientation of the features to the specified DRF. Since datum A is referenced in the lower segment, both the location and parallelism of this zone is controlled relative to datum A. The .004 tolerance zone is located and oriented to datum A but, otherwise, may slide back and forth within the confines of the .010 tolerance zone. The axes of the holes must lie in both zones simultaneously. The basic dimensions to the datums are not unlocked for the lower segment.

THREE SINGLE SEGMENT - ROTATION CONTROL

Position tolerancing is very flexible as it allows the designer to add and release datums to state very exacting requirements while still allowing maximum manufacturing tolerance. The example below illustrates a pattern of holes with three specifications. It is sometimes easier to read the callouts from the bottom up. The bottom specification requires the three holes to be located to each other and the face within .003 dia. The middle specification requires the holes to be located to the pilot and the face within .010 dia. The top specification requires the holes to be located to the face, pilot and slot within .040 dia.

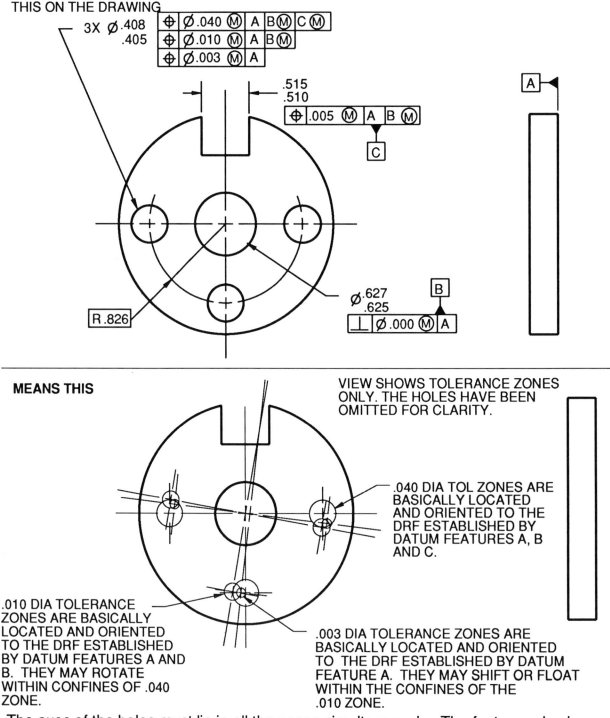

THIS ON THE DRAWING

3X ⌀ .408
       .405

⊕	⌀.040 Ⓜ	A	BⓂ	CⓂ
⊕	⌀.010 Ⓜ	A	BⓂ	
⊕	⌀.003 Ⓜ	A		

.515
.510

| ⊕ | .005 Ⓜ | A | B Ⓜ |

[C]

[A]

R .826

⌀ .627
  .625

[B]

| ⊥ | ⌀ .000 Ⓜ | A |

MEANS THIS

VIEW SHOWS TOLERANCE ZONES ONLY. THE HOLES HAVE BEEN OMITTED FOR CLARITY.

.040 DIA TOL ZONES ARE BASICALLY LOCATED AND ORIENTED TO THE DRF ESTABLISHED BY DATUM FEATURES A, B AND C.

.010 DIA TOLERANCE ZONES ARE BASICALLY LOCATED AND ORIENTED TO THE DRF ESTABLISHED BY DATUM FEATURES A AND B. THEY MAY ROTATE WITHIN CONFINES OF .040 ZONE.

.003 DIA TOLERANCE ZONES ARE BASICALLY LOCATED AND ORIENTED TO THE DRF ESTABLISHED BY DATUM FEATURE A. THEY MAY SHIFT OR FLOAT WITHIN THE CONFINES OF THE .010 ZONE.

The axes of the holes must lie in all the zones simultaneously. The features also have additional tolerance as they and referenced datum features depart from MMC (virtual).

13.24

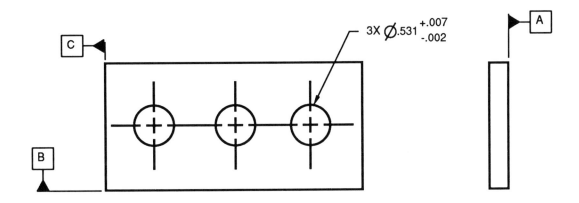

1. On the part above, the three holes can move .020 in the fore and aft direction but only .006 in the up and down direction. Specify a position control to allow this requirement in relation to the DRF established by datum features A, B and C.

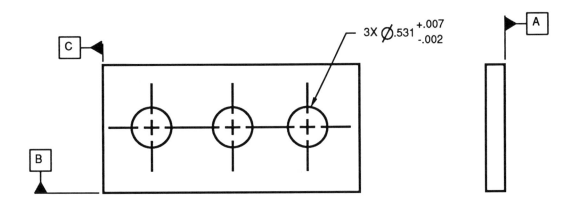

2. On the part above, show the size, shape and direction of the tolerance zones that were defined in the previous part.

3. What is the perpendicularity in relation to datum A for the three holes above? Explain.

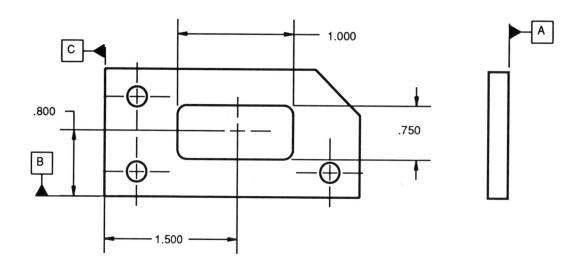

4. On the rectangular hole on the part above, apply a profile of .005 all around in relation to datum A. In addition, position the rectangular hole within a .020 boundary at MMC in relation to datums A, B and C. Assume all dimensions are basic.

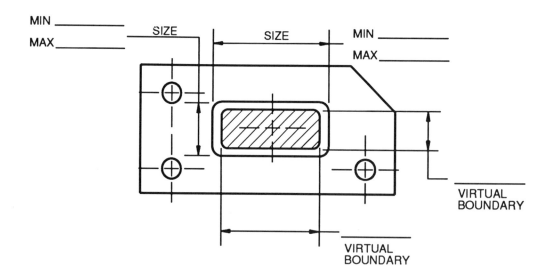

5. On the produced part above, calculate the size of the rectangular hole as well as the size of the virtual position boundary.

6. In the figure below, there is a composite position and a two single segment position specification. The upper entry of the two specifications are identical. There is, however, a difference between the lower segments of the two specifications. In the space provided next to the feature control frames, explain the difference.

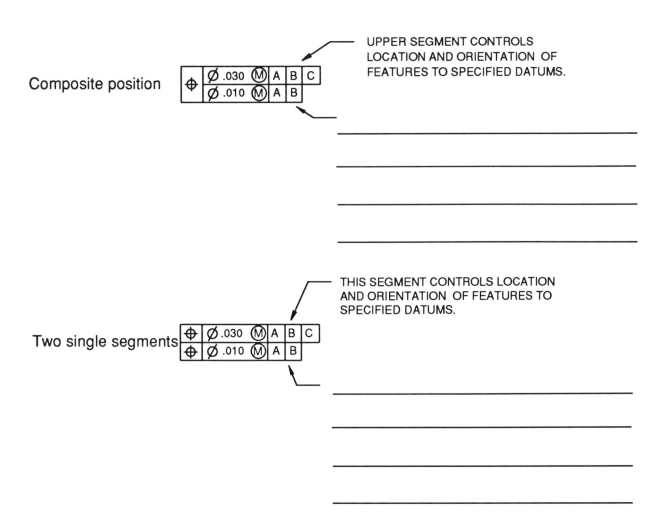

Composite position

UPPER SEGMENT CONTROLS LOCATION AND ORIENTATION OF FEATURES TO SPECIFIED DATUMS.

_____

_____

_____

_____

THIS SEGMENT CONTROLS LOCATION AND ORIENTATION OF FEATURES TO SPECIFIED DATUMS.

Two single segments

_____

_____

_____

_____

7. Turn back the pages to the geometric matrix chart on the last page in unit 10. Complete the chart, filling in the required information on position tolerancing.

8. The three functional gages below can be used to verify the three segment position specification that is shown earlier in this unit. In each case, calculate the size of the functional gage pins or 3D solids. See product drawing earlier in this unit.

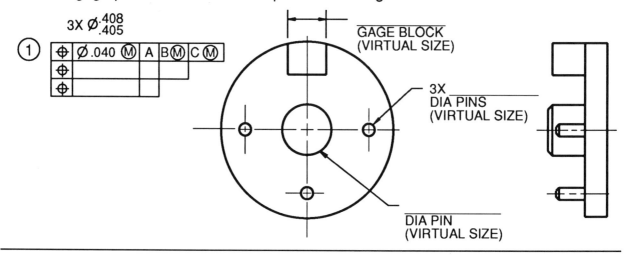

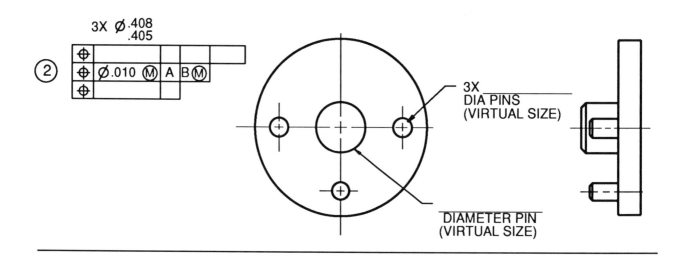

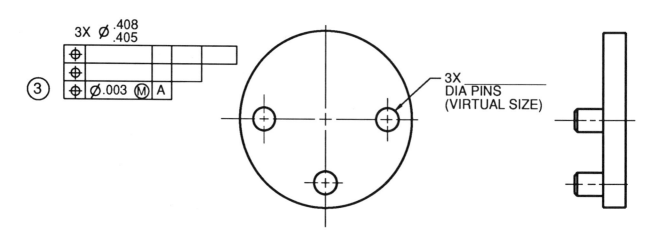

The top drawing is a plate with composite position tolerancing applied. The lower drawing is the produced part. Evaluate the dimensions on the produced part to verify conformance to the position tolerances. Remember there are two specifications. The upper segment and the lower segment. The upper segment can be calculated by using the conversion chart. The lower segment must be completed using the paper gage. Try completing the entire exercise using the paper gage. The answers may vary slightly as the numbers will be estimates. The final outcome, however, will be identical. Use the chart below to record your calculations.

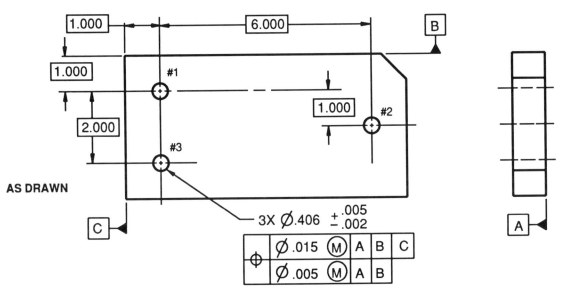

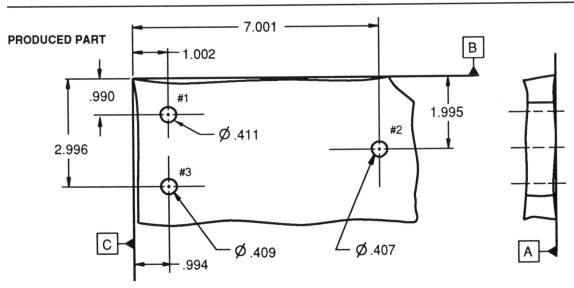

HOLE NO.	HOLE MMC	HOLE ACTUAL SIZE	PATTERN TOLERANCE ALLOWED	PATTERN TOLERANCE ACTUAL	ACC	REJ	HOLE TO HOLE TOLERANCE ALLOWED	HOLE TO HOLE TOLERANCE ACTUAL	ACC	REJ
1										
2										
3										

The top drawing is a plate with composite position tolerancing applied. The lower drawing is the produced part. Evaluate the dimensions on the produced part to verify conformance to the position tolerances. Remember there are two specifications, the upper segment and the lower segment. The upper segment can be calculated by using the conversion chart. The lower segment must be completed using the paper gage. Try completing the entire exercise using the paper gage. The answers may vary slightly as the numbers will be estimates. The final outcome, however, will be identical. Use the chart below to record your calculations.

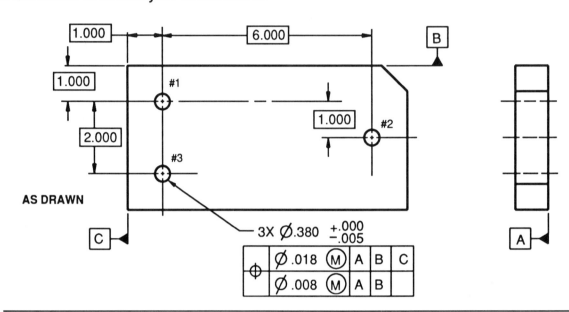

AS DRAWN

3X Ø.380 +.000 -.005

⌖	Ø .018 Ⓜ	A	B	C
	Ø .008 Ⓜ	A	B	

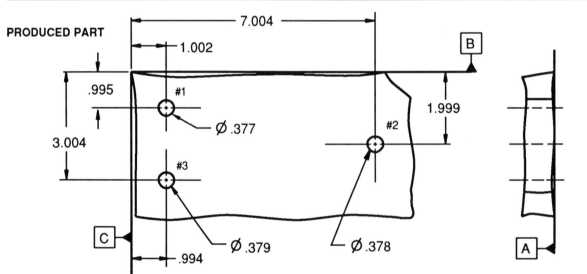

PRODUCED PART

UPPER SEGMENT CALCULATIONS ⟋   LOWER SEGMENT CALCULATIONS ⟋

HOLE NO.	HOLE MMC	HOLE ACTUAL SIZE	PATTERN TOLERANCE ALLOWED	PATTERN TOLERANCE ACTUAL	ACC	REJ	HOLE TO HOLE TOLERANCE ALLOWED	HOLE TO HOLE TOLERANCE ACTUAL	ACC	REJ
1										
2										
3										

UNIT 14

COAXIAL AND NON-CYLINDRICAL CONTROLS

RUNOUT, CONCENTRICITY, POSITION, PROFILE AND SYMMETRY

SHAFT - COAXIAL DATUMS

The shaft below mounts in the assembly on the two journals. Both journals share the same importance. The following geometric call out is a very common procedure for establishing a single datum axis from two datum features.

THIS APPLICATION

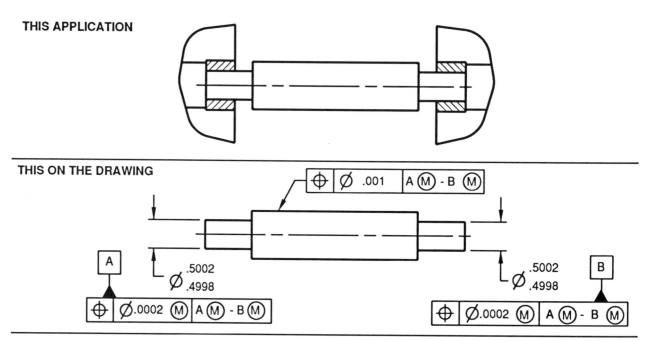

THIS ON THE DRAWING

MEANS THIS   First establish the simulated datum axes A and B.

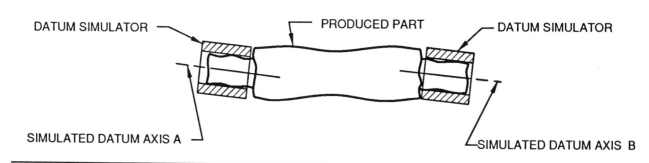

DATUM SIMULATOR — PRODUCED PART — DATUM SIMULATOR

SIMULATED DATUM AXIS A   SIMULATED DATUM AXIS B

Then, collapse the smallest cylinder around the datum axes A and B. The axis of that cylinder is the simulated A - B axis. The A - B axis is the average of the two datum axes.

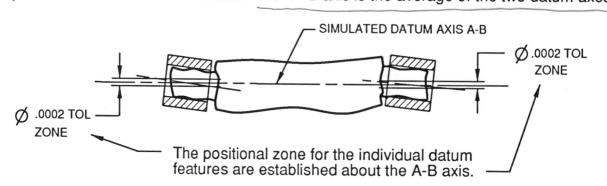

SIMULATED DATUM AXIS A-B

Ø .0002 TOL ZONE

Ø .0002 TOL ZONE

The positional zone for the individual datum features are established about the A-B axis.

VIRTUAL SIZE BOUNDARIES FOR COAXIAL DIAMETERS

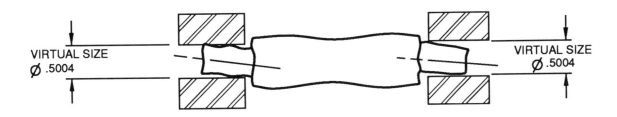

VIRTUAL SIZE
Ø .5004

VIRTUAL SIZE
Ø .5004

When designing the virtual size opening for the mating part, make sure consideration is given to the virtual sizes on the shaft. The above datum features are referenced at MMC. This allows the datum features to tilt or twist within the virtual size boundary as they depart from MMC. If bearings are pressed on the journals, the datums are usually referenced at RFS.

If the size of the journals, and the tilt/twist of the journals are all combined in one value, consider the use of zero position tolerancing.

The above part may be set-up and verified with two V blocks or on a CMM. If a CMM is used, a cylinder alignment is set on both journals simultaneously, this will establish the A - B axis. The individual journals are verified for position conformance to the to the A - B axis.

COAXIAL CONTROL USING A SINGLE DATUM

**THIS ON THE
DRAWING**

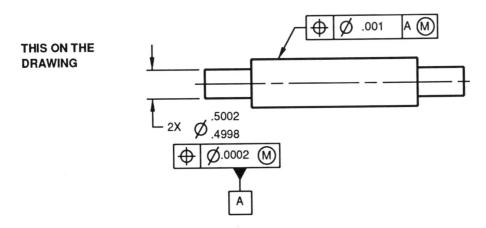

The drawing above is another way to control the relationship between two coaxial diameters. The end result of the above geometric control is identical to the result of the A - B call out. Either specification may be used. If the diameters are of different size, the A - B specification may be more appropriate. See the seat latch part in the datum section for another similar example.

COAXIAL FEATURES - WHICH CHARACTERISTIC TO USE

There are four main geometric characteristics that can be used to control a coaxial relationship between features: position, runout, concentricity and profile tolerances. Each characteristic has a different meaning and will yield different results when applied to control coaxiality relationships. The goal for applying geometric tolerancing is obtaining maximum manufacturing tolerance while still maintaining functional requirements. The geometric callout that is used depends on the application.

COMPARISON OF COAXIAL CONTROLS

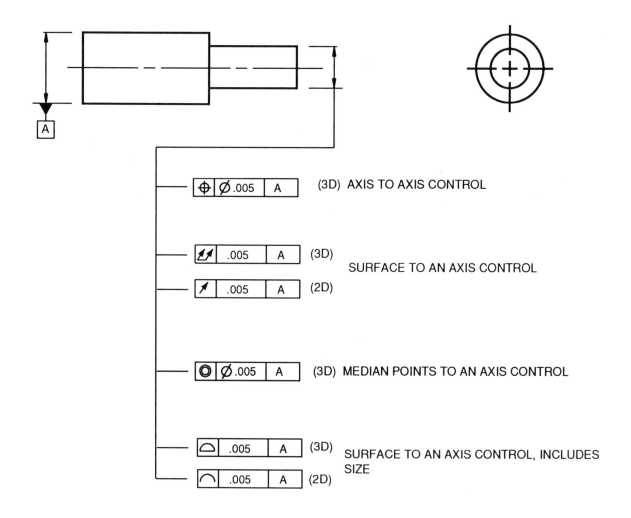

The part above illustrates the application and comparison between the four possible coaxial callouts. In order to offer some guidance as to when to apply a characteristic, this same part is shown four individual times with a position, profile, runout and concentricity tolerance applied. These examples can be found in their individual section in this text. It is suggested the reader study and compare the differences between these coaxial controls. In addition to the detailed interpretation of each characteristic, there is a sample practical application of each callout in each sectio⌐

RUNOUT TOLERANCES

RUNOUT TOLERANCES CONTROL THE FORM, ORIENTATION AND POSITION OF INDIVIDUAL FEATURES
DATUM AXIS REQUIRED

SYMBOL	TYPE OF TOLERANCE	SHAPE OF TOLERANCE ZONE	2D OR 3D	APP OF FEATURE MODIFIER
 ⟋	CIRCULAR RUNOUT	2 CONCENTRIC CIRCLES ABOUT A DATUM AXIS 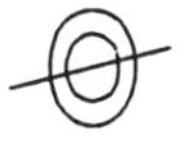 2 CONCENTRIC CIRCULAR LINE ELEMENTS ABOUT DATUM AXIS 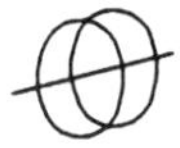	2D	NO
 ⟋⟋	TOTAL RUNOUT	2 CONCENTRIC CYLINDERS ABOUT A DATUM AXIS 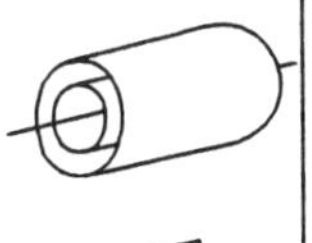 2 PARALLEL PLANES ABOUT A DATUM AXIS 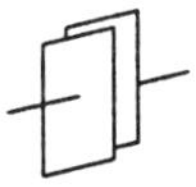	3D	NO

OVERVIEW:

CIRCULAR RUNOUT and TOTAL RUNOUT are both runout tolerances. These tolerances will control the form orientation and location of features. The common shapes of the tolerance zones are shown in the chart above.

Runout tolerances will control surfaces constructed around a datum axis and those constructed at right angles to a datum axis. Runout may or may not have a plane surface referenced as a datum but must always be referenced to a datum axis.

Circular and total runout are identical concepts except that total runout is a 3D control and circular runout a 2D control. On a surface of revolution, total runout will control taper of the feature; but, circular runout will not control taper.

The two runout tolerances are both surface controls. Therefore, the feature modifiers MMC, LMC and RFS are not applicable. Datum modifiers are applied RFS because of the unique nature of the runout callouts, where a feature is verified by rotation about a datum axis.

In the past, runout tolerances were often used on all rotating components. This is not necessarily true and can unnecessarily tighten manufacturing tolerances and increase cost. The other coaxial controls of position and profile of a surface should be considered as well.

Both circular and total runout specifications are axis to surface controls and are often used when a component rides on the surface, such as a wheel or idler pulley in the applications shown in this section. See also the discussion on coaxial controls later in the text.

Runout tolerances can also be applied to surfaces at right angles to a datum axis. If total runout is applied, it will control cumulative variations of perpendicularity, concavity, convexity, flatness and "wobble" of that surface. If circular runout is applied, it will control only the circular elements of the surface. It will not control perpendicularity, concavity, convexity or flatness but will control "wobble".

CIRCULAR RUNOUT

Circular runout is a two dimensional, surface to an axis control. The tolerance is applied independently at each circular cross section. When applied to a surface constructed around a datum axis, circular runout will control the cumulative variations of circularity and coaxiality. Unlike total runout, it does not control taper.

THIS ON THE DRAWING

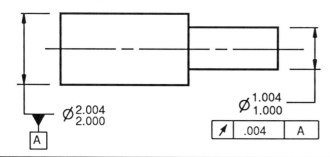

$\varnothing{}^{2.004}_{2.000}$

A

$\varnothing{}^{1.004}_{1.000}$

| ↗ | .004 | A |

MEANS THIS

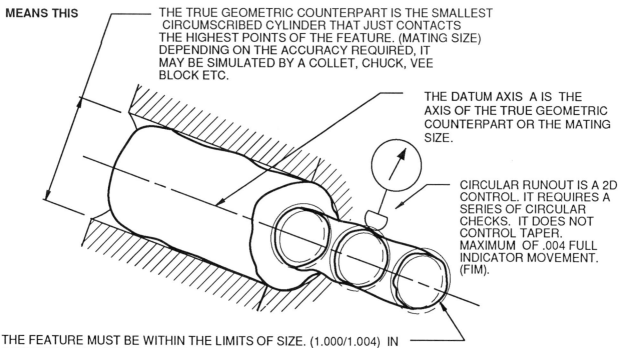

THE TRUE GEOMETRIC COUNTERPART IS THE SMALLEST CIRCUMSCRIBED CYLINDER THAT JUST CONTACTS THE HIGHEST POINTS OF THE FEATURE. (MATING SIZE) DEPENDING ON THE ACCURACY REQUIRED, IT MAY BE SIMULATED BY A COLLET, CHUCK, VEE BLOCK ETC.

THE DATUM AXIS A IS THE AXIS OF THE TRUE GEOMETRIC COUNTERPART OR THE MATING SIZE.

CIRCULAR RUNOUT IS A 2D CONTROL. IT REQUIRES A SERIES OF CIRCULAR CHECKS. IT DOES NOT CONTROL TAPER. MAXIMUM OF .004 FULL INDICATOR MOVEMENT. (FIM).

THE FEATURE MUST BE WITHIN THE LIMITS OF SIZE. (1.000/1.004) IN ADDITION, EACH CIRCULAR ELEMENT OF THE SURFACE MUST LIE BETWEEN TWO CONCENTRIC CIRCLES, ONE HAVING A RADIUS OF .004 LARGER THAN THE OTHER. THE CIRCLES ARE CONCENTRIC TO THE DATUM A AXIS.

NOTE: ON THIS PART THE OUTER BOUNDARY IS 1.008 THE INNER BOUNDARY IS .996.

The circular runout specification may be verified with a dial indicator, CMM or by other methods. If a dial indicator is used, each circular cross section of the surface must lie within the specified runout tolerance (.004 full indicator movement) when the part is rotated 360 degrees about the datum axis. The indicator is reset at every location along the surface in a position normal to the true geometric shape. The feature must also be within the limits of size.

TOTAL RUNOUT

Total runout is a three dimensional, surface to an axis control. Total runout provides a composite control of all surface elements. When applied to a surface constructed around a datum axis, total runout will control the cumulative variations of circularity, straightness, coaxiality, angularity, taper and variations in the surface.

THIS ON THE DRAWING

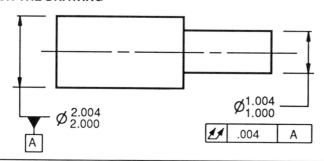

$\emptyset \begin{smallmatrix} 2.004 \\ 2.000 \end{smallmatrix}$ A

$\emptyset \begin{smallmatrix} 1.004 \\ 1.000 \end{smallmatrix}$

| ⟁⟁ | .004 | A |

MEANS THIS

THE TRUE GEOMETRIC COUNTERPART IS THE SMALLEST CIRCUMSCRIBED CYLINDER THAT JUST CONTACTS THE HIGHEST POINTS OF THE FEATURE. (MATING SIZE) DEPENDING ON THE ACCURACY REQUIRED, IT MAY BE SIMULATED BY A COLLET, CHUCK, VEE BLOCK ETC.

THE DATUM AXIS A IS THE AXIS OF THE TRUE GEOMETRIC COUNTERPART OR THE MATING SIZE.

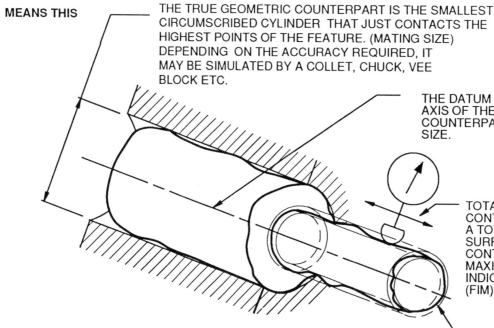

TOTAL RUNOUT IS A 3D CONTROL. IT REQUIRES A TOTAL SWEEP OF THE SURFACE AND CONTROLS TAPER. MAXIMUM OF .004 FULL INDICATOR MOVEMENT. (FIM)

THE FEATURE MUST BE WITHIN THE LIMITS OF SIZE. (1.000/1.004) IN ADDITION, EACH ELEMENT OF THE SURFACE MUST LIE BETWEEN TWO CONCENTRIC CYLINDERS, ONE HAVING A RADIUS OF .004 LARGER THAN THE OTHER. THE CYLINDERS ARE CONCENTRIC TO THE DATUM A AXIS.

NOTE: ON THIS PART THE OUTER BOUNDARY IS 1.008 THE INNER BOUNDARY IS .996.

The total runout specification may be verified with a dial indicator, CMM or by other methods. If a dial indicator is used, the entire surface must lie within the specified runout tolerance (.004 full indicator movement) when the part is rotated 360 degrees about the datum axis. The indicator is placed at every location along the surface in a position normal to the true geometric shape without a reset of the indicator. The feature must also be within the limits of size.

RUNOUT - APPLICATION TO SURFACES AT RIGHT ANGLES

Both total runout and circular runout may be applied to surfaces constructed at right angles to a datum axis as shown below. Total and circular runout are a refinement of size and location tolerances. Total runout and perpendicularity will provide identical results and can be used interchangeably. Circular runout is a 2D specification and controls circular elements. The surface may be convex or concave within the size or location tolerance.

THIS ON THE DRAWING

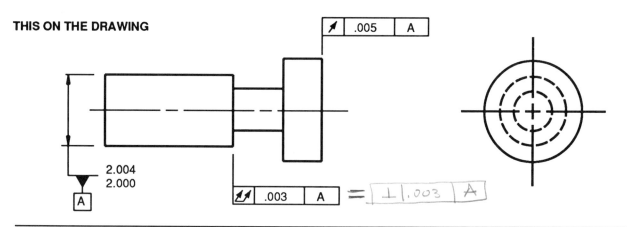

2.004
2.000

MEANS THIS

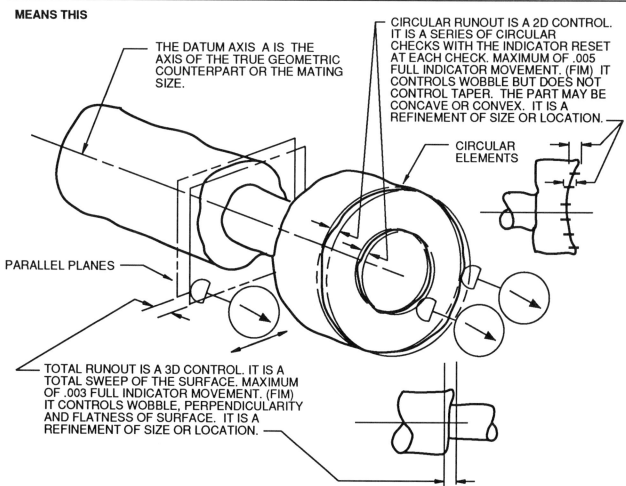

THE DATUM AXIS A IS THE AXIS OF THE TRUE GEOMETRIC COUNTERPART OR THE MATING SIZE.

CIRCULAR RUNOUT IS A 2D CONTROL. IT IS A SERIES OF CIRCULAR CHECKS WITH THE INDICATOR RESET AT EACH CHECK. MAXIMUM OF .005 FULL INDICATOR MOVEMENT. (FIM) IT CONTROLS WOBBLE BUT DOES NOT CONTROL TAPER. THE PART MAY BE CONCAVE OR CONVEX. IT IS A REFINEMENT OF SIZE OR LOCATION.

CIRCULAR ELEMENTS

PARALLEL PLANES

TOTAL RUNOUT IS A 3D CONTROL. IT IS A TOTAL SWEEP OF THE SURFACE. MAXIMUM OF .003 FULL INDICATOR MOVEMENT. (FIM) IT CONTROLS WOBBLE, PERPENDICULARITY AND FLATNESS OF SURFACE. IT IS A REFINEMENT OF SIZE OR LOCATION.

The total and circular runout specifications may be verified with a dial indicator, CMM or by other methods. If a dial indicator is used the specified surface and circular elements must lie within the specified runout tolerance when the part is rotated 360 degrees around the datum axis. In addition, the surfaces must be within the limits of size or location.

TOTAL RUNOUT APPLICATION - IDLER WHEEL

THIS APPLICATION

Idler wheel places tension on the belt and can be adjusted up and down to compensate for size tolerance.

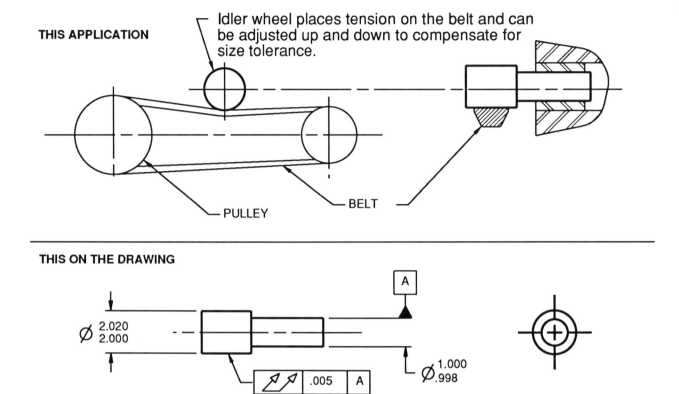

PULLEY

BELT

THIS ON THE DRAWING

$\emptyset$ 2.020 / 2.000

A

.005 | A

$\emptyset$ 1.000 / .998

PRODUCED PART IN ASSEMBLY

RUNOUT IS A SURFACE TO AN AXIS CONTROL

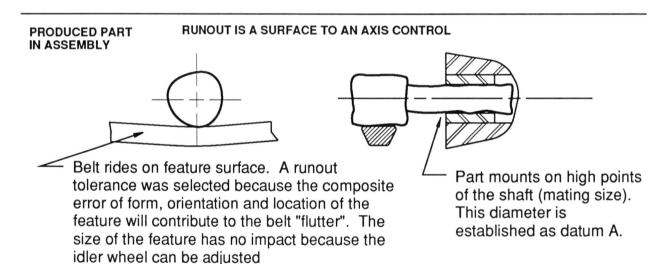

Belt rides on feature surface. A runout tolerance was selected because the composite error of form, orientation and location of the feature will contribute to the belt "flutter". The size of the feature has no impact because the idler wheel can be adjusted

Part mounts on high points of the shaft (mating size). This diameter is established as datum A.

A position tolerance was not selected because the high points of the smallest circumscribing cylinder do not ride on the belt. The ovality of the OD will contribute to the error. Position does not control ovality or form.

A profile tolerance was not selected because the OD of the idler wheel can be adjusted up and down to compensate for the size tolerance. If the idler pulley location was not adjustable, a profile tolerance would be used.

POSITION - COAXIAL FEATURES

Position for coaxial features is a three dimensional, axis to axis control. It defines a cylindrical zone within which the axis of the mating size of the feature must lie. When applied to a feature constructed around a datum axis, position controls only orientation and location. It has no effect on size, form or variations in the surface. If required, the MMC, LMC or RFS modifier may be applied to the feature as well as the datum feature.

THIS ON THE DRAWING

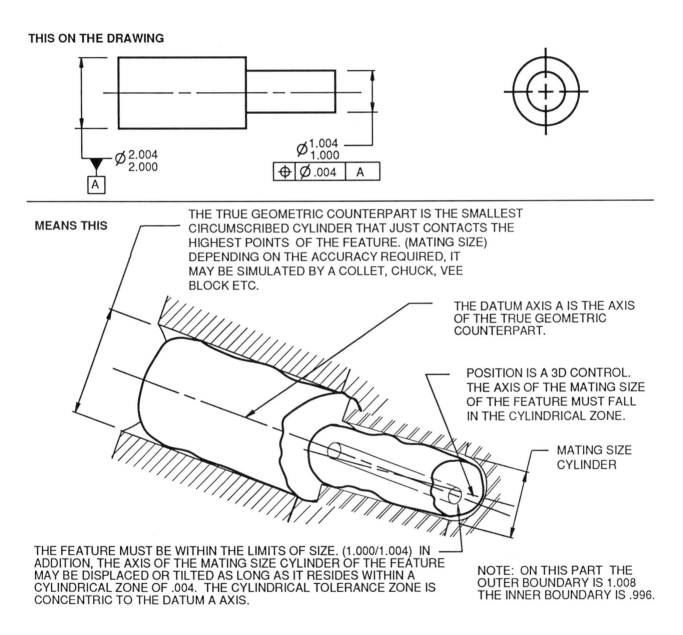

ϕ 2.004 / 2.000

A

ϕ 1.004 / 1.000

| ⌖ | ϕ .004 | A |

MEANS THIS

THE TRUE GEOMETRIC COUNTERPART IS THE SMALLEST CIRCUMSCRIBED CYLINDER THAT JUST CONTACTS THE HIGHEST POINTS OF THE FEATURE. (MATING SIZE) DEPENDING ON THE ACCURACY REQUIRED, IT MAY BE SIMULATED BY A COLLET, CHUCK, VEE BLOCK ETC.

THE DATUM AXIS A IS THE AXIS OF THE TRUE GEOMETRIC COUNTERPART.

POSITION IS A 3D CONTROL. THE AXIS OF THE MATING SIZE OF THE FEATURE MUST FALL IN THE CYLINDRICAL ZONE.

MATING SIZE CYLINDER

THE FEATURE MUST BE WITHIN THE LIMITS OF SIZE. (1.000/1.004) IN ADDITION, THE AXIS OF THE MATING SIZE CYLINDER OF THE FEATURE MAY BE DISPLACED OR TILTED AS LONG AS IT RESIDES WITHIN A CYLINDRICAL ZONE OF .004. THE CYLINDRICAL TOLERANCE ZONE IS CONCENTRIC TO THE DATUM A AXIS.

NOTE: ON THIS PART THE OUTER BOUNDARY IS 1.008 THE INNER BOUNDARY IS .996.

The coaxial position specification may be verified with a dial indicator, CMM or by other methods. If a dial indicator is used, and the full indicator reading does not exceed .004, the part is good. If the full indicator reading exceeds .004, there is a possibility that the part is still good. The reason for this is, as an indicator rides on the surface, it inadvertentaly reports a composite error of position, surface variations and form errors (ovality). Since position is only an axis to axis control, surface variations are not included in the requirement. A "mapping" and further evaluation of the surface might be required to accept features beyond the .004 full indicator movement.

POSITION - COAXIAL GEAR APPLICATION

THIS APPLICATION

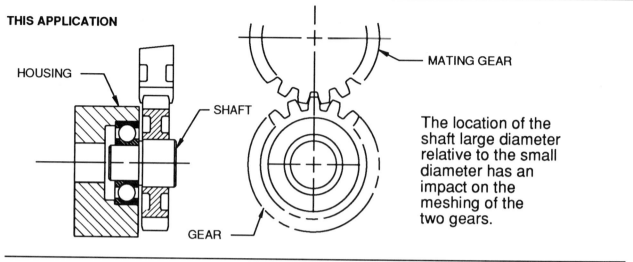

HOUSING

SHAFT

MATING GEAR

GEAR

The location of the shaft large diameter relative to the small diameter has an impact on the meshing of the two gears.

THIS ON THE DRAWING

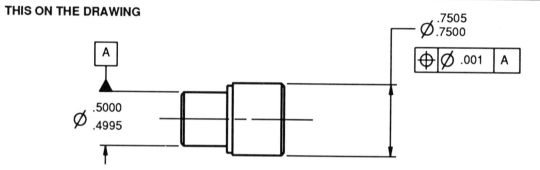

A

$\varnothing$ .7505 / .7500

| ⊕ | $\varnothing$ | .001 | A |

$\varnothing$ .5000 / .4995

PRODUCED PART IN ASSEMBLY

POSITION IS AN AXIS TO AXIS CONTROL

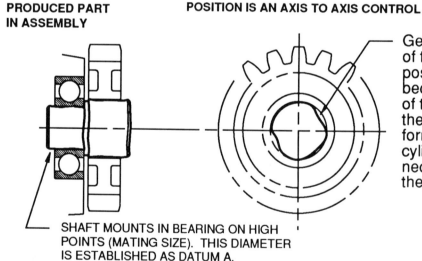

Gear mounts on the high points of the feature (mating size). A position tolerance was selected because the form error (ovality) of the OD does not contribute to the position error. A separate form tolerance of circularity or cylindricity can be applied if necessary to control the fit of the gear to the shaft.

SHAFT MOUNTS IN BEARING ON HIGH POINTS (MATING SIZE). THIS DIAMETER IS ESTABLISHED AS DATUM A.

A runout tolerance was not selected because the form error on the OD of the shaft has nothing to do with the location of the gear center. Runout would include form error in the reading. Position, just like runout, can be verified with an indicator. The form error is just calculated out of the reading. Position will provide maximum manufacturing tolerance while still preserving the functional requirements of the part.

A profile tolerance was not selected because the size of the feature is important for the fit of the gear, and another tolerance is required for the location of the gear. Profile tolerance combines all of these factors in to one tolerance value.

14.12

PROFILE - COAXIAL FEATURES

Profile of a surface for coaxial features is a three dimensional surface to an axis control. It defines a zone of tolerance (.006 wide on each side) that lies between two concentric cylinders that are equally disposed about a basic diameter. When applied to a surface constructed around a datum axis, profile of a surface will control the cumulative variations of size, circularity, straightness, coaxiality, angularity, taper and variations in the surface.

THIS ON THE DRAWING

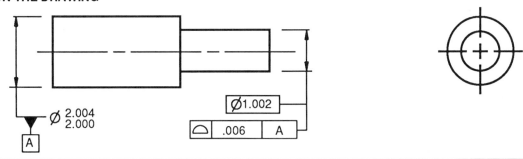

MEANS THIS

THE TRUE GEOMETRIC COUNTERPART IS THE SMALLEST CIRCUMSCRIBED CYLINDER THAT JUST CONTACTS THE HIGHEST POINTS OF THE FEATURE. (MATING SIZE) DEPENDING ON THE ACCURACY REQUIRED, IT MAY BE SIMULATED BY A COLLET, CHUCK, VEE BLOCK, CMM, ETC.

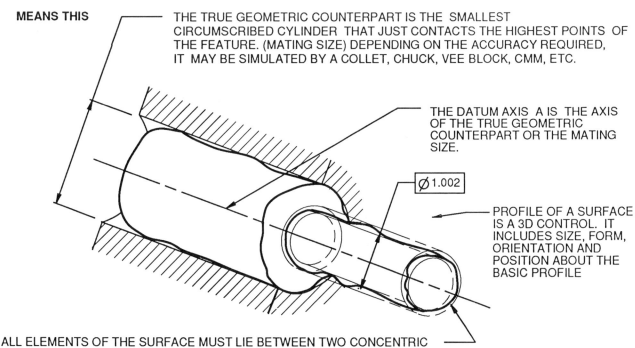

THE DATUM AXIS A IS THE AXIS OF THE TRUE GEOMETRIC COUNTERPART OR THE MATING SIZE.

PROFILE OF A SURFACE IS A 3D CONTROL. IT INCLUDES SIZE, FORM, ORIENTATION AND POSITION ABOUT THE BASIC PROFILE

ALL ELEMENTS OF THE SURFACE MUST LIE BETWEEN TWO CONCENTRIC CYLINDERS. THE TWO CYLINDERS ARE EQUALLY DISPOSED ABOUT A BASIC CYLINDER OF 1.002 DIAMETER. THE OUTER TOLERANCE CYLINDER HAS A DIAMETER OF 1.008, AND THE INNER TOLERANCE CYLINDER HAS A DIAMETER OF .996. THE TOLERANCE CYLINDERS ARE CONCENTRIC TO THE DATUM A AXIS. SINCE THE SURFACE MUST FALL WITHIN THE TOLERANCE ZONE, IT INCLUDES SIZE, FORM, ORIENTATION AND POSITION OF THE FEATURE.

NOTE: ON THIS PART THE OUTER BOUNDARY IS 1.008 THE INNER BOUNDARY IS .996.

The profile of a surface specification may be verified with a dial indicator, CMM or other methods. If a dial indicator is used, the indicator is "mastered" or set at the basic diameter of 1.002. The entire surface, including the size of the feature, must lie within the specified profile tolerance (.006 full indicator movement) when the part is rotated 360 degrees about the datum axis. The indicator is placed at every location along the surface in a position normal to the true geometric shape without a reset of the indicator. The size of the feature is contained within the profile tolerance.

PROFILE - COAXIAL AIR GAP OR CLEARANCE APPLICATION

THIS APPLICATION

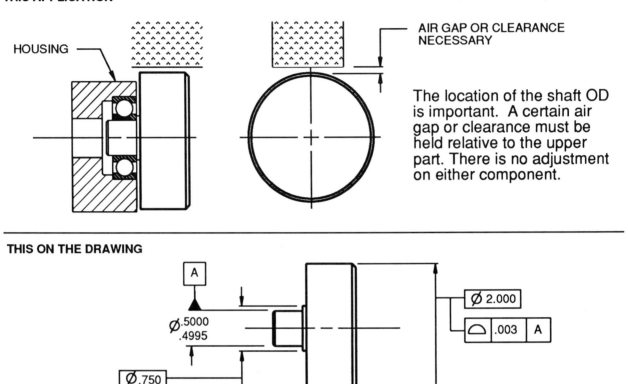

The location of the shaft OD is important. A certain air gap or clearance must be held relative to the upper part. There is no adjustment on either component.

THIS ON THE DRAWING

PRODUCED PART IN ASSEMBLY

PROFILE IS AN AXIS TO SURFACE CONTROL. IT CONTROLS SIZE, FORM, ORIENTATION AND LOCATION.

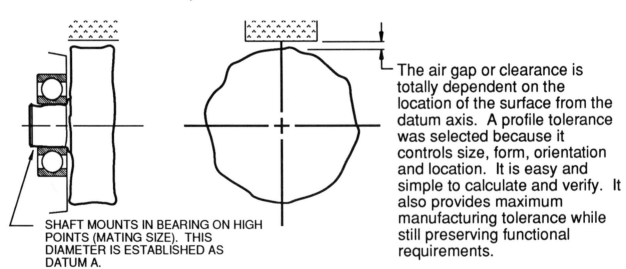

SHAFT MOUNTS IN BEARING ON HIGH POINTS (MATING SIZE). THIS DIAMETER IS ESTABLISHED AS DATUM A.

The air gap or clearance is totally dependent on the location of the surface from the datum axis. A profile tolerance was selected because it controls size, form, orientation and location. It is easy and simple to calculate and verify. It also provides maximum manufacturing tolerance while still preserving functional requirements.

A runout tolerance was not selected because a runout control does not include size. Neither of the two components are adjustable. The size tolerance will also contribute to the air gap or clearance.

A position tolerance was not selected because a position contol does not include size. A separate size and a separate location would have to be selected. The size and the location both contribute to the air gap or clearance.

14.14

CONCENTRICITY AND SYMMETRY TOLERANCES

CONCENTRICITY AND SYMMETRY TOLERANCES CONTROL THE LOCATION OF OPPOSING MEDIAN POINTS OF A FEATURE DATUM AXIS OR MEDIAN PLANE REQUIRED

SYMBOL	TYPE OF TOLERANCE	SHAPE OF TOLERANCE ZONE	2D OR 3D	APP OF FEATURE MODIFIER
◎	CONCENTRICITY	CYLINDRICAL	3D	NO
═	SYMMETRY	2 PARALLEL PLANES	3D	NO

OVERVIEW:

CONCENTRICITY and SYMMETRY are 3D tolerances. The common shape of the tolerance zones are shown in the chart above. The tolerance defines a zone within which the opposing median points of a feature must lie. The concept of concentricity and symmetry are identical except that concentricity is used for controlling opposing median points on cylindrical features, and symmetry is used for controlling opposing median points on non-cylindrical features.

Datums are always required with both concentricity and symmetry. The datums must always be an axis or median plane and are always applied on an RFS basis. The feature tolerance is also always applied on an RFS basis.

Concentricity and symmetry are often misapplied and confusing for the entry level user of geometric tolerancing. The words concentricity and symmetry are very generic sounding terms, but in reality they are very complex requirements. The dilemma is that, often, untrained personnel will apply concentricity and symmetry but really intend to apply position, runout or profile tolerances. If you study the interpretations on the following pages you will find that concentricity and symmetry require a very detailed analysis of the opposing median points of a feature to ensure conformance.

In an assembly, parts will often mount on the high points of the mating feature dictating a position callout. In other cases, they may ride on or clear a surface dictating a runout or profile callout. (See coaxial controls in this text for more information.)

In the prior ANSI Y14.5M, 1982 standard, concentricity was associated with dynamic balance and high speed rotating parts. The concentricity definition has been redefined somewhat in the ASME Y14.5M, 1994 standard, and the association to dynamic balance has been deleted. The application of concentricity to workpieces is very rare. If a coaxial relationship is required between features the user should consider position, runout or profile tolerances.

The symmetry characteristic and definition is new to the ASME Y14.5M, 1994 standard. It was added to provide the same control for non-cylindrical features as concentricity does for cylindrical features. The application of symmetry to workpieces is very rare. If a symmetrical relationship is required between features, the user should consider position or profile tolerances.

Caution:

Although the symbols for concentricity and symmetry in the ASME Y14.5M, 1994 standard are identical to the symbols for these terms in the ISO standard, the interpretation of the specification is not the same.

At present the ISO standard does not recognize the unique interpretation of concentricity and symmetry as defined in the ASME Y14.5M, 1994 standard. The concentricity and symmetry characteristics in ISO have the same interpretations as the position characteristic in the ASME Y14.5M, 1994 standard.

CONCENTRICITY

Concentricity is a three dimensional control. It controls opposed median points to an axis. Concentricity will control location and can have some effect on the form and orientation of a feature. It will not control the form of perfectly oval parts but may have an impact on irregular or "D" shaped features. The application of concentricity is complex and rare.

THIS ON THE DRAWING

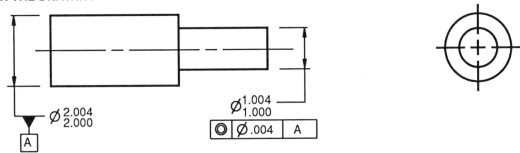

$\phi\,{}^{2.004}_{2.000}$ A

$\phi\,{}^{1.004}_{1.000}$

| ◎ | ϕ .004 | A |

MEANS THIS

THE TRUE GEOMETRIC COUNTERPART IS THE SMALLEST CIRCUMSCRIBED CYLINDER THAT JUST CONTACTS THE HIGHEST POINTS OF THE FEATURE. (MATING SIZE) IT MAY BE SIMULATED BY A COLLET, CHUCK, VEE BLOCK ETC.

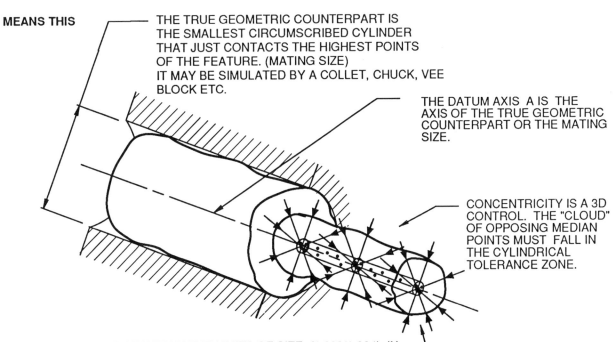

THE DATUM AXIS A IS THE AXIS OF THE TRUE GEOMETRIC COUNTERPART OR THE MATING SIZE.

CONCENTRICITY IS A 3D CONTROL. THE "CLOUD" OF OPPOSING MEDIAN POINTS MUST FALL IN THE CYLINDRICAL TOLERANCE ZONE.

THE FEATURE MUST BE WITHIN THE LIMITS OF SIZE. (1.000/1.004) IN ADDITION, ALL MEDIAN POINTS OF DIAMETRICALLY OPPOSED ELEMENTS OF THE FEATURE (WILL RESULT IN A CLOUD OF POINTS) MUST LIE WITHIN A .004 CYLINDRICAL TOLERANCE ZONE . THE TOLERANCE ZONE IS CONCENTRIC TO DATUM AXIS A.

NOTE: ON THIS PART THE OUTER BOUNDARY IS 1.008 THE INNER BOUNDARY IS .996.

The concentricity specification may be verified with dial indicators, a CMM or by other methods. If dial indicators are used, two diametrically opposed, mastered indicators are placed on either side of the feature and positioned and rotated about the datum axis.

14.17

CONCENTRICITY

If both indicators simultaneously read plus an equal amount or minus an equal amount, the feature's opposed median point at that specific location is perfectly concentric. If one indicator reads plus .001 and the other indicator reads minus .001, the median point of the feature at that specific location is off perfect concentricity by .002 dia. To conform, all opposing median points must fall within a .004 dia zone. The indicators are placed at every location along the surface in a position normal to the true geometric shape without a reset of the indicators. The feature must also be within the limits of size.

CONCENTRICITY VS. RUNOUT VERIFICATION AT ONE CROSS SECTION ON A FEATURE PRODUCED WITHIN SIZE LIMITS AND PERFECTLY OVAL.

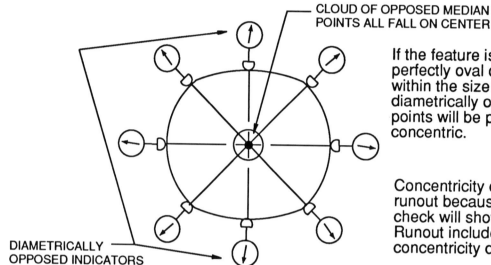

CLOUD OF OPPOSED MEDIAN POINTS ALL FALL ON CENTER

DIAMETRICALLY OPPOSED INDICATORS

If the feature is produced perfectly oval or "symmetric" within the size limits, the diametrically opposed median points will be perfectly concentric.

Concentricity differs from runout because a runout check will show a variation. Runout includes ovality and concentricity does not.

CONCENTRICITY VS. POSITION VERIFICATION AT ONE CROSS SECTION ON A FEATURE PRODUCED WITHIN SIZE LIMITS BUT "D" SHAPED.

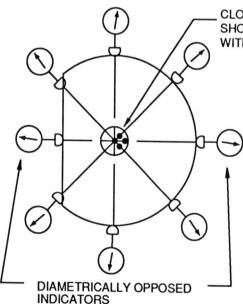

CLOUD OF OPPOSED MEDIAN POINTS ARE SHOWN DISPLACED. THEY MUST ALL FALL WITHIN A CYLINDRICAL ZONE OF .004.

If the feature is produced within size limits but happens to be "D" shaped, the diametrically opposed median points will not be concentric.

Concentricity differs from position. If this part were verified for position it would be reported as perfect. This occurs because position is specified as the axis of the mating size (collapsing cylinder) related to the datum axis. The axis of the mating cylinder is perfectly on center.

DIAMETRICALLY OPPOSED INDICATORS

Concentricity does not control dynamic balance as there is no correlation between the series of cross section checks. All the median points could be disposed to the right on one cross sectional check, and all the median points could be disposed to the left on the next cross sectional check.

14.18

SYMMETRY

Symmetry is a condition where the median points of all opposed elements of a feature are congruent with the axis or centerplane of a datum feature. Symmetry is the same concept as concentricity except that it is applied to non-cylindrical features. Symmetry differs from position in that it controls opposing points (derived median plane) where as position controls the centerplane of the actual mating envelope. The application of symmetry is rare and is commonly misused. Irregularities in the form of an actual feature may make it difficult to establish the location of a feature's median points. Therefore, unless there is a definite need for the control of a features median points, it is recommended that a control of position or profile be used.

Note: Symmetry in ISO standards is not interpreted as shown below. Symmetry in ISO standards is interpreted the same as position.

THIS ON THE DRAWING

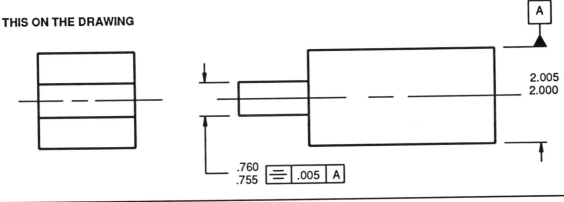

MEANS THIS

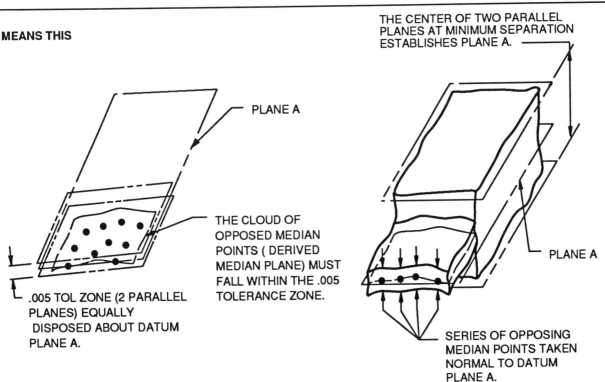

PLANE A

THE CLOUD OF OPPOSED MEDIAN POINTS (DERIVED MEDIAN PLANE) MUST FALL WITHIN THE .005 TOLERANCE ZONE.

.005 TOL ZONE (2 PARALLEL PLANES) EQUALLY DISPOSED ABOUT DATUM PLANE A.

THE CENTER OF TWO PARALLEL PLANES AT MINIMUM SEPARATION ESTABLISHES PLANE A.

PLANE A

SERIES OF OPPOSING MEDIAN POINTS TAKEN NORMAL TO DATUM PLANE A.

Within the limits of size and regardless of feature size, all median points of opposed elements of the feature must lie within two parallel planes .005 apart. The two parallel planes are equally disposed about datum plane A. The specified tolerance and the datum reference can only apply on an RFS basis.

POSITION - NON-CYLINDRICAL FEATURE

The fundamental principles of position tolerancing can be applied to noncylindrical features such as slots and tabs. A position tolerance is shown below locating the center plane of a tab. The tolerance value represents a distance between two parallel planes. This tolerance zone also defines the limits within which variation in attitude or orientation must be confined.

The tolerance on the feature and datum reference below is applied on an RFS basis. If desired the tolerance could also have been applied on a MMC and/or LMC basis as well. In this case, additional tolerance is available as the feature and datum feature depart from the specified material condition.

THIS ON THE DRAWING

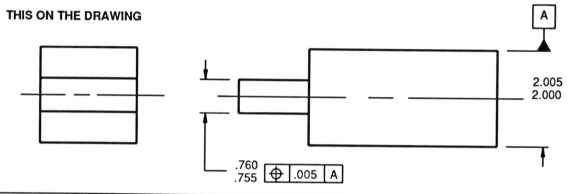

MEANS THIS

THE CENTER OF TWO PARALLEL PLANES AT
MINIMUM SEPARATION ESTABLISHES PLANE A.

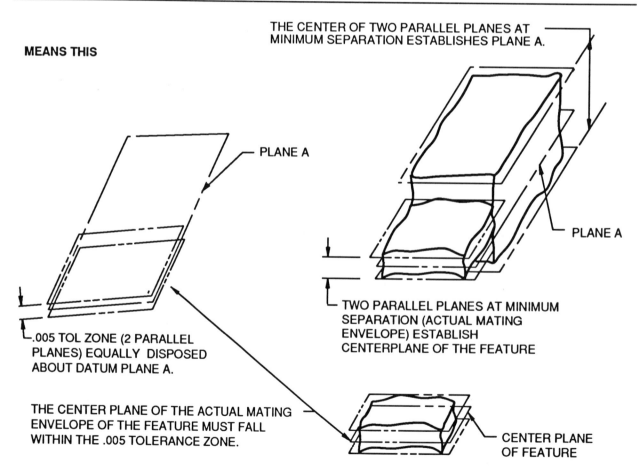

PLANE A

PLANE A

TWO PARALLEL PLANES AT MINIMUM
SEPARATION (ACTUAL MATING
ENVELOPE) ESTABLISH
CENTERPLANE OF THE FEATURE

.005 TOL ZONE (2 PARALLEL
PLANES) EQUALLY DISPOSED
ABOUT DATUM PLANE A.

THE CENTER PLANE OF THE ACTUAL MATING
ENVELOPE OF THE FEATURE MUST FALL
WITHIN THE .005 TOLERANCE ZONE.

CENTER PLANE
OF FEATURE

The center plane of the actual mating envelope of the feature must lie between two parallel planes .005 apart which are equally disposed about the center plane of datum feature A.

THIS ON THE DRAWING

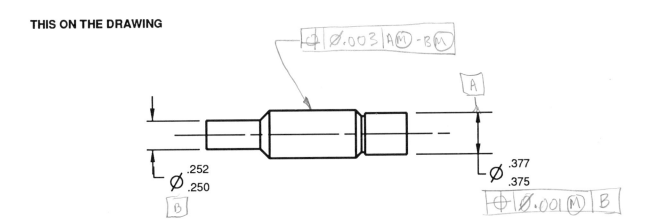

1. What coaxiality is implied between the diameters on the shaft above before any geometric controls are applied?

none

2. On the part above, create a datum axis A - B by using the right and left diameters.

3. Establish a position coaxility between the right and left diameter of .001 diameter at MMC.

4. Position the large centerdiameter within a diameter of .003 RFS in relation to the single datum axis established by the right and left diameters. For this problem, the datum reference for both diameters should be modified at MMC.

5. Calculate the virtual sizes for the right and left diameters and label them on the virtual gage below.

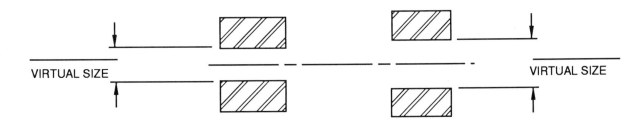

6.   The example below illustrates all the coaxial controls. Next to each feature control frame, identify the best description that identifies the control.   Select from the descriptions below. In addition, state whether the control is a 2D or 3D specification.

MEDIAN POINTS TO AN AXIS CONTROL   AXIS TO AXIS CONTROL

SURFACE TO AN AXIS CONTROL, INCLUDES
SIZE   SURFACE TO AN AXIS CONTROL

COMPARISON OF COAXIAL CONTROLS

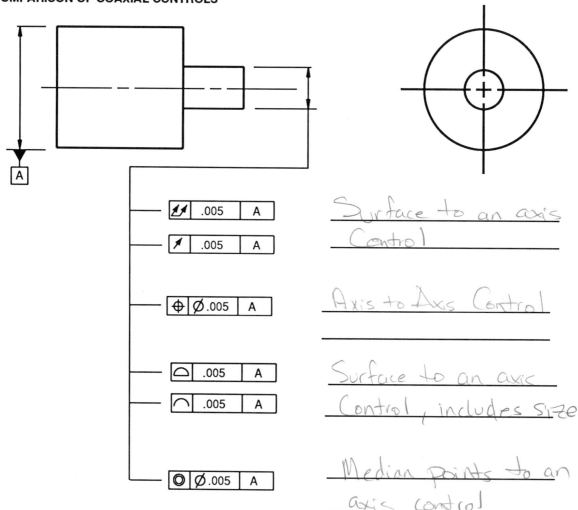

⟋⟋	.005	A

Surface to an axis
Control

⟋	.005	A

⊕	Ø.005	A

Axis to Axis Control

⌓	.005	A

Surface to an axis
Control, includes size

⌒	.005	A

◎	Ø.005	A

Median points to an
axis control

7.   Only one of the controls above are allowed to have the material condition modifiers MMC or LMC applied to the feature tolerance. Which one is it?

8.   Only two of the controls above are allowed to have the material condition modifiers MMC or LMC applied to the datum reference. Which two are they?

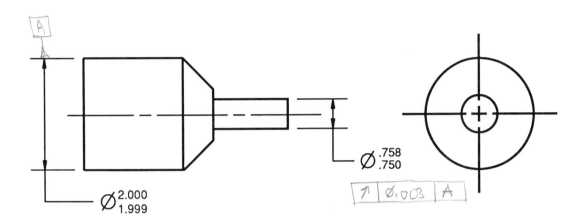

9. On the part above, what coaxiality is implied between the two diameters if no geometric specification is applied?

10. On the part above, identify the large diameter as datum feature A. Specify a .003 circular runout specification to the small diameter in relation to datum A.

11. With the circular runout specification applied, how much taper on a diameter is allowed on the small diameter?

12. If the circular runout specification were replaced with a .003 total runout specification, how much taper is allowed on the small diameter?

13. With the circular runout specification applied, how much circularity, on a radius, is allowed on the small diameter?

14. With the circular runout specification applied, How much position, on a diameter, is allowed on the small diameter relative to datum A?

15. On the part above, specify a runout requirement to make the left face perpendicular to to datum axis within .005 total.

16. On the part above, specify a runout requirement to make the right face circular elements perpendicular within .001.

17. What is the technical difference between the circular runout specification applied to the left face and the total runout specification applied to the right face?

WORKSHOP EXERCISE 14.1

THIS APPLICATION

THE WHEEL MOUNTS ON THE SHAFT AND CAN BE ADJUSTED UP AND DOWN TO COMPENSATE FOR THE VARIATION IN SIZE. THE "RIDE" OR MOVEMENT OF THE SHAFT AXIS MUST BE CONTROLLED TO MOVE NO MORE THAN .005 TOTAL.

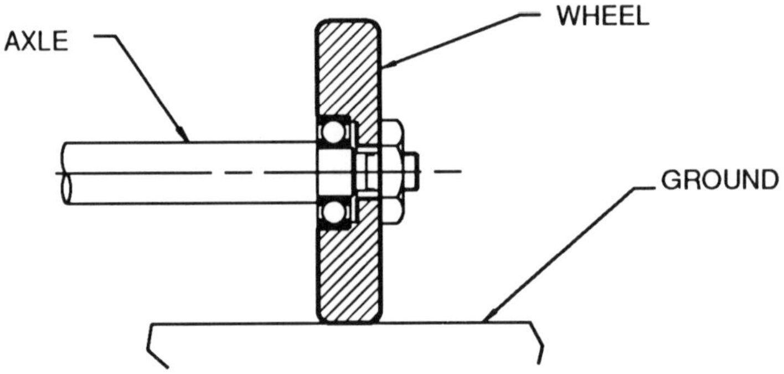

AXLE
WHEEL
GROUND

THIS ON THE DRAWING

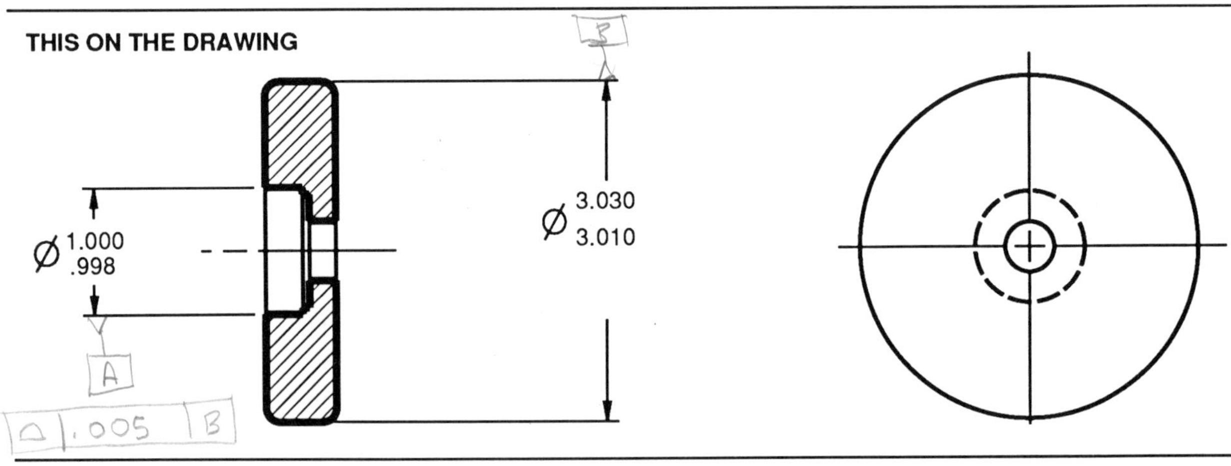

Ø 1.000 .998
A
.005
B
Ø 3.030 3.010

18. Read and study the functional requirements of the application above. Afterwards, on the product drawing, select the datum feature and apply the proper coaxial control to meet the functional requirements. Explain your decision below. Make any necessary sketches to prove your point.

THIS APPLICATION

THE WHEEL MOUNTS ON THE SHAFT AND
CAN NOT BE ADJUSTED UP AND DOWN.
THE AIR GAP OR CLEARANCE TO THE
MATING PART MUST BE FROM .010 TO
.020.

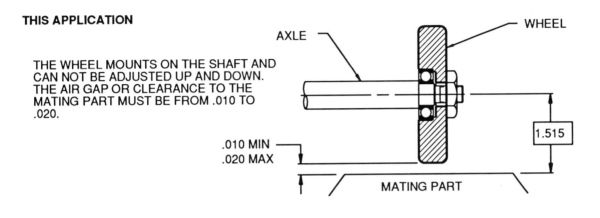

THIS ON THE DRAWING

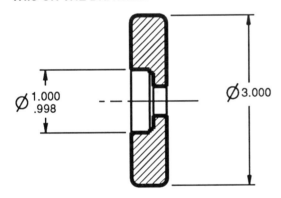

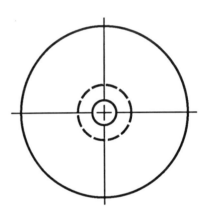

19. Read and study the functional requirements of the application above. Afterwards,
on the product drawing, select the datum feature and apply the proper coaxial control
and tolerance to meet the functional requirements.

20.   Consider another example. Suppose there was a mating fit requirement on the
3.000 OD above and then a separate locaion requirement to be met. Which coaxial
control would you consider using?

21. Turn back the pages to the geometric overview chart on the last page in unit 10.
Complete the chart filling in the required information on runout, concentricity and
symmetry.

UNIT 15

FASTENER FORMULAS AND SCREW THREADS

Floating Fastener Formula
Fixed Fastener Formula
Projected Tolerance Zone
Hatch - Application
Workshop Exercise 15.1

FLOATING FASTENER FORMULA

The floating fastener formula is an important design tool. It is used by the designer for calculating position tolerance between mating parts. A floating fastener assembly is one in which both parts have clearance holes. This assembly is called a floating fastener because the fastener can "float " around in the holes and is not fixed in either part. A common floating fastener example is an assembly that uses bolts and nuts or rivets.

The two flat plates on the following page fasten together using four screws and nuts. This type of assembly is categorized as a floating fastener application. The designer must decide the size of the bolts and how many are required. The dimensions between the holes and the dimensions from the datums are designated as basic dimensions.

In order to calculate the position tolerances, the designer will calculate the difference between the largest (MMC size) bolt and the smallest (MMC size) hole. The difference between these two sizes is the allowable tolerance that can be applied in the feature control frame of each part. The MMC modifier is applied to the feature tolerance which allows additional positional tolerance as the features depart from MMC.

Use standard hole sizes In the following floating fastener assembly, the example uses four 1/4-20 UNC screws. To accommodate these screws, the designer has selected clearance holes for the fasteners at .279/.287 diameter. The size of clearance holes were selected based on standard manufacturing drill sizes. This will help reduce cost in the manufacturing process as special tools will not be required.

If you will notice, standard drilled holes usually have a nominal size, with the plus tolerance being larger than the minus tolerance. As the drill wears it will make the hole larger rather than smaller. As the drill wears or after it is sharpened, it will tend to wander around the hole a little making the hole larger. If the designer uses a standard hole size and tolerance, it will give more life to the tools and, therefore, reduce product manufacturing cost.

FLOATING FASTENER FORMULA

Where two or more parts are assembled with fasteners, such as bolts and nuts or rivets, and all parts have clearance holes, it is termed a floating fastener case.

This formula is used for determining the required position tolerances of mating features to ensure that the parts will assemble. This formula is valid for all types of features or patterns of features and will give a "no interference, no clearance" fit when features are at maximum material condition with their locations in the extreme of position tolerance. Consideration must be given for additional geometric conditions that could affect functions not accounted for in the following formula.

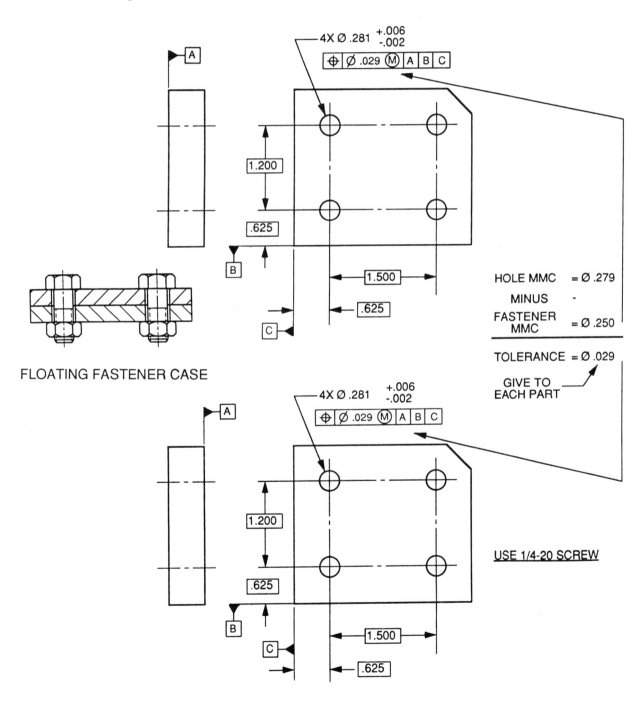

FLOATING FASTENER CASE

HOLE MMC	= Ø .279
MINUS	-
FASTENER MMC	= Ø .250

TOLERANCE = Ø .029

GIVE TO EACH PART

USE 1/4-20 SCREW

FIXED FASTENER FORMULA

A fixed fastener assembly is one in which one of the holes is a clearance hole, and the other hole is not a clearance hole but a press fit or a tapped hole. It is called a fixed fastener assembly because the fastener is "fixed" in the assembly. A common fixed fastener example is an assembly that uses dowels or tapped holes.

The two flat plates on the following page are fastened together using four screws that assemble in the mating part with tapped holes. The dimensions between the holes and the dimensions from the datums are designated as basic dimensions.

To determine the positional tolerances between the parts, the designer will calculate the difference between the largest (MMC size) bolt and the smallest (MMC size) hole. The difference between these two sizes is the allowable tolerance. This tolerance must be divided between the two parts in any combination that does not exceed the allowable tolerance.

The tapped hole is usually given a larger share of the tolerance for two possible reasons. First, the tapped hole is usually two operations - the tap drill and then the tap. This is a more difficult manufacturing operation and therefore usually needs more tolerance.

Second, and probably the most important reason for allotting the tapped hole more tolerance is that, if the MMC modifier is applied to the entire assembly, it will allow additional positional tolerance as the features depart from MMC. According to our rules, unless otherwise specified, positional tolerance for a tapped hole applies to the pitch diameter of the thread. There is very little available additional positional tolerance available as the size of the thread pitch diameter departs from MMC.

The clearance hole usually has much more size tolerance which results in more positional tolerance at LMC for the clearance hole than the tapped hole. To make the positional tolerance more equal at LMC, more positional tolerance is given to the tapped hole. As a general rule of thumb, split the tolerance 60% and 40%. The tapped hole receives 60%, and the clearance hole receives 40%.

To ensure interchangeability, a projected tolerance zone is usually applied to a fixed fastener assembly. A projected tolerance zone tolerance is applied to the tapped holes in the assembly on the following page. The projected tolerance zone is equal to the thickness of the mating part.

The pitch diameter of a screw thread has size; and, therefore, will have a feature modifier applied. As the screw is torqued down in the assembly, there is a centering effect. It is for this reason that, functionally, the position tolerance on a screw thread applies at RFS. In some cases, a designer may apply MMC to a screw thread. If this is the case, consideration should be given to the centering effect and the tolerance on the pitch diameter.

FIXED FASTENER FORMULA

Where one of the parts to be assembled has restrained fasteners, such as screws in tapped holes, studs or press fit pins, it is termed a fixed fastener case.

This formula is used for determining the required position tolerances of mating features to ensure that the parts will assemble. This formula is valid for all types of features or patterns of features and will give a "no interference, no clearance" fit when features are at maximum material condition with their locations in the extreme of position tolerance. Consideration must be given for additional geometric conditions that could affect functions not accounted for in the following formula. This formula does not provide sufficient clearance when threaded holes or holes for press fit pins are out of square. To provide for this condition, a projected tolerance zone was added to the tapped holes below to ensure interchangeability.

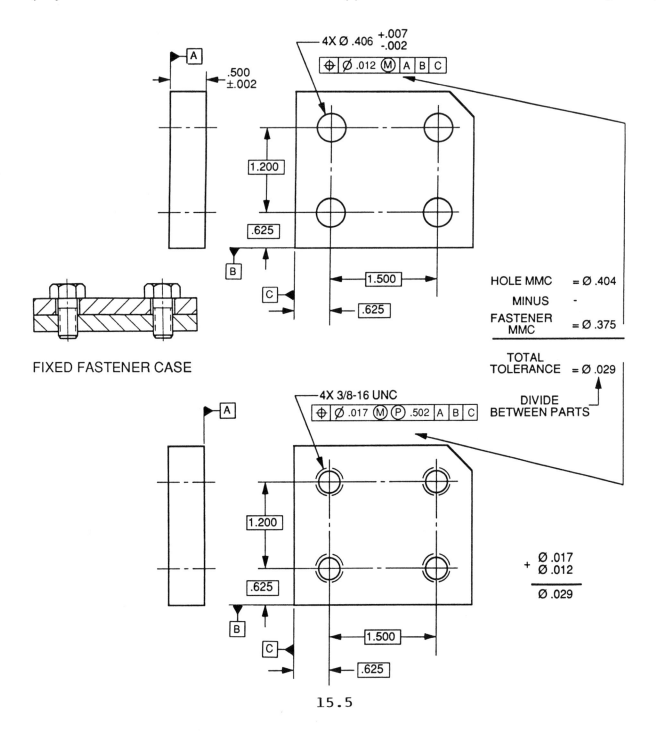

FIXED FASTENER CASE

HOLE MMC	= Ø .404
MINUS	-
FASTENER MMC	= Ø .375

TOTAL TOLERANCE = Ø .029

DIVIDE BETWEEN PARTS

$$+ \begin{array}{c} \varnothing\ .017 \\ \varnothing\ .012 \end{array}$$
$$\overline{\varnothing\ .029}$$

PROJECTED TOLERANCE ZONE

All geometric tolerances applied to features extend for the full length and depth of the feature. In some cases it may be necessary to "project" a tolerance zone. The projected tolerance zone concept will "project" the tolerance zone out of the feature by a specified amount.

Unlike the floating fastener application involving only clearance holes, the attitude of a fixed fastener (for example, screw thread or press fit dowel) is governed by the inclination of the produced hole in which it assembles. The fixed fastener formula does not ensure complete interchangeability between parts. The formula does not take into consideration a possible interference condition if the threaded or press fit holes were to tilt or be out of perpendicular within the allotted position tolerance zone. The projected tolerance zone concept is recommended where we wish to limit the variation of perpendicularity on threaded or press fit holes to ensure interchangeability between parts.

The location and perpendicularity of the threaded holes are only of importance in so far as they affect the extended portion of the engaging fastener. The projected tolerance zone will project the tolerance out of the hole to the extent of the engaging fastener.

The height of the projected tolerance zone is usually the thickness of the mating part including any gaskets or shims. The thickness of washers are usually not included in the calculations as the washers may float in any direction and will not cause interference. If a stud or pin is toleranced as an assembly with the pin or stud installed, a projected tolerance zone is not necessary, as the tolerance extends for the full length of the feature. If the hole in the detail is toleranced without the studs or pins installed, then the hole should have a projected zone applied equal to the height of the studs or pins.

Where the direction of the projected tolerance zone is clear, only a symbolic call-out and height dimension is required. The symbolic call-out is placed in the feature tolerance compartment of the feature control frame following the feature tolerance and any modifier. The height of the tolerance zone follows the projected tolerance zone symbol.

Where the direction of the projected tolerance zone is not clear, the symbolic call-out for the projected tolerance zone is still placed in the feature control frame following the feature tolerance and any modifier. The dimensioned height value of the projected zone is included in the drawing view with a heavy chain line that is drawn closely adjacent to the centerline of the hole.

PROJECTED TOLERANCE ZONE APPLICATION

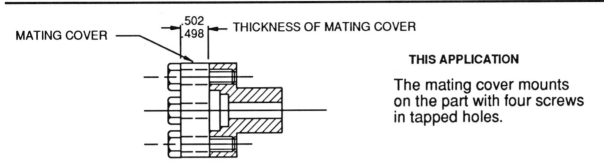

MATING COVER

.502
.498 ← THICKNESS OF MATING COVER

THIS APPLICATION

The mating cover mounts on the part with four screws in tapped holes.

HEIGHT OF PROJECTED TOLERANCE SHOWN IN FEATURE CONTROL FRAME.

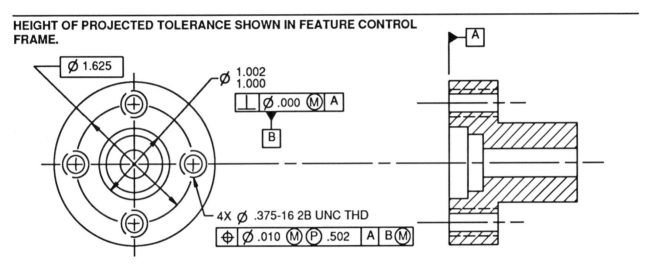

Where the direction of the projected tolerance zone is clear, only a symbolic call-out and height dimension is required. The symbolic call-out is placed in the feature control frame following the feature tolerance and any modifier. The height of the projected tolerance zone follows the projected tolerance zone symbol.

HEIGHT OF PROJECTED TOLERANCE SHOWN IN DRAWING VIEW.

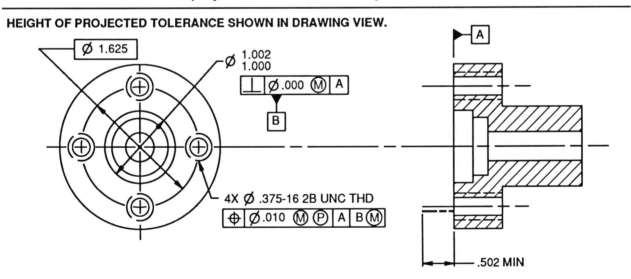

Where the direction of the projected tolerance zone is not clear, the symbolic call-out for the projected tolerance zone is still placed in the feature control frame following the feature tolerance and any modifier. The dimensional height of the projected tolerance zone, however, is shown in the drawing view with a heavy chain line that is drawn closely adjacent to the centerline of the hole.

The projected tolerance zone concept projects the tolerance zone out of the feature to the height indicated. This projection will ensure the assembled screws or press fit pins will be in proper location in the installation.

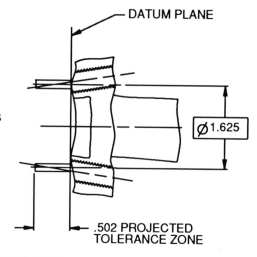

DATUM PLANE

⌀1.625

.502 PROJECTED TOLERANCE ZONE

FUNCTIONAL GAGE CHECK AT MMC FOR PROJECTED TOLERANCE ZONE

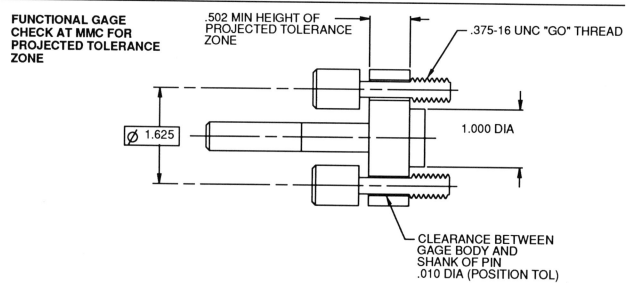

.502 MIN HEIGHT OF PROJECTED TOLERANCE ZONE

.375-16 UNC "GO" THREAD

⌀ 1.625

1.000 DIA

CLEARANCE BETWEEN GAGE BODY AND SHANK OF PIN .010 DIA (POSITION TOL)

OPEN SET-UP TECHNIQUE

In an open set-up, the pitch dia location and the projected tolerance zone may be checked with threaded gage pins. Position at MMC may be checked with "go" threads and are allowed to be "wiggled" (tol on pitch dia) into position. RFS may be checked with gage threads that have slits that expand into the thread allowing no wiggle.

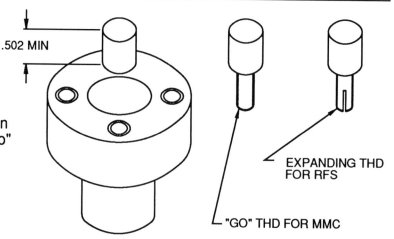

.502 MIN

EXPANDING THD FOR RFS

"GO" THD FOR MMC

GEOMETRIC TOLERANCING APPLICATION - HATCH

The drawing below is a sample application of geometric tolerancing. The part has two complete datum reference frames. The datum features are all related and qualified, and the two datum reference frames are related to each other. There are also applications of form, orientation, position and profile tolerances.

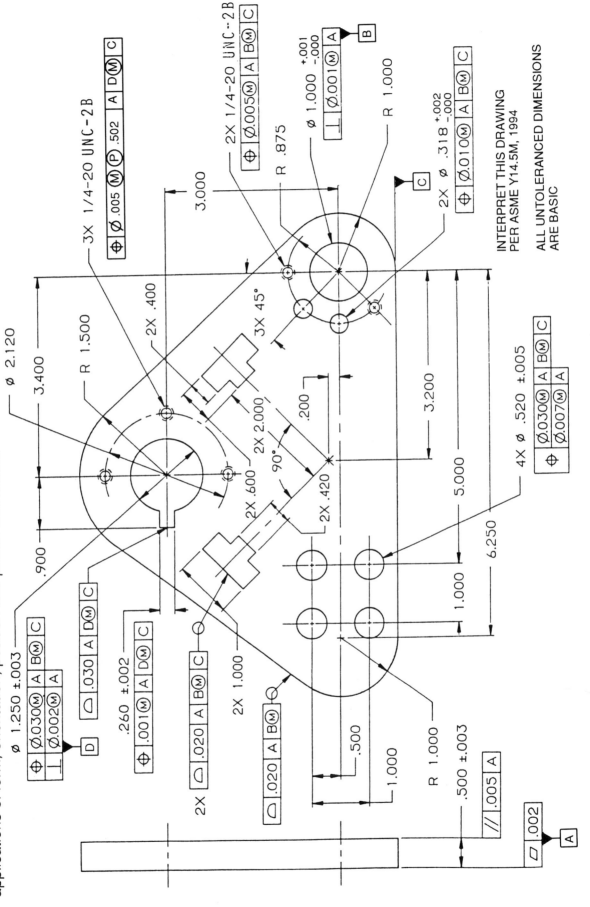

INTERPRET THIS DRAWING
PER ASME Y14.5M, 1994

ALL UNTOLERANCED DIMENSIONS
ARE BASIC

WORKSHOP EXERCISE 15.1 - POWDER CASE

1. The two parts below are mating parts. Select datums and apply any necessary geometric tolerancing to insure functional requirements.

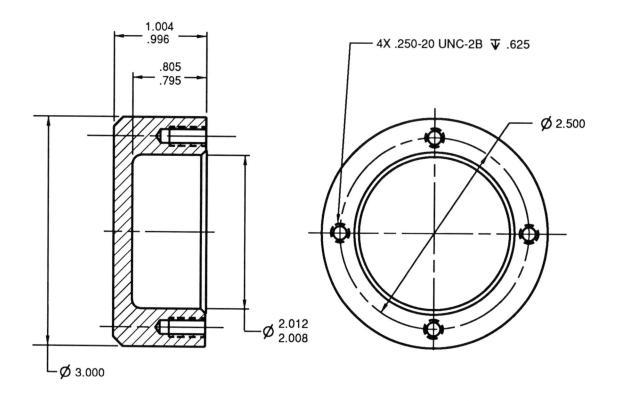

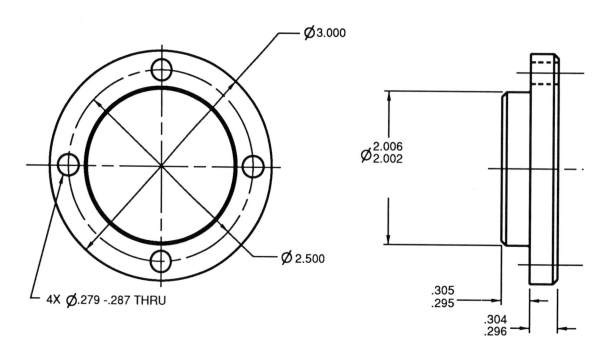

WORKSHOP EXERCISE 15.1 - REINFORCEMENT

2. Study the functional requirements of this assembly. Select datums and apply any necessary geometric tolerancing to insure functional requirements. Assume .020 dia. position on mating part screws.

THIS APPLICATION

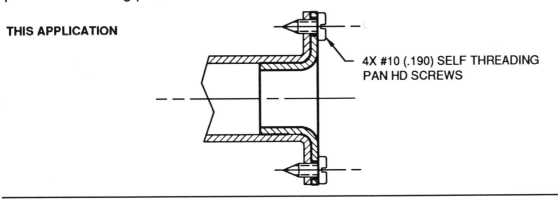

4X #10 (.190) SELF THREADING PAN HD SCREWS

THIS ON THE DRAWING

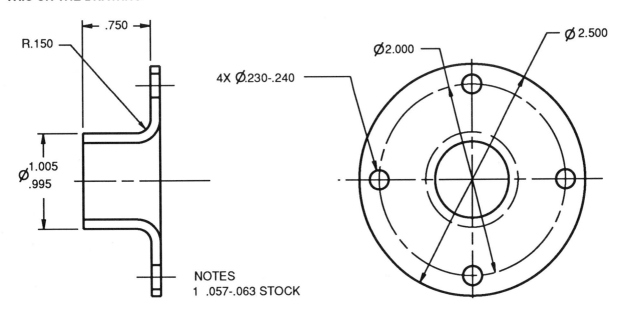

.750

R.150

$\emptyset \begin{smallmatrix} 1.005 \\ .995 \end{smallmatrix}$

4X $\emptyset$.230-.240

$\emptyset$2.000

$\emptyset$ 2.500

NOTES
1 .057-.063 STOCK

Draw a simple functional gage for the toleranced part above.

UNIT 16

IB AND OB STACK-UP CALCULATIONS

INNER & OUTER BOUNDARY AND VIRTUAL & RESULTANT CONDITION

The terms outer and inner boundary and virtual and resultant condition are used in geometric tolerancing to calculate fits and stack-ups. The technical definitions and graphical descriptions of these terms are shown on the following pages.

Inner Boundary (IB) - A worst case boundary generated by the smallest feature (MMC for an internal feature and LMC for an external feature) minus the stated geometric tolerance and any additional geometric tolerance (if applicable) from the feature's departure from its specified material condition.

Outer Boundary (OB) - A worst case boundary generated by the largest feature (LMC for an internal feature and MMC for an external feature) plus the stated geometric tolerance and any additional geometric tolerance (if applicable) from the feature's departure from its specified material condition.

Virtual Condition (VC) - A constant boundary generated by the collective effects of a size feature's specified MMC or LMC material condition and the geometric tolerance for that material condition.

Resultant Condition (RC) - The variable boundary generated by the collective effects of a size feature's specified MMC or LMC material condition, the geometric tolerance for that material condition, the size tolerance, and the additional geometric tolerance derived from the feature's departure from its specified material condition.

The outer boundary and inner boundary concepts are very powerful tools in understanding the effects of geometric tolerancing. The IB and OB are graphical aids to help us understand what happens to the features when geometric tolerancing is applied to features. The terms inner boundary and outer boundary are perfect words to describe the concept.

The outer boundary describes the worst case outside boundary of how much a feature may move around after all the applicable geometric tolerance is applied. The inner boundary describes the worst case inside boundary of how much a feature may move around after all the applicable tolerance is applied.

The terms give us a graphical picture of how much a feature may grow, shrink, shift or move around as the geometric tolerancing is applied. The IB and OB terms can be used with all of the geometric tolerances such as position, form, orientation, runout and profile tolerances.

I am sure that if you have used geometric tolerancing in the past there have been questions in your mind of exactly how much a feature could move around after all the geometric tolerance has been applied.

For example, if a position tolerance at MMC were applied to a hole, we already know that the larger that hole was produced the more the hole could move off center. This is a useful concept if we have a clearance hole where all we want to do is get the material out of the way to clear the bolt or pin that we are inserting in the hole.

The designer will need to calculate the worst case inner boundary to insure clearance to the bolt. This worst case inner boundary is called the virtual condition. To simplify things, this worst case inner boundary can be simply called the inner boundary.

In some cases, it also may be necessary for the designer to calculate the worst case outer boundary of the hole for purposes of wall thickness or rattle in assembled parts. This outer boundary can be referred to as the resultant condition or just simply called the outer boundary.

The terms virtual and resultant condition are only used in the specific cases when geometric tolerances are applied in conjunction with the MMC and LMC principle. The use of the virtual and resultant condition terms are not appropriate when using RFS, runout, profile etc. The terms inner and outer boundary are much more general terms and can be used with all the geometric tolerances such as profile, runout , position, etc.

The general, all encompassing terms IB and OB can also substitute for the terms virtual and resultant conditions. Since the terms virtual and resultant conditions are so specific, they can only be used as substitute terms for IB and OB when used in conjunction with the MMC or LMC principle.

It is important to study the definitions of virtual condition and resultant condition very carefully. These definitions can be a little confusing, especially when working with the MMC and LMC concept. For example, if a hole is located with a position tolerance at MMC the worst case inner boundary is called the virtual condition. The worst case outer boundary is called the worst case resultant condition.

Conversely, if a hole is located with a position tolerance at LMC, the worst case outer boundary is called the virtual condition, and the worst case inner boundary is called the worst case resultant condition. This is often very difficult concept to understand especially when comparing pins and holes at MMC and LMC.

There are graphical examples of the inner and outer boundary concepts and the virtual and resultant condition concepts shown on the next few pages. These terms may take some careful study at first, but after a while, the inner and outer boundary and the virtual and resultant condition concepts will become second nature.

INNER AND OUTER BOUNDARIES RFS CONCEPT - INTERNAL/EXTERNAL

Graphical representations of the inner and outer boundary and the virtual and resultant condition for the RFS concept are shown below. These are important tools to understand and evaluate the "worst case" boundaries of a feature after geometric tolerancing has been applied. The technical definitions of the terms are shown on the preceeding pages.

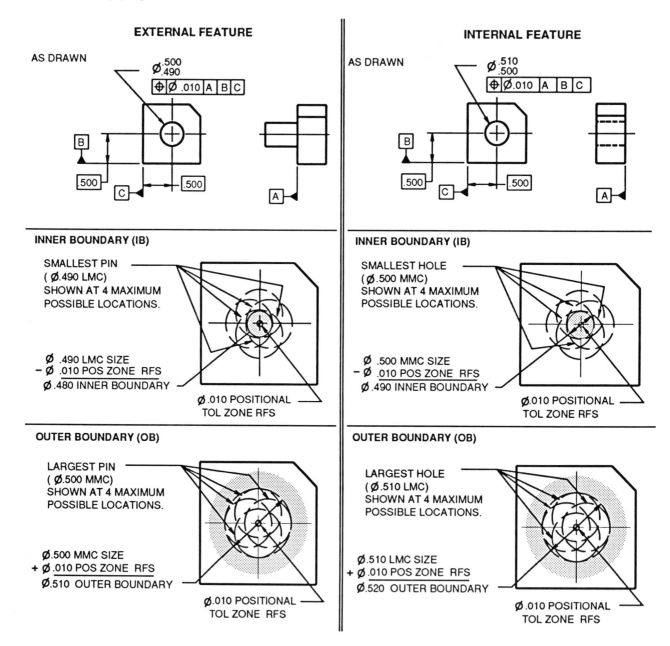

EXTERNAL FEATURE

AS DRAWN

⌀ .500
.490

⊕ | ⌀ .010 | A | B | C

INNER BOUNDARY (IB)

SMALLEST PIN
(⌀.490 LMC)
SHOWN AT 4 MAXIMUM
POSSIBLE LOCATIONS.

⌀ .490 LMC SIZE
− ⌀ .010 POS ZONE RFS
⌀.480 INNER BOUNDARY

⌀.010 POSITIONAL
TOL ZONE RFS

OUTER BOUNDARY (OB)

LARGEST PIN
(⌀.500 MMC)
SHOWN AT 4 MAXIMUM
POSSIBLE LOCATIONS.

⌀.500 MMC SIZE
+ ⌀ .010 POS ZONE RFS
⌀.510 OUTER BOUNDARY

⌀.010 POSITIONAL
TOL ZONE RFS

INTERNAL FEATURE

AS DRAWN

⌀ .510
.500

⊕ | ⌀ .010 | A | B | C

INNER BOUNDARY (IB)

SMALLEST HOLE
(⌀.500 MMC)
SHOWN AT 4 MAXIMUM
POSSIBLE LOCATIONS.

⌀ .500 MMC SIZE
− ⌀ .010 POS ZONE RFS
⌀.490 INNER BOUNDARY

⌀.010 POSITIONAL
TOL ZONE RFS

OUTER BOUNDARY (OB)

LARGEST HOLE
(⌀.510 LMC)
SHOWN AT 4 MAXIMUM
POSSIBLE LOCATIONS.

⌀.510 LMC SIZE
+ ⌀.010 POS ZONE RFS
⌀.520 OUTER BOUNDARY

⌀.010 POSITIONAL
TOL ZONE RFS

INNER AND OUTER BOUNDARIES MMC CONCEPT - INTERNAL/EXTERNAL

Graphical representations of the inner and outer boundary and the virtual and resultant condition for the MMC concept are shown below. These are important tools to understand and evaluate the "worst case" boundaries of a feature after geometric tolerancing has been applied. The technical definitions of the terms are shown on the preceeding pages.

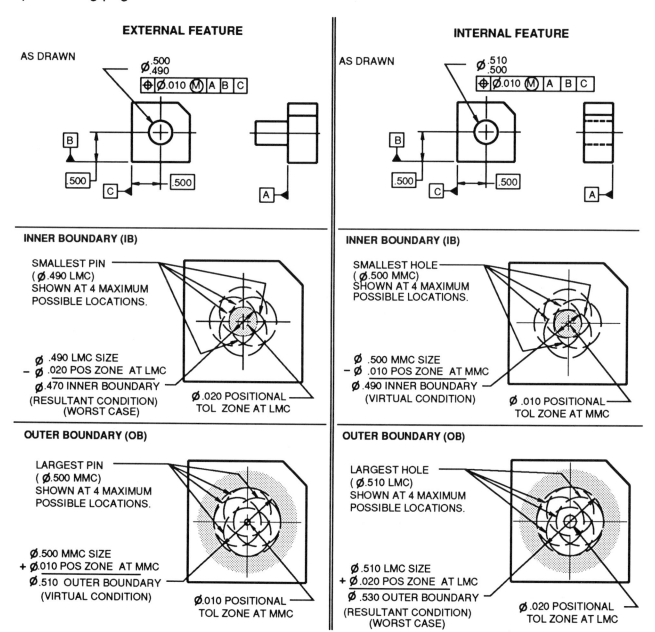

EXTERNAL FEATURE

AS DRAWN

Ø .500 / .490

⊕ | Ø.010 Ⓜ | A | B | C

B

.500   .500

C   A

INNER BOUNDARY (IB)

SMALLEST PIN
(Ø.490 LMC)
SHOWN AT 4 MAXIMUM
POSSIBLE LOCATIONS.

Ø .490 LMC SIZE
- Ø .020 POS ZONE AT LMC
Ø.470 INNER BOUNDARY
(RESULTANT CONDITION)
(WORST CASE)

Ø .020 POSITIONAL
TOL ZONE AT LMC

OUTER BOUNDARY (OB)

LARGEST PIN
(Ø.500 MMC)
SHOWN AT 4 MAXIMUM
POSSIBLE LOCATIONS.

Ø.500 MMC SIZE
+ Ø.010 POS ZONE AT MMC
Ø.510 OUTER BOUNDARY
(VIRTUAL CONDITION)

Ø.010 POSITIONAL
TOL ZONE AT MMC

INTERNAL FEATURE

AS DRAWN

Ø .510 / .500

⊕ | Ø.010 Ⓜ | A | B | C

B

.500   .500

C   A

INNER BOUNDARY (IB)

SMALLEST HOLE
(Ø.500 MMC)
SHOWN AT 4 MAXIMUM
POSSIBLE LOCATIONS.

Ø .500 MMC SIZE
- Ø .010 POS ZONE AT MMC
Ø .490 INNER BOUNDARY
(VIRTUAL CONDITION)

Ø .010 POSITIONAL
TOL ZONE AT MMC

OUTER BOUNDARY (OB)

LARGEST HOLE
(Ø.510 LMC)
SHOWN AT 4 MAXIMUM
POSSIBLE LOCATIONS.

Ø.510 LMC SIZE
+ Ø .020 POS ZONE AT LMC
Ø .530 OUTER BOUNDARY
(RESULTANT CONDITION)
(WORST CASE)

Ø .020 POSITIONAL
TOL ZONE AT LMC

16.5

INNER AND OUTER BOUNDARIES LMC CONCEPT - INTERNAL/EXTERNAL

Graphical representations of the inner and outer boundary and the virtual and resultant condition for the LMC concept are shown below. These are important tools to understand and evaluate the "worst case" boundaries of a feature after geometric tolerancing has been applied. The technical definitions of the terms are shown on the preceeding pages.

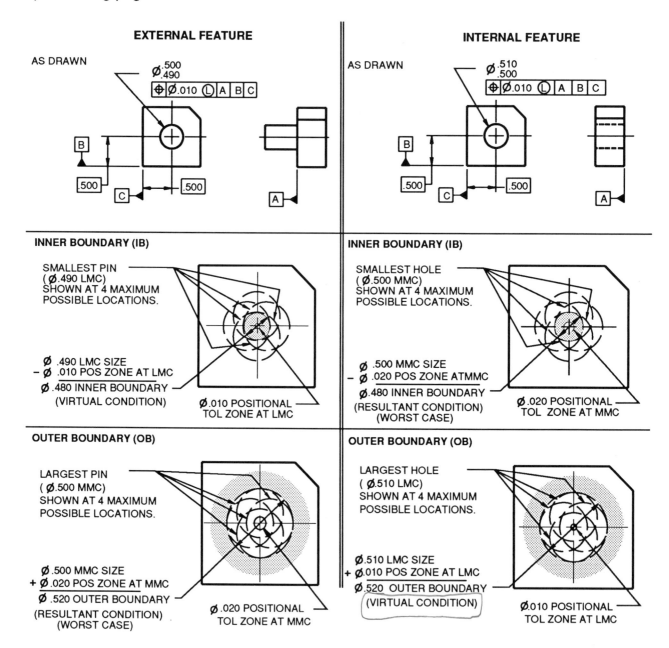

EXTERNAL FEATURE

AS DRAWN

Ø .500 / .490

⊕ | Ø.010 Ⓛ | A | B | C

INNER BOUNDARY (IB)

SMALLEST PIN
(Ø.490 LMC)
SHOWN AT 4 MAXIMUM
POSSIBLE LOCATIONS.

  Ø .490 LMC SIZE
− Ø .010 POS ZONE AT LMC
  Ø .480 INNER BOUNDARY
  (VIRTUAL CONDITION)

Ø.010 POSITIONAL
TOL ZONE AT LMC

OUTER BOUNDARY (OB)

LARGEST PIN
(Ø.500 MMC)
SHOWN AT 4 MAXIMUM
POSSIBLE LOCATIONS.

  Ø .500 MMC SIZE
+ Ø .020 POS ZONE AT MMC
  Ø .520 OUTER BOUNDARY
  (RESULTANT CONDITION)
  (WORST CASE)

Ø .020 POSITIONAL
TOL ZONE AT MMC

INTERNAL FEATURE

AS DRAWN

Ø .510 / .500

⊕ | Ø.010 Ⓛ | A | B | C

INNER BOUNDARY (IB)

SMALLEST HOLE
(Ø.500 MMC)
SHOWN AT 4 MAXIMUM
POSSIBLE LOCATIONS.

  Ø .500 MMC SIZE
− Ø .020 POS ZONE AT MMC
  Ø.480 INNER BOUNDARY
  (RESULTANT CONDITION)
  (WORST CASE)

Ø .020 POSITIONAL
TOL ZONE AT MMC

OUTER BOUNDARY (OB)

LARGEST HOLE
(Ø.510 LMC)
SHOWN AT 4 MAXIMUM
POSSIBLE LOCATIONS.

  Ø.510 LMC SIZE
+ Ø.010 POS ZONE AT LMC
  Ø.520 OUTER BOUNDARY
  (VIRTUAL CONDITION)

Ø.010 POSITIONAL
TOL ZONE AT LMC

1. Calculate the inner and outer boundaries for the following problems. Consult the IB and OB formulas in the text if you have any problems.

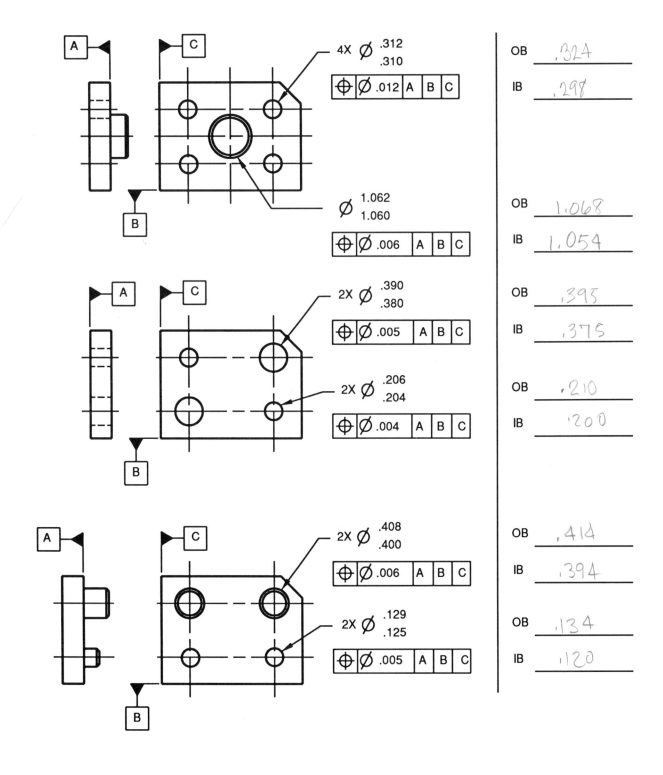

4X ⌀ .312 / .310 ⊕ ⌀ .012 A B C	OB .324
	IB .298
⌀ 1.062 / 1.060 ⊕ ⌀ .006 A B C	OB 1.068
	IB 1.054
2X ⌀ .390 / .380 ⊕ ⌀ .005 A B C	OB .395
	IB .375
2X ⌀ .206 / .204 ⊕ ⌀ .004 A B C	OB .210
	IB .200
2X ⌀ .408 / .400 ⊕ ⌀ .006 A B C	OB .414
	IB .394
2X ⌀ .129 / .125 ⊕ ⌀ .005 A B C	OB .134
	IB .120

2. Calculate the inner and outer boundaries for the following problems. Consult the IB and OB formulas in the text if you have any problems.

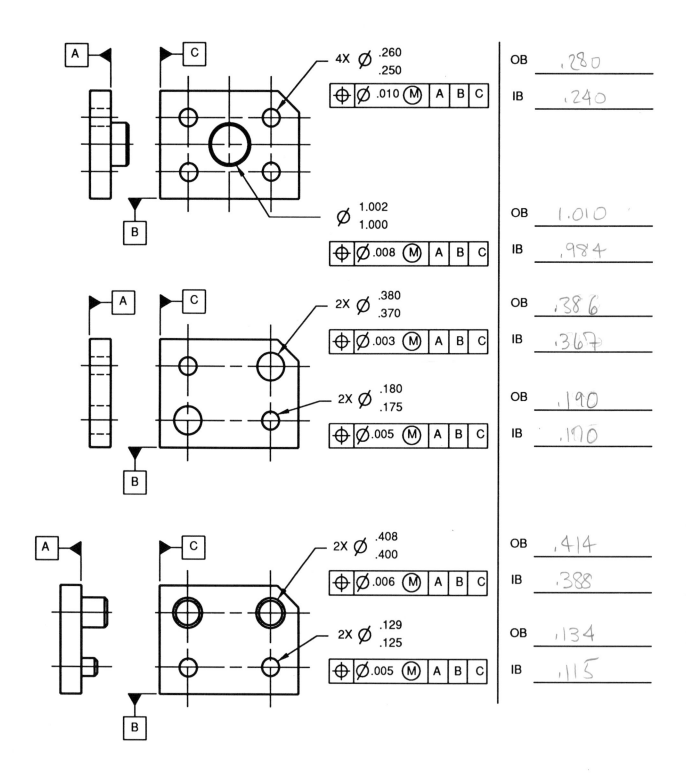

4X ⌀ .260 / .250 ⊕ ⌀ .010 Ⓜ A B C	OB .280 IB .240
⌀ 1.002 / 1.000 ⊕ ⌀.008 Ⓜ A B C	OB 1.010 IB .984
2X ⌀ .380 / .370 ⊕ ⌀.003 Ⓜ A B C	OB .386 IB .367
2X ⌀ .180 / .175 ⊕ ⌀.005 Ⓜ A B C	OB .190 IB .170
2X ⌀ .408 / .400 ⊕ ⌀.006 Ⓜ A B C	OB .414 IB .388
2X ⌀ .129 / .125 ⊕ ⌀.005 Ⓜ A B C	OB .134 IB .115

3. Calculate the inner and outer boundaries for the following problems. Consult the IB and OB formulas in the text if you have any problems.

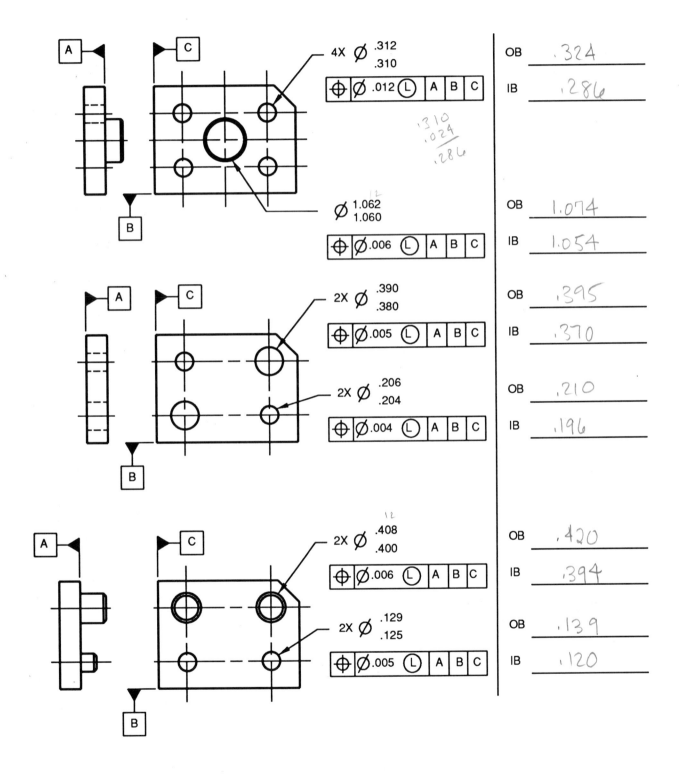

4. Calculate the inner and outer boundaries for the following problems. Consult the IB and OB formulas in the text if you have any problems.

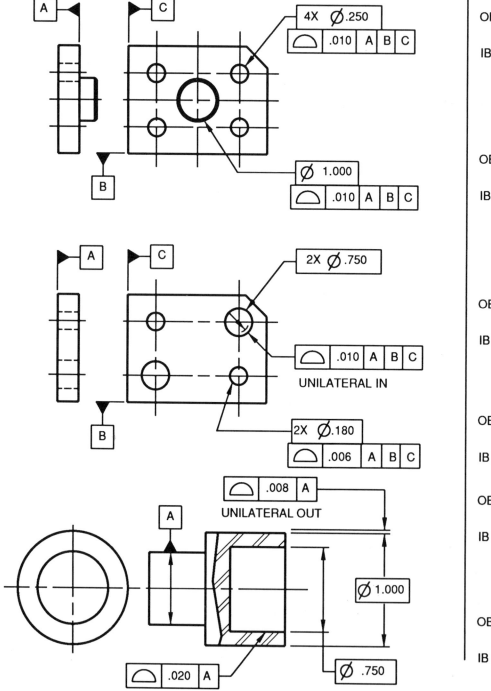

4X ⌀.250	
⌓ .010 A B C	OB .260
	IB .240
⌀ 1.000	
⌓ .010 A B C	OB 1.010
	IB .990

2X ⌀.750	
⌓ .010 A B C	OB .750
UNILATERAL IN	IB .740
2X ⌀.180	
⌓ .006 A B C	OB .186
	IB .174
⌓ .008 A	OB 1.008
UNILATERAL OUT	IB 1.000
⌀ 1.000	
⌓ .020 A	OB .770
⌀ .750	IB .730

5. Calculate the inner and outer boundaries for the following problems. Consult the IB and OB formulas in the text if you have any problems.

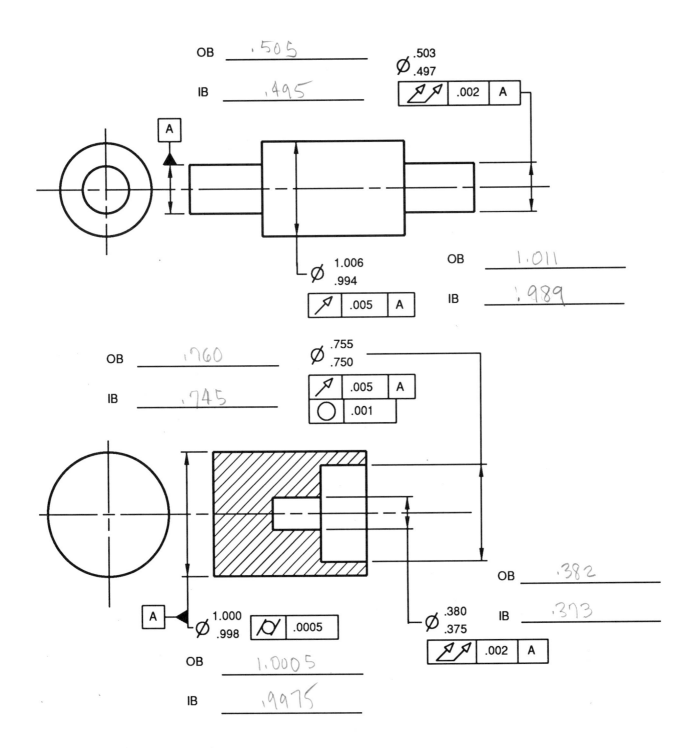

OB _____.505_____

IB _____.495_____

$\emptyset$ .503 / .497

⤢⤢	.002	A

A

$\emptyset$ 1.006 / .994

⤢	.005	A

OB _____1.011_____

IB _____.989_____

OB _____.760_____

IB _____.745_____

$\emptyset$ .755 / .750

⤢	.005	A
◯	.001	

A ◀ $\emptyset$ 1.000 / .998

⟋	.0005

OB _____1.0005_____

IB _____.9975_____

$\emptyset$ .380 / .375

⤢⤢	.002	A

OB _____.382_____

IB _____.373_____

.9980
.0005
.9975

16.11

WALL THICKNESS CALCULATION

The upper drawing is the product drawing. The bottom drawing illustrates a simple min/max wall calculation using the IB/OB and considerations of the form or orientation tolerances on datums.

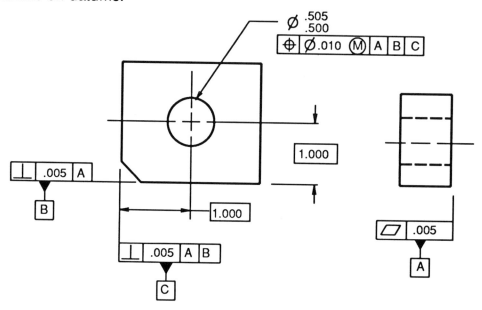

MIN/MAX WALL CALCULATIONS TO A DRF VS. A DATUM FEATURE

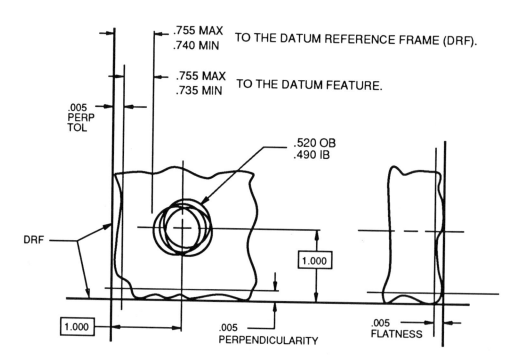

In making min/max wall calculations, remember that the basic dimensions originate from the DRF. The high points of the datum feature make contact with the DRF. The relationship between the datum feature and the DRF (datum qualification) is the flatness or perpendicularity tolerance shown above. When making min/max wall calculations, consideration should be given to the applicable flatness or orientation tolerances.

IB AND OB STACK-UP CALCULATIONS

The upper drawing is the product drawing. The bottom drawing illustrates some simple stack-ups between the features. The IB and OB for each feature is calculated. The center of the IB/OB is defined by the basic dimension. To find the min and max between the features is a simple matter of addition and subtraction of the numbers.

THIS ON THE DRAWING

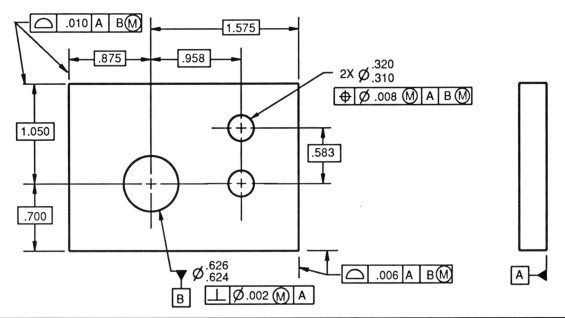

SIMPLE STACK-UPS BETWEEN THE FEATURES

MEANS THIS

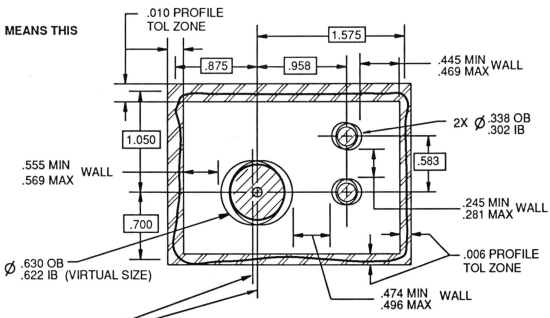

The axis of datum feature B at MMC (IB or Virtual) establishes the axis of the DRF. All basic dimensions originate from this point.

The axis of datum feature B is allowed to be displaced relative to the axis of datum feature B at MMC (IB or Virtual). It may displace in a zone equal to the difference between the virtual size (IB) and the actual virtual size a maximum of .004 dia. Consequently, if a .626 dia hole can displace by .004 dia., its OB is .630. The IB (virtual) is .622 dia.

16.13

1. The top drawing is the product drawing. The bottom drawing requires the calculation of the min/max between the designated features.

THIS ON THE DRAWING

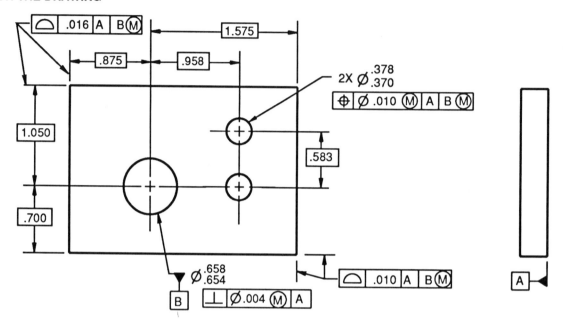

COMPLETE THE SIMPLE STACK-UPS BETWEEN THE FEATURES

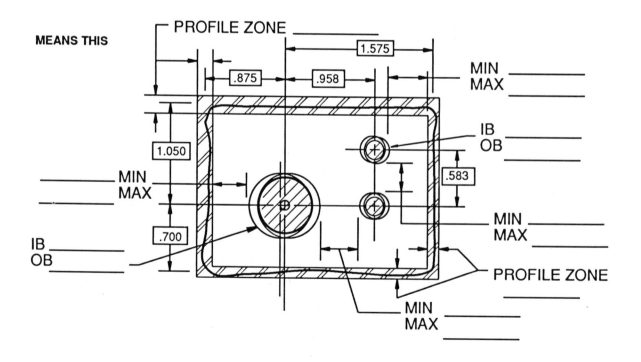

WORKSHOP EXERCISE 16.2

2. Find the IB and OB or min and max as required on this drawing and record your answers on the worksheet. For this example, make all calculations from the DRF or the datum features as specified. Generally, the higher the number the harder the problem. After number 9, watch for multiple DRF's.

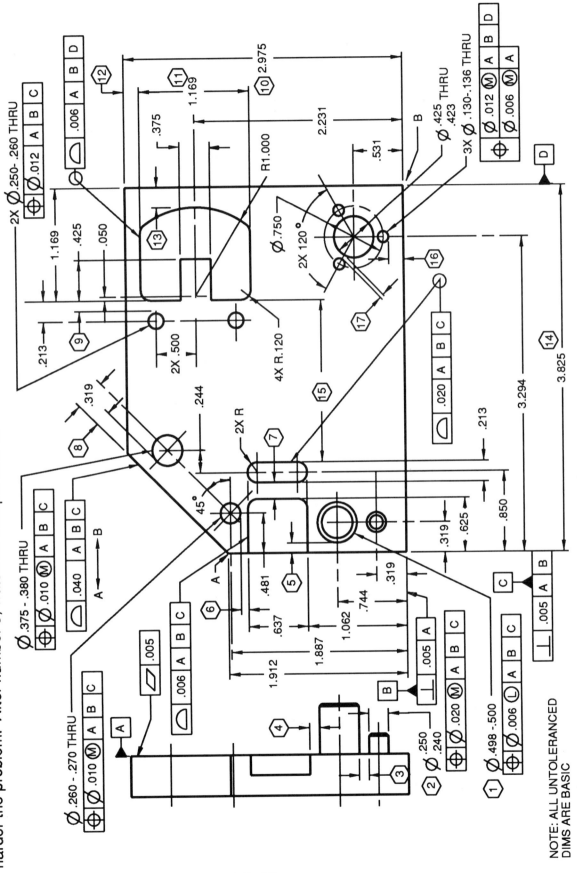

NOTE: ALL UNTOLERANCED DIMS ARE BASIC

16.15

WORKSHOP EXERCISE 16.2 - PLATE WORKSHEET

(1)
OB _____
IB _____

(2)
OB _____
IB _____

(3)
MAX _____
MIN _____

(4)
MAX _____
MIN _____

(5)
MAX _____
MIN _____
MIN TO DATUM
FEATURE _____

(6)
MAX _____
MIN _____

(7)
MAX _____
MIN _____

(8)
MAX _____
MIN _____

(9)
MAX _____
MIN _____

(10)
MAX _____
MIN _____
MIN TO DATUM
FEATURE _____

(11)
MAX _____
MIN _____

(12)
MAX _____
MIN _____

(13)
MAX _____
MIN _____
MIN TO DATUM
FEATURE _____

(14)
MAX _____
MIN _____
MIN TO DATUM
FEATURE _____

(15)
MAX _____
MIN _____

(16)
MIN _____
MIN TO DATUM
FEATURE _____

(17)
MIN _____
MAX _____

WHAT IS WRONG ?

The drawing below is so bad, it looks like a ransom note. See if you can find any mistakes. Look for missing symbols, mis-used symbols, mis-applied datums, incorrect feature control frames etc. If the drawing looks good, take the course again.

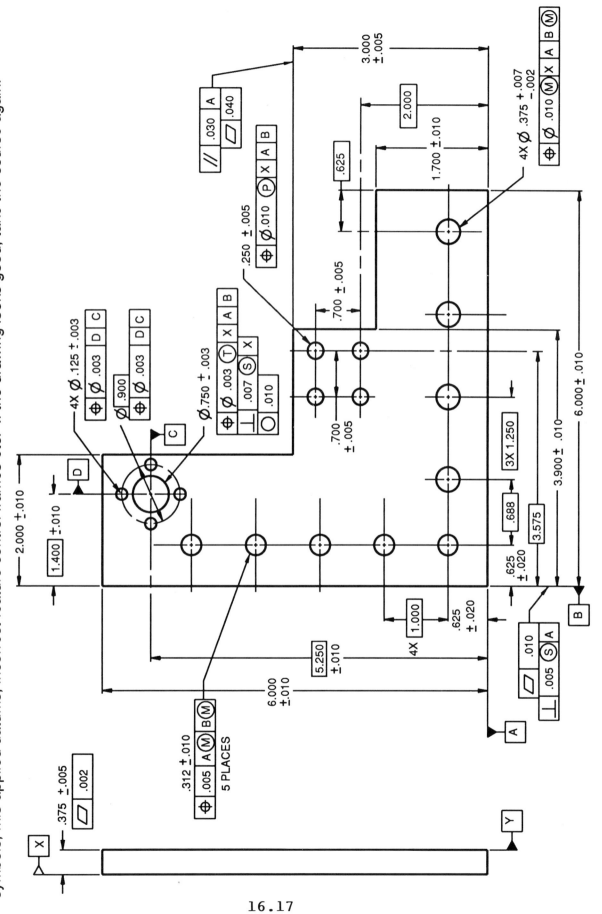

16.17

APPENDIX A

PRINCIPLE CHANGES, IMPROVEMENTS AND ENHANCEMENTS FROM THE ANSI Y14.5M-1982 TO THE ASME Y14.5M-1994 STANDARD

APPENDIX A

PRINCIPLE CHANGES, IMPROVEMENTS AND ENHANCEMENTS FROM THE ANSI Y14.5M-1982 STANDARD

The new updated ASME Y14.5M-1994 standard on dimensioning and tolerancing reflects a culmination of effort extending over 12 years. It is a revision of the ANSI Y14.5M-1982 standard. In the new ASME Y14.5M-1994 standard there are extensive changes, improvements and enhancements from the 1982 standard.

Probably the most significant change is the release of the ASME Y14.5.1M-1994 standard on Mathematical Definitions of Dimensioning and Tolerancing Principles. It is the sister document to the ASME Y14.5M-1994 standard. This new math standard provides mathematical rigor to the dimensioning and tolerancing principles. This new standard creates explicit definitions in three major areas: 1) the mathematical definition of the tolerance zone; 2) the mathematical definition of conformance to the tolerance; 3) the mathematical definition of the actual value. These definitions are especially important in such areas as Computer Aided Design (CAD), Coordinate Measuring Machines (CMM's) and Computer Aided Manufacturing (CAM). The new math standard provides the geometric tolerancing principles with a solid mathematical foundation.

Many of the changes are simple word or definition changes that can have a major impact throughout the entire document or the expansion of individual concepts. The reader must understand that only the major concepts are outlined. Many of the changes, expansions, revisions and term redefinitions are woven through the fabric of the document. This is especially true with the introduction of the new ASME Y14.5.1M-1994 math standard and redefinition of size. (Taylor Principle)

In this short outline it is impossible to address the implication of all the changes. In order to understand the impact of these changes, it is assumed that the reader has knowledge of the old 1982 standard. In the outline below, reference page numbers from this text are noted. The reader can find detailed information on the changes at these locations in the text. It is suggested that the reader study the material in the workbook as it will provide a better general over all feel for extension and expansion of the principles. In addition, in appendix A of the ASME Y14.5M-1994 standard, there is also a listing of the principle changes.

PRINCIPLE CHANGES, IMPROVEMENTS AND ENHANCEMENTS FROM THE ANSI Y14.5M-1982 STANDARD

New and Revised Common Symbols - page 1.6
      Radius - page 1.10
      Controlled radius - page 1.10
      Statistical tolerance - page 1.9
      Datum feature symbol - pages 1.17, 5.11 thru 5.15
            Former practice - pages 1.16, 5.12
      Free state condition (non-rigid parts) - pages 12.21 thru 12.24
      Tangent plane - page 11.18
      Feature control frames - pages 1.15 thru 1.17

New and Revised Terms
      Feature of size - page 1.11
      Actual size - pages 4.7, 4.8
      Actual local size - pages 4.7, 4.8
      Actual mating envelope - pages 4.7, 4.8
      True geometric counterpart - page 5.15
      Datum simulator - pages 5.3, 5.4, 5.16. 5.17
      Simulated datum - page 5.4
      Virtual condition - pages 4.20, 16.2 thru 16.6
      Resultant condition - pages 16.2 thru 16.6

Tolerance Analysis Calculations - pages 16.1 thru 16.14
          Inner boundary - pages 16.2 thru 16.6
          Outer boundary - pages 16.2 thru 16.6

New and Revised Rules
Changes under rules #2 and #3 (applicability of modifiers) - pages 2.17, 2.18
      RFS symbol elimination - page 2.18
      Expansions of modifier applications - pages 2.20 thru 2.30

Rule I, Envelope (Taylor) Principle - Expansions
      Feature of size - redefinition and impact - pages 4.1 thru 4.16

Datum Referencing - Expansions
      Datum feature symbol application - pages 5.11 thru 5.15
      Inclined datum features - pages 8.4, 8.5
      Mathematically defined surfaces - pages 8.6, 8.7
      Multiple datum features (hole patterns) - pages 9.3 thru 9.7
      Multiple datum reference frames - pages 9.9 thru 9.11
      Simultaneous position and profile tolerances - pages 6.10 thru 6.12

Form Tolerances - Expansions
   Line element, orientation control - pages 10.9, 10.10
   Circularity clarification - page 10.15, 10.16

Orientation Tolerances - Expansions
   Multiple datums - pages 11.9 thru 11.15
   Line element, orientation control - page 11.17
   Tangent plane - page 11.18

Profile Tolerancing - Expansions
   Surface versus axis control - page 13.3
   Bilateral unequal distribution - page 12.5
   Composite profile - datum, orientation only control - pages 12.16 thru 12.19
   Location of irregular features - page 12.14, 12.15
   Multiple, single segment requirements - page 12.20

Position Tolerancing - Expansions
   Composite position - datum, orientation only control - pages 13.1 thru 13.7, 13.22, 13.23
   Multiple, single segment requirements - pages 13.8, 13.9, 13.22, 13.23, 13.24
   Boundary position tolerancing - pages 13.9, 13.10
   Projected tolerance zone
      Symbology and applications - pages 15.6 thru 15.8

Concentricity and Symmetry - New and Expansions
   Concentricity - pages 14.15 thru 14.18
   Symmetry - pages 14.15, 14.16, 14.19